IP Convergence:
The Next Revolution in
Telecommunications

For a listing of recent titles in the *Artech House Telecommunications Library,* turn to the back of this book.

IP Convergence:
The Next Revolution in
Telecommunications

Nathan J. Muller

Artech House
Boston • London

Library of Congress Cataloging-in-Publication Data
Muller, Nathan J.
 IP convergence : the next revolution in telecommunications /
Nathan J. Muller.
  p.   cm. — (Artech House telecommunications library)
 Includes bibliographical references and index.
 ISBN 1-58053-012-5 (alk. paper)
 1. Internet telephony.   2. TCP/IP (Computer network protocol)
 I. Title.   II. Series.
TK5105.8865.M85   1999
621.382′12—dc21   99-045837
                                                   CIP

British Library Cataloguing in Publication Data
Muller, Nathan J.
 IP convergence : the next revolution in telecommunications. —
 (Artech House telecommunications library)
 1. TCP/IP (Computer network protocol)   2. Computer networks
 I. Title
 004.6

 ISBN   1-58053-012-5

Cover design by Igor Valdman

© 2000 ARTECH HOUSE, INC.
685 Canton Street
Norwood, MA 02062

International Standard Book Number: 1-58053-012-5
Library of Congress Catalog Card Number: 99-045837

10 9 8 7 6 5 4 3 2 1

To Shirley Tyke

Contents

Contents

Preface

The combination of the telephone and the Internet in recent years has unleashed a torrent of innovation that promises to redefine telecommunications in the twenty-first century. The principles of transporting voice over the Internet—as exemplified in the "voice funnel" of the 1980s—were rediscovered in the 1990s when a handful of vendors started selling software to enable those with a multimedia computer and an Internet connection to converse with each other as if by telephone. Word spread fast that for the cost of an Internet account, users could talk for as long as they want—to anyone, anywhere in the world—without having to pay long-distance call charges.

However, establishing a connection entailed cumbersome procedures. Voice quality was hampered by the inherent variable delay of the Internet. Dropped packets caused speech segments to be clipped. Constant adjustments had to be made in an effort to eliminate echo, get the right volume, and optimize performance by choosing an appropriate compression rate that would work best with the modem's current operating speed. There was always confusion about whose turn it was to talk at any given time. These problems prompted many industry analysts to doubt the viability of Internet telephony as anything more than a hobbyist's toy.

The relentless pace of technology improvement changed all that. Advancements in digital signal processor (DSP) technology improved voice quality, new compression techniques squeezed the digitized voice segments to levels that could be more easily routed through the Internet, and standards emerged for enabling different voice products to interoperate. Special gateways started appearing, which joined the Internet to the public switched telephone network (PSTN), enabling ordinary phones to be used for calls over managed IP nets. In fact, the technology has improved to such an extent that today every major telco equipment, PBX, and interconnect manufacturer has incorporated IP telephony into their existing offerings, while competitors are emerging all the time to bring innovative products to this new market.

Regional telephone companies, long-distance carriers, national Internet service providers, and so-called next-generation service providers now offer commercial IP-based voice services over their high-speed fiber backbones. Corporations are adding IP telephony service to leverage their investments in LANs and intranets and save on long-distance call charges between far-flung offices. This and other activity are the result of growing recognition that IP is a more efficient method of transporting voice than the traditional circuit-switched environment of the PSTN.

According to Forrester Research, the consumer market for Internet telephony services will be worth $2.5 billion by the year 2004. However, given the level of capital investment by hardware manufacturers, service providers, and application developers, this is rather a conservative projection. In 2004, the consumer market for Internet telephony services can easily exceed $20 billion. By that time, consumers will not know or care that most of their conversations are being transported over routed IP networks rather than conventional circuit-switched networks. What they will know and appreciate is that they will have access to a range of integrated voice, data, and video services that will cost less than buying them separately today. They will deal with only one service provider, get only one bill, and be able to take the equipment with them when they move and use it at their new location. In essence, the use of proven, standardized, and ubiquitously available Internet technologies for mainstream voice and value-added voice-enabled applications will change the whole communications paradigm in the next century.

A key ingredient in this new paradigm is "convergence." This is the combination of voice, video, and data in the same application or different applications running over the same connection simultaneously. Although this is possible with other technologies such as ISDN, frame relay, and ATM, it can be done more efficiently, economically, and ubiquitously on IP-based networks. With IP, the distinct traffic types are reduced to a single stream of binary ones and zeros, which can be carried by any delivery platform and traverse any environment from LAN to WAN—and at much less cost and complexity than ISDN, frame relay, or ATM.

Internet protocols are flexible enough to overcome the traditional boundaries between voice and data services. Developers can come up with innovative new services that combine different content formats and immediately load them onto the existing IP infrastructure. Because particular services are no longer locked into specific forms of infrastructure, convergence creates new markets and new efficiencies. The competition spawned from these new markets and efficiencies lowers the cost of communications and fuels the continuous cycle of innovation.

This book lays out the blueprint for the new communications paradigm of the twenty-first century. It not only deals with the technologies that make the new paradigm possible, but delves into the benefits, applications, vendor approaches, and regulatory issues that will shape the paradigm. This book is not intended to be an exhaustive overview of every product, service, and technology that has implications for voice-data convergence—the pace of innovation makes that impossible. Instead, this book focuses on mainstream topics that are relevant now and in the near term to provide readers with the "how and why" of voice-data convergence so they will be better prepared to deal with these issues as they arise as possible solutions for the office, in the home, and on the road.

The information contained in this book—especially as it relates to specific vendors, products, and services—is believed to be accurate at the time it was written and is, of course, subject to change with continued advancements in technology and shifts in market forces. Mention of specific products and services is for illustrative purposes only and does not constitute an endorsement of any kind by either the author or the publisher.

Nathan J. Muller

CHAPTER

1

Contents

IP: The once and future protocol

The core suite of TCP/IP (Transmission Control Protocol/Internet Protocol) protocols upon which the Internet is built is 30 years old, yet stands poised to dramatically extend and enhance communications in the twenty-first century. The capabilities that were built into IP to allow the easy exchange of data between different computers interconnected over low-speed lines in an unmanaged environment are even more advantageous when used over well-managed high-speed internets. These advanced networks are comprised of ATM-based switches and routers that are capable of supporting data transmission at gigabits-per-second over self-healing SONET-based fiber optic backbones and feeder links.

Since new applications continually are being added to the public Internet and private intranets—many of which were not envisioned by the original developers of TCP/IP—standards have had to evolve and protocols have had to be developed to keep

1

pace with user demands. The Internet community, notably the Internet Engineering Task Force (IETF), is responsible for accepting, circulating, and facilitating consensus on technical proposals that strive to enhance TCP/IP so it can handle new and emerging applications.

Today, TCP/IP nets are already capable of converging voice and data to handle information distribution, telephone calls, videoconferencing, and interactive gaming, as well as streaming audio/video and multicast delivery. The use of IP makes the global distribution of information and services more efficient and economical than using traditional public switched telephone networks (PSTN), while permitting their integration with legacy data and applications as well. With continued improvements in the performance of Internet protocols and the availability of higher-speed backbones and managed infrastructures, multimedia messaging, telemedicine, distance learning, and on-demand entertainment will become ubiquitously available via TCP/IP nets.

As the vehicle of choice for the convergence of voice, video, and data traffic, IP will characterize the next generation of public networks. Some industry analysts predict that 99% of the world's bandwidth capacity will be used for Internet traffic and all voice traffic will be carried over IP by 2004. This convergence will lead to new economic models for the data and telecommunications industry and will determine which mergers, acquisitions, and partnerships will be formed to exploit the new opportunities that will arise in the emerging networked economy.

1.1   The big deal about IP

IP is the underlying protocol for routing packets over the Internet and other TCP/IP-based networks. Within the context of next-generation networks, IP is important for several reasons. It is the unifying protocol between computing and communications. It provides a solution for merging new technology with legacy infrastructures. It is nonproprietary and open, and it offers efficient, cost-effective ways to merge voice and data traffic on a common platform—even allowing the creation of new applications, such as "shared space," which allows people to collaborate on projects using only a Web browser.[1] IP also provides seamless integration between wireline and wireless networks. Although there are other protocols that offer compelling advantages of their own—notably,

LibQUAL+™

Thank you for your participation!

Questions about the LibQUAL+™ survey? Email us: librarysurvey@utsa.edu

Can't complete the survey online? Stop by the Reference Desk for a paper survey.

Tell us what you think of the quality of our services, facilities, and collections so that we can serve you better.

Library

LibQUAL+™

How are we doing?
UTSA Library wants to
hear from YOU!

Log on to:
http://survey.libqual.org/
index.cfm?ID=330150

Oct 23rd - Nov 17th

Enter your email
address at the end of
the survey to

Win an iPOD!

Or one of these prizes:

- $50-$100 Gift Certificate
- Popular novel from Barnes & Noble
- Bowling & Popcorn party at University Bowl
- 4 tickets to the Botanical Garden
- McDonald's Big Mac
- Chipotle burrito

Survey takes approximately
10 minutes to complete.

Asynchronous Transfer Mode (ATM)—none is capable of matching IP in terms of economy, efficiency, and global reach.

The incumbent carriers still maintain separate voice and data networks. In the new era of voice-data convergence, this is not only costly, it is noncompetitive. It has been estimated by various industry analysts that AT&T, for example, stands to lose between $620 million and $950 million in international calls to IP telephony by 2001 alone, which would account for more than 4% of its total revenue.

Shared IP networks also are more efficient than circuit-switched PSTNs, which rely on dedicated connections that are set up between endpoints. These connections are idle much of the time. During a typical voice conversation, as much as 40% of circuit-holding time is wasted with silent periods. In an IP network, packets are sent out onto the network only when there is voice to be conveyed. This frees the network to handle much more traffic over the available bandwidth. The packetization scheme allows the network to handle any mix of voice and data at the same time over the same links.

IP can be run over other services, such as ATM. Like IP, ATM is highly scalable, and both are capable of supporting mixed traffic—data, voice, video, and other multimedia. Unlike IP, however, ATM offers built-in quality of service (QoS), which improves the delivery of delay-sensitive traffic. For QoS to work over IP, other protocols must be added. Several protocol standards exist—such as RSVP (Resource Reservation Protocol) and RTP (Rapid Transfer Protocol)—but none is universally implemented, except on private intranets where an organization has more control of the routers and switches within its domain. There also are proprietary techniques for dealing with the latency, but again these are better suited for private nets.

The performance problem of IP nets has not gone unnoticed by the IETF. There are new proposals being considered, such as differentiated services (Diff-Serv) and simple integrated media access (SIMA), which define IP classes of service (CoS), but these are not likely to be agreed

---

1. In partnership with Changepoint Corp., U S West offers this capability as a free service called involvFree Web Teaming. It provides an easy way for people to collaborate on projects, regardless of their location using only a Web browser. The basic service provides step-by-step instructions that help users instantly create and utilize an interactive team space on the Web. It includes a personal work space for each team member to create memos, post calendar items, comment in discussions, add Web links, assign tasks, and take polls. More information about this service can be found at http://www.involv.net/.

upon or widely deployed any time soon. Currently, next-generation networks are running IP over ATM on synchronous optical network (SONET) fiber optic links. This architecture provides multiple advantages:

▶ IP provides near-universal connectivity at the lowest cost.

▶ ATM improves network performance by prioritizing traffic (even IP traffic) according to QoS parameters.

▶ SONET provides integral monitoring features that can be used to trigger automated link restoral functions without disrupting service.

A growing number of these so-called next-generation carriers have emerged in recent years that have built high-performance networks that offer all of these advantages. They not only offer more bandwidth than AT&T, MCI WorldCom, and Sprint, they offer advanced voice-data services at a lower price than those charged by the incumbents. Among these next-generation carriers are ITXC, Level 3 Communications, and Qwest LCI. Depending on the specific carrier, they may offer customers leased lines, dark fiber,[2] VPNs (virtual private networks), and IP-based telephone and facsimile services. Eventually, as market demand increases, these networks will support the most advanced integrated voice-data services as well. IP is the only protocol capable of doing it all, and the incumbent long-distance carriers have taken note.

1.2   Incumbent carriers adopt IP

Sensing that the networking landscape is about to change in a big way, AT&T, MCI WorldCom, and Sprint have all established high-speed internets. AT&T and British Telecom (BT) have even teamed up in a $10 billion joint venture to build an intelligent, managed IP-based global network that will provide telephony, electronic commerce, call centers, and new Internet-based applications in support of international organizations and executives on the move. The two companies will spend another $1 billion

---

2. Dark fiber is leased from the provider as is; it is the leasing organization's responsibility to supply the equipment at each end and manage the facility. For example, many companies use their own SONET equipment to implement disaster recovery; that is, move traffic from a cut or failed line to a spare or underutilized fiber link.

in the United States to acquire companies that offer products and services that can be plugged into this new IP infrastructure.

AT&T's interest in the Internet most dramatically manifested itself in the $48 billion acquisition of cable television operator TCI.[3] The long-distance company is using TCI's cable lines to deliver high-speed Internet, IP-telephone, and television services. Many cable operators already offer Internet access and some plan to offer IP telephone service as well—both through the same cable modem. Among other things, the use of cable for telephone calls could give companies low-cost entry into the local services market, enabling them to compete more quickly and effectively against the incumbent telephone companies.

The regional Bell telephone companies are also looking to get into the IP telephony business in some form or another. In September 1998, Bell Atlantic became the first regional telephone company to sign a termination contract with an IP telephony wholesaler. Under the agreement, Bell Atlantic will start accepting calls from ITXC Corp., giving it a "last-mile" connection for international IP calls coming into the United States. Terminating gateways operated by Bell Atlantic translate the IP calls back into the traditional voice format for completion over the local telephone network. Bell Atlantic currently has 40 million access lines to homes and businesses throughout the East Coast. Bell Atlantic is trying to establish a presence in the IP telephony market because it sees voice and data networks rapidly converging and expects Internet-based calls to nearly double every year for the next 5 years.

IP nets may provide the means for smaller providers to compete effectively against the large telephone companies in the provision of local telephone service, without having to invest in local loop infrastructure. In mid-1998, a regional Internet service provider became the first ISP anywhere to receive certification to provide local telephone service as a competitive local exchange carrier (CLEC). The Colorado Public Utilities Commission granted Rocky Mountain Internet Inc. status as a CLEC, enabling it to compete with U S West in the provision of local telephone

---

3. At this writing, AT&T had acquired MediaOne, the third largest cable company in the U.S. If approved by regulatory authorities, the $58 billion acquisition of MediaOne gives AT&T a total of 26 million cable customers, and expands its ability to provide local phone service and Internet access in major markets across the country. Comcast had also tried to buy MediaOne, but abandoned its bid under a separate agreement with AT&T, under which Comcast acquired 2 million cable subscribers from AT&T. Both companies also agreed to eventually cooperate in providing telephony service to their cable subscribers, much like an earlier agreement between AT&T and Time Warner, Inc.

service. The company provides local exchange services in Colorado including private line, switched and dedicated access service, intraLATA toll, and advanced features under the name Rocky Mountain Broadband. The company already provides long-distance IP telephony service to clients throughout the United States through strategic partnerships with such companies as Frontier Corp. (long-distance services) and Vienna Systems (IP telephony).

By mid-1999, every major telco equipment, private branch exchange (PBX), automatic call distributor (ACD), and interconnect manufacturer has added the voice-over-IP (VoIP) capability to its products. The market has also attracted numerous smaller firms and startup companies that offer innovative VoIP products and applications for businesses and consumers. There are about dozen manufacturers of IP-PSTN gateways alone, some of which are addressing the market for IP-based call centers as well. The big three router manufacturers—Cisco Systems, Bay Networks (now owned by Nortel Networks), and 3Com—also support voice over IP. As discussed in Chapter 8, Cisco has even evolved VoIP into an elaborate concept it calls "multiservice networking," which entails a single network infrastructure that converges voice and data services using IP as the common internetworking protocol. The infrastructure combines the best features of the IP and PSTN worlds for substantial savings in both capital expenditure and operational costs.

1.3 Impact of utility companies

With passage of the Telecommunications Act of 1996, utility companies were given the freedom to enter the competitive telecommunications marketplace. With their extensive infrastructures for power distribution, which touch every home and business, they can attract partners for just about any kind of service they want to offer their customers. Their impact to date has not been enough to worry incumbent telephone companies because market entry entails a huge investment of capital.

However, through a new technology—called digital power line (DPL), which was developed in the United Kingdom—utility companies in the United States may soon be able to offer high-speed Internet access to their customers via power transmission lines. DPL delivers data at speeds of up to 1 Mbps—several times faster than current digital services such as basic rate Integrated Services Digital Network (ISDN).

The technology turns the low-voltage signals going between houses and local electricity substations into a local area network (LAN). The substations are then linked by fiber optic circuits to Internet switching points.

Multiple applications can be supported by DPL. In addition to basic Internet access, DPL can support IP telephony, multimedia, smart applications/remote control, home automation and security, home banking and shopping, data backup, information services, telecommuting, and entertainment. DPL also offers permanent, "always-on," connection to online services with the potential for lower charges.

At the customer premises, a standalone unit connects to the power supply by standard coaxial cable. The unit is then connected to the computer by a standard Ethernet cable. It supports laptops as well as PCs and Macs and supports a wide variety of platforms. It also supports the Universal Serial Bus (USB), enabling several machines to be connected to a single DPL box simultaneously to create a LAN in the home or office.

The nature of this technology allows utility companies to roll out service in discrete, targeted phases very economically until consumer demand builds. Utilities not wishing to operate data services themselves have the option of charging a right-to-use fee to an experienced IP service provider for accessing their plant. Regardless of how they participate—actively or passively—the electric utility industry represents the dark horse in the new era of IP convergence. The utility industry now has a powerful technology at its disposal to assume a leading position among next-generation carriers seeking to build market share with advanced IP-based services.

1.4 A closer look at IP nets

A next-generation network is one which provides integrated voice-data services over an economic IP-based infrastructure and also meets the requirements for integration, scalability, reliability, mediation, manageability, security, and global reach. Although the PSTN comes close to meeting many of these requirements, an IP-based next-generation network takes these requirements to a higher level—and at only 20% of the cost of a conventional circuit-switched network of comparable capacity.

1.4.1 Economy

The cost savings on network infrastructure can be accrued in a number of ways. Since many next-generation service providers own the rights of way for their fiber routes, they save on permit fees. Since they are building their networks from scratch, they can deploy the latest dense wave division multiplexing (DWDM) equipment with new, higher-quality optical fiber (discussed later). This combination results in far more network capacity than incumbent carriers have, and at a fraction of the cost to build.

Building networks from scratch has another advantage—it is possible to minimize fiber installation costs and even eliminate them entirely. The Williams Company, which has its roots in running oil and gas pipelines, has installed much of its fiber in abandoned pipelines at a cost far below that of burying it underground. Qwest LCI has done even better. The company started out in 1992 as a construction company that built fiber optic lines for others, most notably GTE Corp., WorldCom, and Frontier Corp. Not only did Qwest use a patented rail-plow fiber deployment process that enabled it to lay down fiber deeper and faster than anyone else, but for every fiber Qwest laid for a customer, it installed another for itself, much of it along rights of way that it owned. In effect, the company's customers—who are also its long-distance competitors—have helped Qwest eliminate most of its network construction costs.

Finally, tremendous cost savings can be achieved by using routers for moving traffic across the network, rather than the central office switches used by local telephone companies and interexchange carriers. Routers are much less complex than switches, yet can be as effective in handling real-time applications such as voice over an IP-based fiber optic network, especially one that is specifically designed to minimize end-to-end delay.

Businesses also can build their own IP network infrastructures and reap substantial cost savings. Instead of relying on leased lines or expensive public services to tie remote offices and mobile employees to the backbone network, a VPN can be carved out of a carrier's managed IP network. In many cases, the performance of the VPN in terms of reliability and availability can be guaranteed by the carrier under a service level agreement (SLA), which will be discussed later. Special monitoring and reporting tools enable network managers to check performance to ensure the carrier is living up to the SLA.

1.4.2 Integration

The integration capabilities of IP are amply illustrated by its ability to provide interoperability between wireline and wireless networks, and its ability to unify traditionally separate applications through plug-in protocols.

1.4.2.1 Network integration

Network integration is not a new concept, but consolidating various traffic types over IP nets offers new economies of scale. Conventional integrated access solutions consolidate voice, data, Internet, and video wide area network (WAN) services through a single access system and shared high-speed access lines. Integrated access systems are essentially WAN gateways that unify the functions of different types of equipment such as channel service units (CSUs), data service units (DSUs), multiplexers, and inverse multiplexers. These WAN gateways optimize the use of the physical infrastructure of the PSTN, enabling users to quickly and economically take advantage of new telecommunication services.

What is new is the use of IP-based services for integrating diverse traffic types for realizing even more cost savings (see Figure 1.1). Sprint's Integrated On-Demand Network (ION), for example, is an IP over ATM solution that carries voice, data, and video traffic for a fraction of the cost

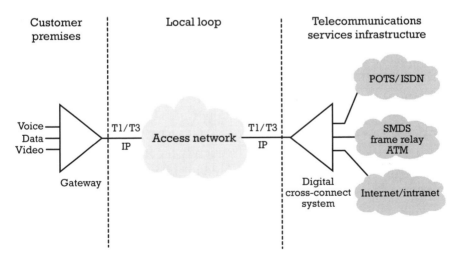

Figure 1.1 An IP-based integrated access topology.

of dedicated services. In addition, ION offers unlimited bandwidth on demand and a single bill for both telephone and data networking. ION eliminates the need for multiple service providers to handle specific applications like voice, data networking, and Internet access. Sprint claims that the use of ATM's cell-based technology lowers the cost of a typical voice call by over 70%. AT&T offers a similar service called Integrated Network Connect (INC), which also concentrates voice and data traffic directly onto an ATM premises device, before pushing it out over broadband local and long-distance links.

A different perspective on network integration involves community antenna television (CATV) and plain old telephone service (POTS). It is the goal among some cable operators to combine television programming, Internet access, and telephone service over the same cable system. Over 60 million households in the United States subscribe to CATV. Since cable connections have higher bandwidths than telephone lines, being able to combine different services on the same medium offers CATV operators the opportunity to offer new services and at less cost.

For example, the CATV infrastructure offers great potential for Internet TV, which combines Web content with TV programming through a set-top box connected to the television. This enables viewers of a news program, for example, to call up related content on the Web via a keyboard for simultaneous display on the television screen through a picture-in-picture (PIP) feature. With some services, such as WebTV, a small icon pops up in the corner of the TV screen when there is relevant information on the Web about the show the viewer is watching—whether it is a sitcom, a documentary, or a soap opera.

Some systems, WebTV included, support chat rooms and e-mail as well. In fact, many news programs routinely inform viewers where they can go on the Web for more information related to their news stories and invite viewers to e-mail their opinions or take part in polls. They also have chat rooms on current events and frequently report on what is being said there. The same technology can be used to enhance the home shopping experience.

One of the best known services of this type is provided by WebTV Networks, Inc., a subsidiary of Microsoft. The service is not a fully integrated offering, however. Programming is received via cable, over-the-air broadcast, or direct broadcast satellite. Web content is requested and delivered through a phone line connected to a 56-Kbps modem.

A few years ago, it looked as if telephone service, television programs, and Internet access might be available as an integrated service provided over the same cable used to bring TV programming to the home. The reason such integration is not generally available today is that the CATV networks must be upgraded to support two-way communication. Of course, this is a very expensive proposition. Several architectures have been tried in an effort to support integrated services more economically, including fiber-to-the-curb (FTTC), hybrid fiber/coax (HFC), and numerous iterations of the two. To date, however, not many cable operators have made commitments to any of these approaches.

The idea of providing integrated services over cable is not dead yet. It may get a shot in the arm from the PacketCable initiative, a project conducted by Cable Television Laboratories (CableLabs) and its member companies. The project is aimed at identifying, qualifying, and supporting Internet-based voice and video products over cable systems. These products will represent new classes of services utilizing cable-based packet communication networks. New service classes include telephone calls and videoconferencing over cable networks and the Internet. The services would be delivered using IP.

Unlike the PSTN, IP voice will be delivered through a cable modem connected to an HFC network. Subscribers would receive an audible dial tone when placing a call. A busy signal or message would be returned if the receiving station is offline or engaged in another call. Audible ringing would be provided once the private virtual circuit has been established to the receiving station. In instances where the terminal device is a standard telephone set, the system will be able to produce ringing for incoming calls.

The IP voice product envisioned by CableLabs should perform with round trip network latencies not to exceed 200 ms. The backbone connection for the IP voice product would be a private ATM backbone functioning as a private virtual network capable of originating and terminating calls to stations similarly connected on the shared medium and to and from stations connected to the PSTN.

The IP voice system would support call features such as call waiting, call forward, call hold, call transfer, conference, and speed dial as well as such ancillary services as inbound and outbound facsimile. It is envisioned that these and other features would be provided by application servers used in conjunction with directory and connection servers in the network.

While the CableLabs approach appears very promising, an alternative is available right now. Digital subscriber line (DSL) technologies offer telephone companies and their competitors a very inexpensive way to increase the bandwidth capacity of existing local loops. One local loop upgrade technology in particular, asymmetrical digital subscriber line (ADSL), allows for the transmission of about 6 to 7 Mbps downstream (central office to customer) over existing twisted-pair copper wiring and 640 Kbps upstream (customer to central office). This is enough bandwidth to support multimedia—video, audio, graphics, and text—over the same lines that already provide POTS.

ADSL does this by carving up the local loop bandwidth into several independent channels suitable for any combination of services. The electronics at both ends compensate for line impairments, increasing the reliability of high-speed transmissions and reducing the carriers' support costs.

Several regional telephone companies have announced their intention to offer integrated services over DSL-enhanced lines. U S West, for example, intends to use several DSL technologies to offer its customers high-speed Internet access and digital cable television service delivered over their phone lines and fully integrated with their phone service within the next five years. Customers will even be able to have telephone Caller ID information flash on the bottom of their TV screens when someone calls them on the phone. Data transmissions will be taken off the voice network at the wire center switch and carried over U S West's separate dedicated ATM backbone.

Network integration can also be viewed from the perspective of uniting wireline and wireless networks. Here again, IP plays a key role. Wireless IP, also known as cellular digital packet data (CDPD), is a standard for providing data service over cellular voice networks at speeds of up to 19.2NKbps. The media used to transport data consist of the idle radio channels typically used for Advanced Mobile Phone System (AMPS) cellular service.

The CDPD infrastructure employs existing cellular systems to access a backbone router network that uses IP to transport user data. Personal digital assistants, palmtops, and laptops running applications that use IP can connect to the CDPD service and gain access to other mobile users via messaging services or to corporate computing resources that rely on wireline connections. Typical applications of CDPD include mobile office, financial transactions, telemetry, public safety, transportation, and Internet access.

CDPD blends digital data transmission, radio technology, packetization, channel hopping, and packet switching. With encryption and authentication procedures built into the specification, CDPD offers effective protection against eavesdropping. As with wireline networks, users can also customize their own end-to-end security for added levels of protection.

CDPD employs digital modulation and signal processing techniques, making transmissions more reliable and more resistant to radio interference than analog transmission technologies. The digital signals are broken down into a finite set of bits, rather than transmitted in a continuous waveform. When signal corruption occurs, error-detection logic at the receiving end can reconstruct the corrupted digital signal using error-correction algorithms. Digital technology also enables the use of processing techniques that compensate for signal fades without requiring any increase in power.

In accordance with IP, the data is packaged into discrete packets of information for transmission over the CDPD network. In addition to addressing information, each packet includes information that allows the data to be reassembled in the proper order at the receiving end and corrected if necessary.

Channel hopping automatically searches out idle channel times between cellular voice calls. Packets of data select available cellular channels and go out in short bursts without interfering with voice communications. Alternatively, cellular carriers may also dedicate voice channels for CDPD traffic to meet high traffic demand. This situation is common in dense urban environments where cellular traffic is heaviest and the channel-hopping mode is difficult to use.

The CDPD overlay network may use either OSI's Connectionless Network Protocol (CLNP) or IP at the network layer. These protocols have virtually the same functionality: they both interpret device names to route packets to remote locations. The inclusion of IP in the CDPD specification is intended to accommodate the vast number of networked devices already using it. IP support also ensures that existing applications software can be used in CDPD networks with little or no modification.

The drawback to CDPD service is that it is not as ubiquitous as standard analog cellular services. However, a group of six companies has come up with a specification that expands coverage nationwide by creating a gateway between CDPD and circuit-switched cellular. The group consists of Ameritech Cellular, AT&T Network Systems, Bell Atlantic Mobile, GTE Personal Communications Services, Isotel Research, and Nynex Mobile Communications. Essentially, the gateway allows customers to access the CDPD network from anywhere in the country—either through the cellular network or via a landline connection.

Linking the two types of cellular networks could prove critical to the future of CDPD. It effectively gives cellular service providers a way to deliver nationwide CDPD—even though the digital packet transport is now up and running in 3,000 cities in the United States. Once users access the CDPD system, they can take advantage of all its special features, including security, customized billing, and access to the Internet.

Another feature offered by CDPD is mobility management, which routes messages to users regardless of the location or the technology. With the gateways in place, the cellular network will have the capability to recognize when subscribers move out of the CDPD coverage area and transfer messages to them via circuit-switched cellular.

Another aspect of wireless-wireline network integration comes in the form of satellite networks that support IP. In striving to reach the goal of true interactive, multimedia television through the Internet, Hughes Network Systems, a unit of Hughes Electronics, offers DirecPC, a service that provides satellite-to-PC Internet access. The DirecPC package includes a 21-inch satellite dish, 100 feet of coaxial cable for connecting the dish to a PC, a 16-bit ISA bus card adapter for the PC, and Windows access software. The user connects to the ISP with a modem and issues requests in the normal fashion, but data is returned via satellite at 400 Kbps, a rate 14 times faster than that of a 28.8 Kbps modem connection.

When the user requests the desired file from a remote Web site, the return address of the user's PC is automatically attached, telling the remote server where to send the requested file. At this point, the DirecPC software intercepts the request for the file before it leaves the user's computer and adds an address header for the DirecPC Operations Center. The request is then received by the DirecPC Operations Center and forwarded to the remote site's server. The remote server then sends the requested file to the operations center instead of to the user's computer. The Operations Center then receives the file and transmits it to the DirecPC satellite. A satellite transponder receives the signal and, in turn, beams it back to earth, where it is received by the user's satellite antenna. The signal then travels from the antenna to the DirecPC adapter installed in the user's PC. Since the file contains the return address of the PC, only that PC is capable of receiving the requested file.

Other IP-based satellite services are emerging. Among the most eagerly anticipated satellite networks is Teledesic, which is being built by Boeing. Teledesic is among several emerging low earth orbit (LEO) satellite networks that will provide ubiquitous personal communications services, including telephone, pager, two-way messaging, and Internet access. In addition to supporting direct wireless communication between users, many of these satellite networks will also interconnect with next-generation IP-based fiber optic networks for linking enterprise LANs, providing broadbandet access, and supporting videoconferencing and interactive multimedia.

LEO satellites circle the earth at about 600 miles up. Since their low altitude means that they have non-stationary orbits and they pass over a stationary caller rather quickly, calls must be handed off from one satellite to the next to keep the session alive. The omnidirectional antennas of these devices do not have to be pointed at a specific satellite. There is also very little propagation delay. And the low altitude of these satellites means that earthbound transceivers can be packaged as inexpensive low-powered, handheld devices.

When fully deployed in 2002, the Teledesic's network will consist of 288 LEO satellites, divided into 12 planes, each with 24 satellites. Teledesic will have a pair of 500 MHz channels that will operate in a portion of the high-frequency Ka band—28.6 to 29.1 GHz on the uplink and 18.8 to 19.3 GHz on the downlink. Gateways and earth terminals on the ground will provide interconnection with wireline networks, enabling the satellite network to interconnect with carrier networks, including next-generation IP networks.

1.4.2.2 Applications integration

The application integration capabilities of IP are illustrated by the concept of universal messaging, whereby all voice and data messages are consolidated in one inbox, regardless of source. This obviates the need to continually check separate devices, applications, and services to retrieve all messages. In addition to user convenience, universal messaging simplifies back-office administration, which otherwise entails multiple databases, platforms, and training. The success of IP-based universal messaging depends on standards so the products of different vendors can interoperate. The Voice Profile for Internet Mail (VPIM), developed by a working group of the Electronic Messaging Association (EMA), fulfills this role.

The VPIM specification is basically a Multipurpose Internet Mail Extensions (MIME) profile that enables voice-mail servers to encode and exchange messages via Simple Mail Transfer Protocol (SMTP) with any other SMTP/MIME-capable server, over an IP network.

With VPIM-enabled products, users can send and receive messages with the same ease and simplicity as electronic mail. Instead of having an e-mail address, however, users are found through their normal phone number. A voice-mail user, for example, can record a message and enter the recipient's phone number. The voice-mail system on a corporate LAN (or carrier network) recognizes that the address is not one of its internal users and performs a directory lookup to a public directory, using the Lightweight Directory Access Protocol (LDAP) to find the Internet e-mail address assigned to handle voice messages for that individual. The system converts the message into a common MIME attachment and routes it through SMTP to the receiving voice-mail system, which delivers it to the recipient.

The VPIM standard also offers easier integration between the different forms of messaging—voice mail, video clips, e-mail, and fax. Because these types of messages are sent as a MIME attachment, all kinds of messaging systems, such as conventional e-mail clients, can accept it. E-mail clients and messaging servers that are equipped with audio/video capabilities can decode and play the message without having to first send it to a voice-mail system or a computer's audio/video player application.

Several voice-mail system vendors already provide mechanisms for combining the delivery of e-mail, voice-mail, and fax from a single mailbox. With Centigram Communications' VPIM-compliant OneView, for example, an InBox window collects all voice, fax, and e-mail messages in one place, showing the sender's name, the subject of the message, the time it was sent, and whether it is urgent (see Figure 1.2). It allows users to make, send, receive, play, forward, and store messages with a click of the mouse—without having to pick up a phone or head for a fax machine. To increase productivity, the visual interface allows users to quickly prioritize messages and respond to the most important ones first. Voice

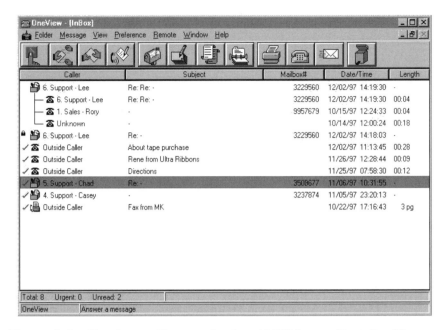

Figure 1.2 Centigram Communications' VPIM-compliant OneView unifies all types of messages—fax, e-mail, and voice—on a single computer.

messages are converted to .WAV files for playback on multimedia desktop or notebook computers.

The inclusion of video and images in messages also is addressed in the VPIM standard. The EMA offers a Common Messaging Call (CMC) application programming interface (API) that can be used by developers to build video-enabled messaging applications. Stream functions are included in the CMC interface to facilitate the reading and writing of large-content information, including video in the MPEG (Motion Picture Experts Group) format. The API can also be used to build messaging applications that incorporate images in the form of JPEG (Joint Photographic Experts Group) and GIF (Graphics Interchange Format). MPEG, JPEG, and GIF are all standard MIME types supported on IP networks, including the services that run over them, such as the World Wide Web.

Next-generation networks based on IP are superior to the PSTN for voice and data convergence because they include all the advantages of interoperability and integration afforded by internationally recognized Internet standards. Although PSTNs also adhere to internationally recognized standards for the transmission of voice, as well as for the signaling systems used for circuit management, the standards do not address applications other than voice to the extent that the Internet community does.

This may change in the near future as more carriers seek to leverage their investments in intelligent network (IN) platforms to do such things as interface e-mail, voice-mail, and wireless systems. Among the fast-sticking and addictive applications that are possible now are:

▶ Call center with rebound lets an agent call from within a voice-mail system to the person who left the message.

▶ Seamless messaging lets a caller reply to messages via a voice system.

▶ Single-number retrieval lets a caller access messages via one number.

Even with the sophisticated features possible over IN platforms, next-generation service providers eventually could match these offerings, including the wireless component, at less cost and better pricing.

1.4.3 Scalability

A network that easily scales has become a fundamental business requirement in recent years. It entails the ability to expand (or contract) the network as business needs change without making extensive changes to hardware and software. From a network perspective, there are several related scalability issues that are being addressed by next-generation service providers: users, gateways, routers, and capacity. Businesses wishing to deploy IP-based PBXs (private branch exchanges) and ACDs (automatic call distributors) must be concerned with scalability as well. In addition, scalability applies to certain applications that are run over public and private IP networks, such as OLAP (online analytical processing).

1.4.3.1 Users

In terms of users, scalability is illustrated by the expansion potential of a VPN. The closed user group capability provided with the VPN service lets businesses limit incoming or outgoing voice and data calls to only members of the group. At the same time, businesses can quickly and efficiently add (or remove) users to their IP services without having to install remote systems such as access servers and modem pools.

Furthermore, the built-in security features of the VPN service—usually packet filters applied to the source address—allow for dial-in and dedicated access without the need for expensive firewalls on company premises. When a data packet reaches a VPN access router, it is checked against a table of authorized source addresses. If there is no match, the packet is discarded. Similar filters are applied to connection points between the VPN service provider's IP backbone and the public Internet to ensure that only authorized traffic originates and terminates on the service provider's IP backbone.

The same packet filtering capabilities of the VPN can enable businesses to extend their reach to allow users within other organizations to have access to certain data sources and applications. With an IP-based infrastructure, companies can seamlessly go from being an intranet to an extranet using the same development tools, systems, and services.

For example, a business can offer extranet service to its strategic partners which gives their users access to its warehouse inventory and an electronic data interchange (EDI) application that allows them to order and pay for selected items online, as well as track the status of order fulfillment. Assuming that the VPN provides effective security, extending this extranet further to include other strategic partners is basically a matter of providing them with an appropriate IP address and reconfiguring the packet filters to grant them access to specific network resources.

Adding and dropping phone users, remapping numbers, and dynamically assigning phone numbers are all easier on an IP network than on traditional voice switching systems. IP address assignment and configuration may be accomplished using the Dynamic Host Configuration Protocol (DHCP), a standard in the data network environment for configuring and managing various IP devices. DHCP can also be applied to IP phones connected over the LAN, making setup and address management virtually automatic.

For example, DHCP allows IP phones to be moved and plugged in anywhere on the IP network without manual database adjustments or wiring changes. The phone's RJ45 connector is simply removed from the wall jack in one office and inserted into the wall jack at the new location. The IP phone will automatically boot and reregister with the server running the VoIP management application, at which time the phone's IP address is assigned from the pool of available addresses.

While the IP phones can communicate with each other directly over the LAN, the call management server comes into play only when supplementary or enhanced services are required, such as call forward, multiparty conference, multiple line appearances, automatic route selection, speed dial, and last-number redial. New features added to the server are automatically available to all IP phones on the network. In terms of scalability, the number of possible IP phones increases as the network infrastructure grows. Call management servers can be added and internetworked to keep pace with user growth and provide redundancy.

1.4.3.2 Gateways

Scalability also applies to gateways, which must accommodate incremental upgrades to handle increasing traffic and processing demands. Until relatively recently, one significant barrier to offering telephony services to a mass market via next-generation IP networks had been gateway scalability. There is the requirement for IP-PSTN gateways that are comparable in capacity to today's central office switches, which are capable of supporting not merely tens or hundreds of concurrent sessions, but tens of thousands.

The gateway scalability problem is being addressed by the major central office switch vendors as well as customer premises equipment (CPE) vendors. Nortel's IPConnect product line, for example, is a family of IP gateways that provides an interface between a managed IP network and the traditional voice network. IPConnect supports the delivery of business-class enhanced services over a variety of deployments, ranging from small-office 24-port installations to carrier-class installations capable of carrying 100,000 VoIP connections.

1.4.3.3 Routers

Of course, scalability applies to more than just IP-PSTN gateways; it applies to any type of network device, especially routers. Not only are next-generation service providers using fiber optic networks for high speed and reliability, they are increasing router capacity to handle more traffic and adding more routers to eliminate excessive router hops that contribute to delay.

Current router capacity is becoming CPU-limited. In other words, the current capacity is less than 500 kilopackets per second, and three to six OC-3 trunks running at 155 Mbps each is enough to congest a router. This leads to excessive packet loss and, of course, this diminishes throughput. The problem gets worse as more router hops are involved in a TCP/IP session. As networks scale in size and logically interconnect, the span of the network also increases, which magnifies performance problems.

The elimination of excessive router hops can be achieved by employing high-speed switching or routing technologies in the core of the network and invoking the routing function only at its edges—that is, at the entry and exit points of the network. By reducing the number of hops to a fixed number—perhaps as few as two—performance gets much more predictable, regardless of the size of the network. The high-speed technologies used to accomplish this will include ATM, emerging gigabit/terabit switch-router technologies, and the use of new routing techniques such as tag or label switching, which speed up routers by compressing the information they need to determine where to send traffic. Combined, these methods can result in a 10 to 20 times increase in packet throughput, effectively eliminating the router CPU bottleneck and overcoming the effects of network scaling.

1.4.3.4 Bandwidth capacity

With regard to capacity, the network must be able to handle baseline and peak load traffic, as well as accommodate long-term increases in traffic. Next-generation IP networks meet the capacity needs of businesses through the use of optical fiber and by deploying capacity-enhancing technologies.

For example, when a photonic technology called DWDM is applied to SONET links or rings, transport capacity can be increased by a factor of 10, without the carrier having to lay any additional fiber optic cable. The use of SONET and DWDM is important for another reason—they go a long way in overcoming a point of contention about ATM's inordinately high overhead relative to its cell size.

ATM's cell structure is fixed at 53 octets, of which 5 are reserved for the header and 48 for data. This translates into roughly 10% of the traffic being devoted to overhead functions, which becomes a cost issue with incumbent carriers. By contrast, IP over SONET incurs only 2% overhead.

With DWDM to increase fiber's already high capacity, ATM's overhead is no longer a serious issue. Instead, the focus is on ATM's unique ability to provide an appropriate quality of service for any given application, even IP-based applications, if IP is carried over ATM.

Interestingly, DWDM only works on the newer grade of optical fiber being installed by the next generation network service providers. These carriers all rely on nondispersion-shifted fiber, which can support up to 80 channels on a single hair-thin strand. The incumbent service providers—AT&T, MCI WorldCom, and Sprint—use mostly dispersion-shifted fiber, which suffers from the problem of signal bleeding. This problem limits the older fiber to eight channels per strand, which are derived from the older wavelength division multiplexers (WDM).

1.4.3.5 PBXs/ACDs

The same scalability problem that confronted first-generation IP-PSTN gateways applies to the first generation of IP-based PBXs and ACDs as well—in addition to providing comparable features and functionality as conventional systems, they must be able to accommodate incremental upgrades to handle increasing traffic and processing demands.

Vendors such as Selsius, Inc. (acquired by Cisco Systems in 1998) offer scalable IP-based PBXs that can transport intraoffice voice over an Ethernet LAN and wide-area voice over the PSTN or a managed IP network. Full-featured digital phone sets link directly to the Ethernet LAN via a 10BaseT interface, without requiring direct connection to a desktop computer. Phone features can be configured using a Web browser. Existing analog devices, such as phones and fax machines, can be linked to the LAN via a gateway. In addition to IP nets, calls can be placed or received using T1, primary rate interface (PRI) ISDN, or traditional analog telephone lines.

All the desktop devices have access to the calling features offered through the IP PBX management software running on a LAN server. The call management software allows up to 1,500 client devices on the network, like phones and computers, to perform functions such as call hold, call transfer, call forward, call park, and calling party ID. Even sophisticated PBX functions such as multiple lines per phone or multiple phones per line are performed by the call management software.

In addition to the scalability of the system, which spans from 2 to 1,500 users with a single server running the call management software, the Selsius IP PBX is rated to deliver in excess of 75,000 busy hour call completions. This provides a truly linear PBX that can keep pace with corporate growth through the deployment of additional phones and gateway ports, while accommodating substantial increases in traffic.

Other vendors, such as Lucent Technologies and Nortel, offer IP interfaces to their conventional PBX systems. Lucent, for example, offers an IP trunk interface for certain models of its DEFINITY product line. The IP interface supports 24 ports and allows businesses to integrate least cost routing and class of service features, giving network managers the ability to add the Internet and intranets as alternative routes for voice and fax services. The ability to add the IP trunk interface directly into the DEFINITY also reduces the cost of obtaining Internet telephony capabilities by eliminating the need to buy a separate IP-PSTN gateway. Nortel also offers a 24-port IP interface for its Meridian communication system, enabling the routing of real-time voice and fax calls over IP data networks, rather than the PSTN.

Another approach to interconnecting PBXs over IP nets is provided by PSINet's Voice iPEnterprise, a fully managed service that delivers internal voice traffic over the same connection as corporate Internet data (see Figure 1.3). This service leverages existing investments in PBXs and uses a T1 dedicated connection to the PSINet network, allowing businesses to carry voice services between their offices.

PSINet's managed service is equal in quality to traditional tie-line or dedicated voice services between PBXs. The service is aimed at companies with branch offices, especially those that must scale rapidly due to internal growth or operate in widely dispersed geographic locations. The service could save corporate customers 50% over traditional carrier services. IP desktop fax, conference calling, and unified messaging capabilities can be added to the service.

IP-based ACDs are scalable as well—usually able to serve up to 50 agents in a call center environment. These devices allow call center agents to handle and manage Internet/intranet calls with traditional tools such as hold, retrieve, transfer, and conference. One vendor, PakNetX, offers an IP-based ACD that integrates audio, video, and data as one contact, offering a new level of personalized customer service over the Web (see Figure 1.4).

Callers reach the customer service agent by clicking on a "Connect Me" button on the company Web site, by calling the company call center

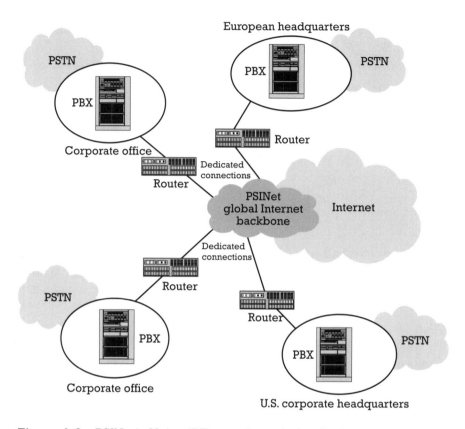

Figure 1.3   PSINet's Voice iPEnterprise solution for interconnecting corporate PBXs over the carrier's managed IP backbone.

with a standard Internet phone, or by calling with a traditional phone connected to an Internet telephony gateway. The PNX ACD software queues and delivers the incoming call to the next available agent who can then provide the customer with video, Web browser sharing, text chat, file transfer, and data collaboration services using H.323/T.120 compliant tools. For PSTN callers, an Internet telephony/PSTN gateway can be used in conjunction with the PakNetX solution to bridge the callers to the PNX ACD.

While a caller is waiting to be connected to a live agent, streaming audio and video are pushed to the company's Web page, which provides opportunities for cross-selling and education. In addition, PNX ACD provides Internet telephony/PSTN gateway support for mixed network

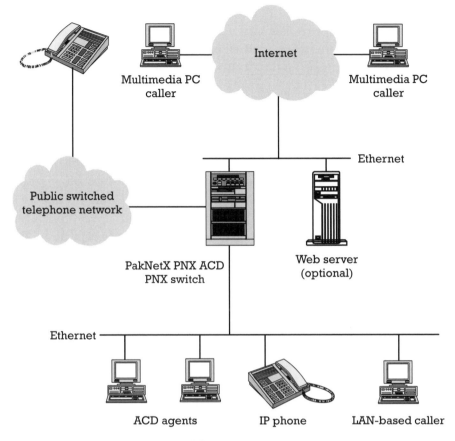

Figure 1.4   PakNetX's PNX ACD topology.

topologies, integrated firewall functions, and computer-telephony integration (CTI) interfaces for use with existing telephony environments.

1.4.4   Reliability

There are at least two aspects of reliability that are being addressed by
next-generation networks: the reliability of applications and the reliability of the physical infrastructure.

1.4.4.1   Applications reliability

Voice and other real-time applications have their own unique set of
challenges that impact reliability. While it is technologically easy to pipe

real-time applications over IP networks, it does not mean that this will result in a viable service. The service provider must be able to design and manage the network to control such performance parameters as latency, throughput, and availability. Of these, controlling latency is the most problematic.

Latency is the amount of delay that affects all types of communications links. Delay on telecommunications networks is usually measured in milliseconds (ms), or thousandths of a second. A rule of thumb used by the telephone industry is that the round-trip delay for a telephone call should be less than 100 ms. If the delay is much more than 100 ms, participants think they hear a slight pause in the conversation and begin speaking. But by the time their words arrive at the other end, the other speaker has already begun the next sentence and feels that he or she is being interrupted. When telephone calls go over satellite links the round-trip delay is typically about 250 ms, and conversations become full of awkward pauses and accidental interruptions.

Latency affects the performance of applications on data networks as well. On the Internet, for example, excessive delay can cause packets to arrive at their destination out of order, especially during busy hours. The reason packets may arrive out of sequence is that they can take different routes on the network. The packets are held in a buffer at the receiving device until all packets arrive and are put in the right order. While this does not affect e-mail and file transfers, which are not time sensitive, excess latency does affect real-time multimedia applications and certain legacy applications such as Systems Network Architecture (SNA). With a videoconference, for example, too much delay causes voice and video components to arrive out of synchronization.

If the packets containing voice or video do not arrive within a reasonable time, they are dropped. When packets containing voice are dropped, a condition known as clipping occurs, which is the cutting off of the first or final syllables in a conversation. Dropped packets of video cause the image to be jerky. Excessive latency also causes the voice and video components to arrive out of synchronization with each other, causing the video component to run slower than the voice component; for example, in a videoconference, a person's lips will not match what he or she is really saying.

The problem of latency on IP networks can be addressed by resource management protocols, such as RSVP. This Internet Engineering Task

Force (IETF) standard improves the QoS by making enough bandwidth available on a priority basis end-to-end to support telephony or multimedia applications.

There are also many proprietary techniques for dealing with latency. Among them is priority output queuing, a technique used by Cisco Systems in its routers. With this technique, network managers can classify traffic into four priorities and provide the available bandwidth to the queues in the order of their priority. The highest priority queue gets as much bandwidth as it needs before lower priority queues get serviced.

Alternatively, IP can be run over ATM with a QoS assigned to the application, which identifies it as being time sensitive and requiring priority over other, less time sensitive, data types. QoS can be handled by the network (i.e., routers, hubs, switches), the operating system, or a combination of both hardware and operating system working together.

Potentially, IP-based next-generation networks are able to offer higher quality voice than traditional circuit-switched networks, which are engineered for 64 Kbps signals within a frequency range of about 3.1 KHz. By contrast, the bit rate and frequency range of next-generation networks are limited only by the capabilities of the devices and transducers at the communications endpoints. These inherent advantages can be leveraged to deliver two-way voice communications within a frequency range of about 7 KHz, which is more than two octaves better than that supported by conventional circuit-switched voice (see Figure 1.5).

This level of audio quality also greatly enhances integrated audio-video applications, such as distance learning, telemedicine, videoconferencing and collaborative work sessions, the playback of archived television programs, online advertising, and interactive gaming. It could also be useful for certain electronic commerce applications.

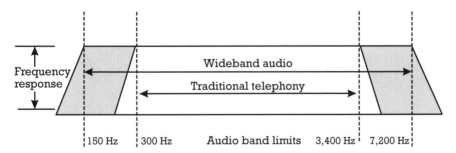

Figure 1.5   A comparison of narrowband and wideband voice.

For example, potential customers could access content from music CDs and video DVDs online before choosing to buy, or preview new movie releases before spending time and money at a local theater. Although such sampling is routinely available today over the Internet, the user experience can be greatly enhanced with a wideband capability that incorporates the latest audio, acoustic, and digital signal processor (DSP) technologies. This would allow the potential customer to evaluate audio-video products based on technical quality as well as content.

1.4.4.2 Infrastructural reliability

The reliability of the physical infrastructure is determined by the quality of the optical fiber and the survivability of the links in case of disaster.

As noted, the use of newer nondispersion-shifted fiber not only supports many more channels per strand than older dispersion-shifted fiber, it offers a lower error rate as well because it does not suffer from the problem of signal bleeding.

The capacity of nondispersion-shifted fiber is illustrated by the Macro Capacity Fiber Network of Qwest LCI. When it becomes fully operational in 1999, this network is able to transmit two trillion bits of multimedia information per second—the equivalent to transmitting the complete contents of the Library of Congress across the country in 20 seconds. In terms of errors, it is estimated that nondispersion-shifted fiber offers less than one bit of error in every quadrillion bits—the equivalent of one grain of sand being out of place on a 20-mile stretch of beach.

However, it is not enough for a network to have great capacity and offer error-free transmission—it must be able to survive virtually any type of disaster. SONET's built-in management functions enable link integrity to be continually monitored in real time and on a nonintrusive basis. Thus, when a link experiences even the slightest performance degradation, a sequence of events is triggered automatically that will result in the transfer of traffic to another link without end users being aware that anything different has happened. Typically, the cut-over takes about 150 ms to implement.

1.4.5 Mediation

Another advantage of IP is that it is able to translate or mediate whatever form of information it receives, regardless of what medium (copper wire, optical fiber, wireless) or service (native IP, frame relay, ATM) it may run on. Some examples of the types of network situations mediation handles include:

- ▶ Incompatible client/server software;

- ▶ Incompatible LAN environments (i.e., Windows and UNIX);

- ▶ Different communications systems (i.e., IP and POTS);

- ▶ Different QoS mechanisms (i.e., IPv4, IPv6, and ATM);

- ▶ Different bandwidth capabilities (i.e., caching, mirroring, load distribution).

This mediation capability was built into the TCP/IP protocol suite from the start. As a platform-independent set of standards, TCP/IP bridges the gap between dissimilar computers, operating systems, and networks. It is supported on nearly every computing platform, from PCs, Macintoshes, and UNIX systems to thin clients and servers, legacy mainframes, and the newest supercomputers. In supporting both local and wide area connections, TCP/IP also provides seamless interconnectivity between the two environments.

Although ATM also provides seamless interconnection between local and wide area networks, it is far more expensive to implement than TCP/IP, especially at the desktop where each machine must be equipped with a special network interface card and middleware to make existing applications "ATM aware." Without this awareness, existing applications will not be able to take advantage of ATM's most powerful feature—standards-based QoS. Of course, the applications can be run and metered at the server, in which case only that device needs to have the middleware and an ATM interface for server-to-server or WAN interconnectivity. This approach enables users to obtain the type of connection that best matches the requirements of their applications, such as:

- ▶ Constant bit rate (CBR)—Provides a fixed virtual circuit for applications that require a steady supply of bandwidth. This QoS is suited for delay-sensitive applications such as videoconferencing and IP telephony.

- ▶ Variable bit rate (VBR)—Provides enough bandwidth for bursty traffic such as transaction processing and LAN interconnection, as long as rates do not exceed a specified average.

▶ Available bit rate (ABR)—Makes use of available bandwidth and minimizes data loss through congestion notification. This QoS is suited for routine applications such as e-mail and file transfers.

A possible drawback to this approach is that it may entail changing the way many users are accustomed to working in that it increases their dependence on the server and forces them to wait if there is no license available for the application they want to use. Despite this potential inconvenience, the client-server approach is gaining in popularity among large organizations because it simplifies management and administration.

On the WAN, carriers will use ATM on their fiber optic backbones. Some carriers, like AT&T, will use ATM for their on-ramps as well. Instead of forcing customers to use separate access lines for different types of traffic, the approach entails placing an ATM switch on the customer premises, which will consolidate voice, frame, and IP traffic on the same access line. This allows multiple traffic types to be fed into the switch in their native formats where they will be assigned an appropriate QoS.

1.4.6 Management

In an effort to match the performance of PSTNs, next-generation service providers rely on managed IP backbones to support time-sensitive applications. The public Internet, which has no central point of management, suffers from delays and congestion that often degrade the performance of real-time applications. To ensure the peak performance of their IP networks end to end—even across different carriers' transport facilities—next-generation service providers as well as incumbent carriers like AT&T are turning to an international standard that enables multivendor network elements to be overseen from a single platform.

To ensure common operational capabilities and a common architecture, the International Telecommunications Union (ITU) Telecommunication Management Network (TMN) standard architecture is being pursued. Among other things, the OSI-based TMN specifies the use of standard interfaces through which internal network management functions are made available. This allows groups of service providers to enter into business-level agreements and deploy resource-sharing arrangements that can be administered automatically through interoperable interfaces. This is of particular value to next-generation network operators because it allows them to extend the reach of their networks globally through strategic partnerships rather than physically lay more fiber optic cable. It also allows them to provide a consistent level of performance end to end so that they can attract new customers through service level guarantees.

By making all internal network management functions available through standardized interfaces, the TMN architecture allows service providers to achieve more rapid deployment of new services, both domestic and international, and make maximum use of automated functions. Network equipment vendors can offer specialized management systems known as element managers which can integrate readily into a service provider's larger management hierarchy. In turn, this architecture allows service providers to align the development of their operations support systems with current and future transport technologies in a way that is disassociated from the vendor-specific aspects of network element implementation.

As networks worldwide become more advanced and as service providers continue to engage in mergers and alliances in an attempt to better serve customer needs, adopting the interoperable management solution offered by TMN standards becomes a key factor in determining their market success.

The use of Simple Network Management Protocol (SNMP) will continue long into the future, but SNMP networks will be integrated into the secure, object-oriented, distributed TMN network when necessary via gateways. Such a gateway is offered by Sun Microsystems, for example. The company's Solstice TMN/SNMP Q-Adaptor software acts as a gateway between the TMN world and the SNMP world, translating OSI protocols to SNMP protocols and vice-versa. To the TMN manager above it, the Solstice TMN/SNMP Q-Adaptor is an OSI Common Management Information Protocol (CMIP) agent. But to the network devices below it, it acts as an SNMP manager. This CMIP/SNMP mapping is defined by standards issued by the Network Management Forum.[4]

---

4. The standards are NMF: Forum 026, Translation of Internet MIBs to ISO/CCITT GDMO MIBs; and NMF: Forum 028, ISO/CCITT to Internet Management Proxy.

1.4.7 Security

All types of networks are vulnerable to attacks of one kind or another. In the voice world, the telecommunications industry has been dealing with security breeches like toll fraud for decades, starting with "phone phreaking," the term given to a method of payphone fraud that originated in the 1960s which employed an electronic box held over the phone's speaker. When a user was asked to insert money, the electronic box played a sequence of tones, indicating to the billing computer that money had been inserted. Since then, theft-of-service techniques have been applied to mobile phones as well. The cost of toll fraud ranges from $3.5 to $5 billion a year in the United States alone.

In the data world, not only is there the potential for theft of service, but also theft of intellectual property. A joint survey by Computer Security Institute, Inc. (CSI) and the Federal Bureau of Investigation (FBI), for example, shows that "cybercrime" is on the rise and that security-related financial losses totaled $137 million in 1997. This figure may actually be as high as $7 to $10 billion a year, since most computer break-ins are not reported. Many companies are afraid that reporting such incidents will expose them to litigation and further financial losses.

Security solutions for the Internet and corporate intranets include encryption, tunneling, certificate services, route authentication, automated intrusion detection and reconfiguration, and virus scanning. These security features are usually implemented by proxy servers, firewalls, routers, and remote access servers—alone or in combination. As the circuit-switched and packet-switched worlds converge in next-generation networks, the best security features from both will likely be applied to the new hybrid environment.

Effective security will be mandatory for companies seeking to take advantage of the economies and efficiencies of IP-based next-generation networks or VPNs for mission-critical applications and processes. Tunneling protocols, such as the IETF's IPSec, provide for packet-by-packet authentication, encryption, and integrity. Authentication positively identifies the sender. Encryption allows only the intended receiver to read the data. Integrity guarantees that no third party has tampered with the packet stream. Voice can be run over VPNs just like any other application, with performance improved by the QoS parameters that can be set for all applications on the VPN. The privacy of voice conversations is also improved on the VPN, because it inherits security properties from IPSec.

Some next-generation and incumbent carriers have chosen to deal with the various security concerns of customers by adopting every industry standard for security as it emerges. This includes the adoption of competing standards for secure electronic commerce. The intent of this "support-all" strategy is to provide customers with a choice of solutions that best meet their needs or which leverage their investments in existing solutions.

1.4.8 Global reach

Because the Internet is ubiquitous, it immediately gives any business worldwide reach. The ability to beam bits across the globe gives businesses the means to tune operations, empower employees with instant access to information, and add flexibility to their lives. It also provides the means to support customers, attract new customers, and penetrate new markets. Although global reach is provided by PSTNs, the current rates for international calls include very high access and settlement charges that greatly inflate the cost of telecommunications and make people think twice about its use for extended periods.

Among the top global communications issues for businesses are remote access and the costs associated with providing global reach. Remote workers, for example, must have the ability to dial into the corporate data network from international locations. And for those employees trying to access sensitive corporate information, a secure service is required. Some next-generation service providers offer VPNs for this purpose. Security is provided by packet filtering, encryption, and other technologies to prevent intrusion and protect sensitive information.

VPNs are an increasingly popular option for extending communication between corporate locations over managed IP nets. VPNs can be local, regional, national, or international in scope. Basically, a VPN allows businesses to carve out their own IP-based WANs within the carrier's managed high-speed IP backbone. The major carriers provide service-level guarantees to overcome concerns about latency and other quality of service issues traditionally associated with the Internet. As noted, certain multipartner applications such as EDI can benefit from the security and global reach of VPNs.

Another application that has recently undergone modification to accommodate the needs of distributed users is OLAP, a type of decision support tool that enables users to access a data warehouse on the corporate network from remote locations over dedicated or dial-in connections to the VPN. OLAP transforms the raw data in a warehouse into strategic information by providing a multidimensional view of aggregate data for further analysis using a simple, standardized query syntax.

Most OLAP vendors now provide access to multidimensional data over IP nets through a Web browser, allowing strategic information to be made conveniently available to all who need it without organizations having to buy, deploy, and maintain hundreds or thousands of client tools. All users need is a network connection and a Java-enabled Web browser, which is available free from companies like Microsoft and Netscape Communications.

OLAP vendors usually base their offerings around a Web gateway that translates between a Web server and an application server, which either holds the multidimensional database or generates the queries against a relational database management system (RDBMS). This gateway acts as a translator between the Web server and OLAP server. Some vendors offer the Web gateway as part of the OLAP engine, while others make it available as an extra-cost option.

To request information, users go to the appropriate Web page set up by their company where a query form is displayed. This form is made of standard and product-specific Hypertext Markup Language (HTML) tags. When the user generates an OLAP query, these tags are passed to the Web server. While the Web server processes the standard HTML tags, the OLAP-specific tags are sent on the Web gateway, which uses a Common Gateway Interface (CGI) script to translate them into queries that can be used to generate the request to the OLAP server. When the OLAP server has retrieved the requested data, it passes the returned data to the gateway, which converts it to HTML and sends it to the Web server. Once the data arrives at the Web server, it is rendered in the proper format by the Web client.

When implementing OLAP over the VPN, a critical issue is the capacity of the OLAP server to deal with the increased number of users. Whatever benefits the organization hoped to achieve by making OLAP tools more widely accessible over the VPN will be negated if users are forced to wait for resources to become available before their query can be handled.

One way to improve performance is to configure multiple Web servers for clustering, enabling them to act as a single unit. This configuration has several advantages, including high availability, load-sharing, and fault-tolerance. In addition, a server feature called demand paging can improve response time by sending one report page at a time to the remote user, thus minimizing network traffic and improving performance for all OLAP users. Application performance also can be improved by other server functions such as prioritizing and scheduling report requests. The server can also regulate how long the reports can run, and how much data they can return.

For companies that do not need a global VPN, a viable alternative comes in the form of a service called IP roaming. With global IP roaming, companies can build a logical network footprint around the world so remote workers can access corporate resources from virtually any location by dialing a local number. There are three major benefits to IP roaming: a business can give employees access to its network anywhere, anytime. Second, a company can run its business globally without huge investments in infrastructure. Third, employees can have access to corporate resources without worrying about paying exorbitant long-distance charges.

IP roaming gives telecommuters, mobile employees, and travelers dial-up access to the Internet with a local call from virtually anywhere in the world. From the Internet, they can access resources on their corporate intranet. They can also access e-mail, transfer files, and browse the Web without having to worry about long-distance call charges. Roaming services also allow users to maintain just one Internet account, and log on with one user name and password.

One of the largest IP roaming alliances is iPass, which operates through a global network of ISPs. Participants install free iPass Roam-Server software on their authentication servers and corporations install it behind a firewall for security purposes. Once installed, ISP subscribers or corporate employees can dial into the Internet through the access points of other iPass partners anywhere in the world. At this writing, iPass has 2,500 active points of presence (POPs) representing over 250 ISPs and corporations in 151 countries. Another worldwide alliance of this type is that of Gric Communications, which has 350 ISPs and telephone companies that provide 2,300 POPs in more than 100 countries.

Corporations benefit from being able to significantly reduce remote access costs by eliminating all of the Internet and intranet access challenges experienced by their IS staff, such as maintaining and upgrading modem banks. Their business travelers and telecommuters no longer require multiple Internet accounts and e-mail addresses, and do not have to rely on long-distance and toll-free services, which can reduce access expenses by as much as 90%.

Individual users install a simple client software tool on their portable computers that contains an international phone book of iPass access numbers. Anywhere users travel, they access the iPass client software to connect to a local Internet access number. They log in with the same user ID and password as they normally use, but with their domain name after the user ID (e.g., yourname@domain).

When users dial into any of the worldwide iPass access numbers, they are actually logging into their own ISP or corporate account, even though they dialed in through a different Internet service provider. Once connected, users have access to all of the services offered by their ISP or company through all of the client software they normally use.

iPass RoamServer communicates with the iPass network through Secure Socket Layer (SSL) technology to securely deliver authentication and accounting records to the authentication systems of the ISP or corporation. At the end of a user session, the remote ISP passes information regarding the length of the session to the iPass server. iPass pays the remote ISP for the time spent using its service. iPass bills the company or ISP for use of the service. iPass produces usage statements very similar to a telephone bill. The statement shows the usage transactions of each individual user.

Next generation networks will also support global roaming through alliances. Even incumbent carriers like AT&T will offer this service. At this writing, AT&T WorldNet currently has a global roaming trial under way, which will enable dial-up from more than 100 countries back to a corporate data network.

1.4.9 Ubiquity

Finally, IP nets are ubiquitous. Just about every country is tied into the public Internet in some way. Only about a dozen countries, mostly in Africa, have no access to the Internet. By year-end 1998, there were close to 40 million hosts visible to the Internet. Millions more are private and are not directly visible to the Internet.

The United States has the largest number of online users. According to a survey by Nielsen Media Research and CommerceNet released in August 1998, 70.5 million (34.9%) of the 202 million United States adults use the Internet. Almost 80 million in the United States and Canada are online, up 21 million over the previous nine months. And for the first time, more than 50% of those in the crucial 16-to-34-year-old advertising bracket are Internet users. Only 18 million were online when Nielsen did its first Internet survey in 1995. This represents an increase of 340% in slightly more than three years.

Because the Internet is ubiquitous, it immediately gives any business worldwide reach. An IP roaming capability enhances this global reach by facilitating personal mobility. As noted, roaming gives telecommuters, mobile employees, and travelers dial-up access to the Internet with a local call from virtually anywhere in the world, allowing them to maintain just one Internet account, and log on with one user name and password.

The Internet offers unparalleled opportunities for organizations to economically extend communication capabilities, applications, and information to office staff, remote offices, mobile professionals, and other companies with which they do business. The main issue for most companies is not whether to establish a TCP/IP network, but how fast it can be established to start reaping the demonstrable benefits. The proven nature of the technologies and protocols involved make the decision to set up a TCP/IP net a risk-free one.

1.4.10 Application protocols

The Internet is a network, but so are AT&T, MCI WorldCom, Sprint, and numerous others. More accurately, the Internet is a network of networks that draws upon diverse facilities, services, and technologies that are available from many sources. In terms of network intelligence, however, IP networks operate in an entirely different manner than traditional networks specifically designed for telephony.

IP is an end-system to end-system network protocol that encourages smart hosts and dumb networks. This is opposite of traditional networks where the network is smart and the terminals (handsets) are dumb. The actual services provided to end users through the Internet are defined not through the routing mechanisms of TCP/IP, but depend instead on higher-level application protocols, such as Hypertext Transfer Protocol (HTTP), File Transfer Protocol (FTP), Network News Transfer Protocol (NNTP), and SMTP.

Because these protocols are not embedded in the Internet itself, a new application-layer protocol can be operated over the Internet through as little as one server computer that transmits the data in the proper format, and one client computer that can receive and interpret the data. The technology of the Internet allows new types of services to be layered on top of existing protocols, often without the involvement or even the knowledge of network providers that transmit those services. The utility of a service to users, however, increases as the number of servers that provide that service increases.

In the traditional phone network, intelligence is embedded in various switches throughout the network. To support a standard set of features throughout the network, or to add new features, each switch must be equipped with the same intelligence. Such networks are very expensive, time-consuming to upgrade, and require a high level of cooperation between vendors and service providers for interoperation. These factors were chiefly responsible for the 10-year time lag in getting ISDN ready for the consumer market.

With the Internet and other IP-based networks, the services are decoupled from the underlying infrastructure to such an extent that new services such as Internet telephony can be introduced without requiring changes in transmission protocols, or making changes to the tens of thousands of routers spread throughout the network. This does not mean, however, that such applications would not benefit from the addition of intelligence in the IP network infrastructure. In fact, more intelligence is indeed being added to private IP networks and VPNs to implement QoS for the various applications that run over them.

This intelligence (e.g., RSVP, RTP), which resides in the routers and switches, can be added to the public Internet, but in the absence of a central administrative authority, progress is slow. However, as ISPs increasingly seek to offer value-added services and applications, they will have more incentive to equip their sites with more intelligence and link up with similarly equipped ISPs and national backbone providers. Those who lag behind in this area will find themselves at a severe competitive disadvantage.

1.5   Internet2

The increasing popularity and uncontrolled growth of the Internet has had an adverse effect on its performance to such an extent that high-level researchers have difficulty using it as a tool. To remedy this situation, a group of 34 leading universities successfully petitioned the Clinton Administration in October 1996 for $100 million to develop a separate Internet, known as Internet2.

The justification for a new Internet was that the privatization of the existing Internet had reduced it to a commodity that no longer served the needs of research and education and threatened implementation of national initiatives in distance learning and telemedicine. A new Internet would offer high bandwidth, reliability, and security. It would not replace the original Internet, but provide a backward-compatible, performance-enhanced alternative for power users.

Internet2 is designed to provide guaranteed bounded delay, low data loss, and high capacity to support a variety of advanced applications such as remote control of high-powered telescopes and electron microscopes, interconnection of distributed supercomputers and digital libraries, advanced computer simulation, multimedia integration, and real-time collaboration. The network itself offers end-to-end QoS, supports different payload types and synchronizes the various data streams in multimedia applications, provides flow and congestion control, enables packet source tracing after arrival, and enhances reliability via correct reproduction of the sent data stream(s).

Internet2 is implemented via high-performance nodes called GigaPOPs, which provide points of interconnection and service delivery between one or more institutional users and one or more service providers. Participating institutions obtain a variety of services over a single high-capacity communications link from the nearest GigaPOP.

Equipment at a GigaPOP site includes packet data switch/routers capable of supporting at least OC-12 (622 Mbps) link speeds and switched data streams, as well as packet data routing. SONET/ATM-compliant multiplexers handle the allocation of link capacity to different services. Physically diverse fiber optic and wireless communications paths protect against network outages. Redundant network operations centers monitor all equipment remotely via both in-band and out-of-band circuits.

Many of the standards and technologies developed and tested on Internet2 will eventually find their way to the public Internet and benefit all users.

1.6 Service-level guarantees

With incumbent and next-generation service providers carefully managing their IP networks to ensure peak performance, they are beginning to offer customers SLAs that specify the performance parameters within which a network service is provided. SLAs are available for all kinds of services, including VPNs, intranets, extranets, and Web hosting. Telecom service providers feel compelled to offer written performance guarantees to differentiate themselves in the increasingly competitive market and to migrate the mission-critical applications of large companies from dedicated leased lines to shared IP nets.

The SLA defines such parameters as the type of service, data rate, and what the expected performance level is to be in terms of delay, error rate, port availability, and network uptime. Response time to system repair and/or network restoral also can be incorporated into the SLA, as can penalties for noncompliance. The increasing use of SLAs has resulted in a proliferation of performance measurement and reporting tools to provide users with the documentation they need to confront service providers when network performance falls below mutually agreed upon thresholds.

SLAs are even becoming popular among ISPs as a means to lure business applications out of the corporate headquarters to outsourced Web hosting arrangements. To do this successfully requires that the ISP operate a carrier-class data center, offer reliability guarantees, and have the technical expertise to fix any problem, day or night. The SLA may also include penalties for poor performance, such as credits against the monthly invoice if network uptime falls below a certain threshold. Some ISPs even guarantee levels of accessibility for their dial-up remote-access customers. IBM, for example, offers an SLA for its remote-access customers that guarantees a 95% success rate on dial-up connections to the IBM Global Network.[5]

GTE Internetworking offers one of the industry's strongest Internet SLAs for its approximately 3,000 Internet Advantage dedicated access customers. The SLA includes credits for network outages, the inability to reach specific Internet sites, and packet losses. The SLA also stipulates a day's credit if a customer has more than a 10% packet loss on GTE's backbone within any 10-minute interval. The SLA covers equipment, wiring, and telephone circuits within and between the carrier's POPs. It does not include equipment on the customer premises, telephone circuits from customer locations to GTE's POPs, or non-GTE equipment or network connections.

Uunet Technologies Inc., a subsidiary of MCI WorldCom, offers service level agreements for its frame relay, dedicated 56 Kbps, T1, T3, and OC-3 Internet access services. The SLAs are available for U.S. customers and U.S. customers purchasing leased line access in its multinational service program. The SLAs include guarantees on network availability and latency as well as offering notification when SLAs are not met and guarantees on the time it takes to install access lines. Specifically, Uunet guarantees:

- ▶ 100% availability of the Uunet backbone as well as the Uunet-ordered customer access circuit.

- ▶ Average monthly latency of no more than 85 ms round-trip within Uunet's backbone in the contiguous United States and of no more than 120 ms between New York and Uunet's international gateway hub in London.

- ▶ Proactive outage notification that guarantees customer notification by Uunet operations within 15 minutes of an outage.

- ▶ Installation by a quoted install date, which will be not more than 40 business days for frame relay, 56 Kbps, and T1 customers and 60 business days for T3 customers in the United States. If these conditions are not met, Uunet will automatically reimburse companies.

Instead of relying on the carriers for performance information, a variety of third-party monitoring and reporting tools are becoming available,

---

5. AT&T acquired IBM's global network in mid-1999 and has incorporated IBM's corporate Internet and IP Remote Access services under its own brand name. Customized SLAs are available, and detailed usage information is provided to customers on a secure Web site accessible only to the customer, with information including most frequently used access numbers, initial modem connection speed, and call failure rate.

many of which allow users to access performance information over the Web in near real-time. Even interconnect vendors are embedding SLA monitoring tools into their network management systems.

Nortel's Bay Networks, for example, has enhanced its Optivity management suite with a throughput measurement tool that can be used to verify SLA performance. The key application in the suite is the Response Time Monitor, which gives network managers the ability to determine whether the network is slow or if it is the client or server that is causing the problem. With another tool, the Bay Ping MIB, a network manager can set up ping sessions within the network. Sending a ping from the probe on the local segment to the default gateway provides response-time results, which can then be used to set color-coded baseline thresholds for various applications. Developing a graphical display makes it easier for network managers to tell at a glance if response time is good, bad, or deteriorating from the norm.

With the growing popularity of the Internet for corporate communications, some vendors offer Internet diagnostic systems to provide real-time data on the "access experience"—revealing just how often actual users are frustrated by busy signals, failed call attempts, and unacceptably slow download times. This information can be used to document network performance so appropriate steps can be taken to improve response.

Inverse Network Technology, for example, offers AccessRamp, a service that enables Internet service providers and managers of corporate dial-up networks to monitor, measure, and diagnose the actual quality of service all users are receiving, rather than having to draw conclusions from a sampling or simulation of users. By providing immediate, actionable data on user access problems, AccessRamp enables customers to reduce user support demands and costs, and offers credible proof that SLAs are being met.

1.7 Analysis

Companies not willing to incur the expense of installing and managing their own intranets should consider subscribing to a carrier-provided service. With local and long-distance companies and next-generation service providers carefully managing their TCP/IP networks to ensure peak performance, they are beginning to offer customers SLAs that specify the performance parameters within which a network service is provided. SLAs are available for all kinds of services, such as VPNs, intranets, extranets, and Web hosting for electronic commerce applications.

While the next-generation carriers may be way out in front of the incumbent local and long-distance carriers when it comes to bandwidth capacity and their use of IP for voice-data convergence, it is important for telecom managers to look at other facets of these networks before bringing their recommendations to top management. Specifically, they must evaluate next-generation carriers in terms of cost savings, local access, geographical coverage, service portfolio, and service-level guarantees.

1.7.1 Cost savings

With regard to cost savings, next-generation carriers beat the big three hands down. For example, Qwest LCI offers a coast-to-coast T1 circuit for about $6,000 a month, while Frontier charges about $9,000. The big three are all in the same range of about $14,000 for the equivalent circuit.

The savings are not nearly as attractive for IP telephony, however. Qwest LCI, for example, prices its IP telephony service at 7.5 to 10 cents per minute. This is more than what large companies pay the big three when volume discounts are applied. ICG Netcom has established breakthrough pricing with its 5.9 cents per minute rate for on-net calls and 8.9 cents per minute for off-net calls. Although next-generation service providers want to eventually get the per-minute charge under 5 cents, none offers it yet. Some industry analysts predict the per-minute charge will eventually reach 3 cents.

1.7.2 Local access

Since the new service providers are not able to match the incumbent interexchange carriers in the number of POPs—which let local customers dial into their backbones—geography will play a critical role in selecting a next-generation network provider. While there is substantial cost savings to be had on long-haul transport, much of the savings can be consumed if the traffic must travel a long distance to get to its destination.

Next-generation carriers address this problem by backhauling, which involves getting the customer's traffic to its destination by whatever route is necessary—regardless of how many route-miles it must traverse. Next-generation carriers like Qwest LCI claim they do this safely because they over-provision their feeder pipes and do not charge customers for the extra distance. But because backhauling can add about 25% to total distance, it is important to check the performance of such networks to determine if the extent of backhauling interferes with real-time mission-critical applications such as SNA, which cannot tolerate much delay.

In addition to having an adequate number of POPs and in just the right locations, the new carriers also must negotiate with the regional Bell operating companies (RBOCs) to get access to the local loop. Some next-generation carriers have been quite successful at this, while others have not. In large part, this is because some RBOCs have been more open to the concept of competition, while others continue to put up obstacles.

Other times, it is the nature of the agreement between the RBOC and next-generation carriers that does not pass legal scrutiny. The agreement between Qwest LCI and U S West to provide U S West customers with a local and long-distance service package, for example, ran into legal trouble because U S West had not yet received regulatory approval to offer long-distance service within its 14-state region.

Although this arrangement has been discontinued because of regulatory constraints, it illustrates the competitive impact next-generation service providers are expected to have in the future as they move in new directions. Their success may hinge on developing such partnerships, enabling them to compete in the long-distance market without having to invest in local loops. It also lets regional telephone companies into the long-distance market without having to invest in state-of-the-art long-haul transport facilities. Such partnerships mean more competition and new programs that provide customers with the kind of low cost, value-added services they have been waiting for and which the Telecom Act of 1996 promised.

1.7.3 Service portfolio

To date, the next-generation service providers have focused primarily on selling wholesale bandwidth and dark fiber to ISPs, CLECs, and RBOCs. This has given them the capital they need to continue building out their networks, but it also means that the service portfolios of next-generation carriers are still quite limited, compared to those offered by the big three.

Some next-generation carriers deliver leased lines, circuit-switched local and long-distance telephone service, frame relay, Internet access,

VPNs, and IP fax. Others include IP telephony in the service mix. While this is a good start, it must be remembered that incumbent carriers have a strong track record when it comes to the range of services offered, their willingness to customize to meet specific needs, the timeliness of service provisioning, management options, and customer support. For many next-generation service providers, these are still unknowns.

Next-generation service providers are still a relatively new element in the competitive telecommunications marketplace. With their state-of-the-art optical fiber networks, they have abundant capacity. What these carriers lack is a broad customer base. For certain kinds of services, such as leased lines, bargains abound, and telecom managers would be wise to check the offerings now. For other types of services, such as telephony, the volume discounts offered by incumbent carriers is probably going to be the best value until next-generation carriers get the cost-per-minute well under 5 cents.

For most telecom managers, a wait-and-see approach might be the best course to follow before making big commitments to next-generation service providers. However, telecom managers should follow the progress of these carriers, noting the large corporate customers they sign up and the partnerships they form to expand services. The problems next-generation carriers experience with hyper-growth and what impact these problems have on their customers also bear watching.

1.8 Conclusion

Numerous complaints continue to surface about the inadequacy of IP to support advanced applications, especially telephony. After all, multimedia applications did not exist 30 years ago and IP was certainly not developed with the idea of someday building telephone networks that could pose a threat to Ma Bell.

As new applications emerged over the years—each requiring a corresponding improvement in network performance—refinements to the original protocol and the development of numerous plug-in protocols have enabled IP to adapt to new requirements. Even when new technologies and protocols were developed to meet different sets of needs—such as LANs, ATM, and frame relay—the advantages of IP could be retained by running it on top of them.

In Arthurian terms, it can truly be said of IP that it is the "once and future" protocol.

More information

The following Web pages contain more information about the topics discussed in this chapter:

AT&T: http://www.att.com/
Bay Networks (see Nortel Networks)
Bell Atlantic: http://www.bellatlantic.com/
British Telecom: http://www.bt.com/
Cable Television Laboratories (CableLabs): http://www.cablelabs.com/
Changepoint: http://www.changepoint.com/
Cisco Systems: http://www.cisco.com/
Computer Security Institute: http://www.gocsi.com/
Electronic Messaging Association: http://www.ema.org/
Federal Bureau of Investigation: http://www.fbi.gov/
Frontier: http://www.frontiercorp.com/
Gric Communications: http://www.gric.com/
GTE: http://www.gte.net/ or http://www.bbn.com/
Hughes Network Systems (DirecPC): http://www.direcpc.com/
IBM Global Network: http://www.ibm.com/globalnetwork
International Telecommunications Union: http://www.itu.org/
Internet Engineering Task Force: http://www.ietf.org/
Internet II documents: http://uu-gna.mit.edu/
Inverse Network Technology: http://www.inverse.net/
iPass: http://www.ipass.com/
ITXC: http://www.itxc.com/
Level 3 Communications: http://www.L3.com/
Lucent Technologies: http://www.lucent.com/
MCI WorldCom: http://www.mciworldcom.com/
National Science Foundation: http://www.nfs.gov/
Nortel Networks: http://www.nortel.com/
PakNetX: http://www.paknetx.com/
Qwest LCI: http://www.qwest.net/
Rocky Mountain Internet: http://www.rmi.net/

Selsius (see Cisco Systems)
Sprint: http://www.sprint.com/
3Com Corp.: http://www.3com.com/
U S West: http://www.uswest.com/
Uunet Technologies: http://www.uu.net/
WebTV Networks: http://www.webtv.com/
WorldCom: http://www.wcom.com/

The concept and ramifications of IP telephony

IP telephony has been around since the 1980s when it was referred to as a "voice funnel." In 1983, both the ARPANET and Internet were being run from the Network Operations Center facility at the offices of Bolt Beranek and Newman (BBN) in Cambridge, Massachusetts. There, among the workstations dedicated to special projects, was one labeled "voice funnel" that digitized voice, arranged the resulting bits into packets, and sent them through the Internet between sites on the East and West Coasts. The voice funnel was part of an ARPA research project concerning packetized audio. ARPA and its contractors used the voice funnel, and related video facilities, to do three-way and four-way videoconferencing, saving travel time and money.

The technology was rediscovered in the 1990s, but did not become popular until 1995 when improvements in microprocessors, digital signal processors (DSP), codec technology, and routing protocols all came together to make feasible products for mainstream use. Since then, IP telephony has emerged as a successful telecommunications service. Not only are new vendors emerging to pursue the market for IP telephony, traditional vendors are developing and implementing strategies for supporting VoIP across their product lines.

Motorola, for example, is taking things as far as they can possibly go. In partnership with NetSpeak, the company intends to develop IP as the multimedia platform of the future—across all of its businesses, which span terrestrial wireless (cellular and private), satellite, copper networks, paging, and HFC (hybrid fiber/coax) broadband systems.

One of Motorola's leading competitors, Nokia, also is in a race to mesh its wireless technologies with IP. Toward that end, it purchased IP telephony vendor Vienna Systems. The acquisition will allow Nokia to develop consumer and business products capable of supporting voice, data, and Internet access. Nokia is positioning itself to be able to transmit data at megabits-per-second speeds through its wireless devices, once new radio frequencies are assigned for that purpose by the World Radio Conference (WRC).

New service providers are offering IP telephony, some intending to compete directly with established long-distance carriers and RBOCs, allowing consumers to use IP telephony services with a look and feel identical to today's phone service. Corporations are adding IP telephony service to leverage investments in private intranets and save on long-distance call charges. They also see the potential of IP telephony for adding value to existing network applications such as call centers, customer support, help desks, and electronic commerce.

2.1 Development background

Internet telephony started to become popular in 1995 with the introduction of client software that enabled computer users to engage in long distance conversation between virtually any location in the world, via the Internet, without regard for per-minute usage charges. All that was needed was an Internet access connection and a multimedia computer equipped with sound card, microphone, and speakers or headset.

Users at each end had to use the same client software to communicate and had to be online at the same moment the call was placed. Users at both ends of the connection spent much of their time tweaking the controls for volume and compression and trying to eliminate echo so they could hear each other better.

The poor quality of the calls limited first-generation Internet telephony software to casual users, hobbyists, and college students. Voice quality was marred by the long pauses induced by the Internet's variable delay, clipped speech that resulted from dropped packets, feedback-induced echo from having the computer's speakers too close to the microphone, and confusion about whose turn it was to talk at any given time. A continuous stream of advancements has addressed these and other problems and, in the process, propelled IP telephony to commercial status.

IP telephony is no longer limited to PC-to-PC communication over the Internet or even PC-to-phone communication. Today, commercial services are offered over managed IP backbones and allow phone-to-phone communication as well. Companies are also leveraging their private IP nets by adding support for telephone service for internal calls—enabling them to save on long-distance call charges.

The success of IP telephony services has had two important ramifications. It has helped spawn a new buzz word that has dominated the telecommunications industry—IP convergence, which entails putting voice, video, and data on a single network. Although ISDN and ATM promised to do the same thing, IP does so more economically and is globally available. The success of IP telephony has also put pressure on the Federal Communications Commission (FCC) to regulate IP telephony as it does the rest of the telecommunications industry, which could lead to higher costs, stifle innovation, and set a precedent for further regulation of the Internet (see Chapter 9).

2.2 First-generation technology

First-generation IP telephony focused on establishing calls over the public Internet between similarly equipped multimedia PCs. Placing calls involved logging on to the Internet and starting up the "phoneware," which provided several ways of establishing a voice link. Users could connect to the vendor's directory server to check the "white pages" for other phoneware users who also were logged on to the Internet. The public directories were organized by user name and topic of interest to make it easy for everyone to strike up a conversation with like-minded and willing participants. The directory was periodically updated, reflecting changes as people entered and left the network.

Alternatively, users could click on a name in a locally stored private phone book. Of course, users could simply enter the IP address or e-mail address to establish a direct user-to-user connection from the start without having to first go through a directory server (see Figure 2.1). Whether a public or private listing, the names corresponded with the static IP addresses of other users. As Internet service providers (ISPs) increasingly adopted dynamic IP addresses,[1] the names in public directories and private address books corresponded with e-mail addresses instead.

2.3   System requirements

To make calls over the Internet, users required a computer equipped with a modem, sound card, speakers (or headset), and a microphone. Sound cards came in two types—half-duplex and full-duplex. Half-duplex works like a citizen's band (CB) radio, where one person can talk at a time and says "over," indicating when he or she is finished talking. With full-duplex audio cards, both parties can talk at once, just like an ordinary telephone call. However, if a full-duplex user connected to a half-duplex user, the conversation defaulted to the half-duplex mode.

In addition to the hardware, three software components were typically required—a TCP/IP dialer program (since most users dialed into the Internet with a modem), Web browser, and the phone software itself.

---

1. With dynamic IP addressing, an address is assigned by the ISP each time the user dials into the server. The ISP has a pool of IP addresses for this purpose. With static IP addressing, the same address is used each time the user connects to the Internet. The proliferation of TCP/IP-based networks, coupled with the growing demand for Internet addresses, makes it necessary to conserve IP addresses. Issuing IP addresses on a dynamic basis provides a way to recycle this finite resource. Even companies with private intranets are increasingly using dynamic IP addresses, instead of issuing unique IP addresses to every machine. The standard that addresses this issue is the Dynamic Host Configuration Protocol (DHCP), developed by the IETF. From a pool of IP addresses, a DHCP server doles them out to users as they establish Internet connections. When they log off the net, the IP addresses become available to other users.

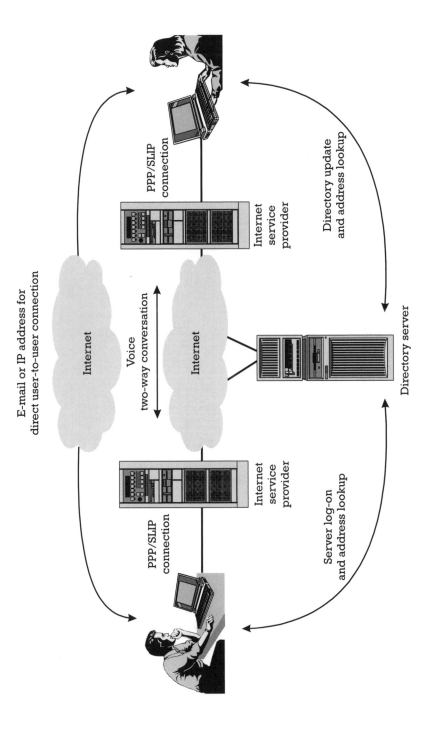

Figure 2.1 Early phoneware products logged users on to a directory server, which enabled them to receive a list of other registered phoneware users. The connection is then user-to-user, bypassing the vendor's directory server.

The critical component is the phoneware, which provides algorithms that compress the recorded speech obtained from the sound card and apply optimization techniques to ensure its efficient delivery over the Internet in the form of data packets.

Phoneware vendors use a variety of compression algorithms to minimize bandwidth consumption over the Internet. A software coder/decoder (codec) utilizes different mathematical strategies to reduce the bit rate requirements as much as possible and yet still provide acceptable reproduction of the original content. Numerous codecs have appeared on the scene in recent years, with more coming all the time.

For example, NetSpeak's WebPhone started out in 1995 using two audio compression algorithms—GSM and TrueSpeech. GSM is the Global System for Mobile Communications and is a worldwide standard for digital cellular communications. It provides close to a 5:1 compression of raw audio with an acceptable loss of audio quality on decompression. True-Speech, a product of the DSP Group, Inc., provides compression ratios as high as 18:1 with an imperceptible loss of audio quality on decompression. NetSpeaks's WebPhone used GSM compression when it was installed on a 486-based computer and TrueSpeech when it is installed on a Pentium-class computer. Offering a high compression ratio, True-Speech is more CPU-intensive than GSM, so it requires a faster processor to compress the same audio signal in real time.

The current strategy of NetSpeak is to deal with codecs in a "plug-and-play" fashion to allow new codecs to be incorporated into the client application with little or no effort on the part of the user. As NetSpeak adds new codecs to its repertoire, they are easily incorporated into the clients. The clients can then negotiate between themselves the best codec to be used for an individual session based on available session bandwidth limitations, network delay characteristics, or individual PC resource limitations.

Other phoneware vendors offer proprietary compression algorithms and support one or more accepted industry-standard algorithms. Most now support the G.7xx international standards (discussed later) as well, which guarantee various levels of speech quality and facilitate interoperability between various H.323-compliant products. H.323 is the international umbrella standard under which the G.7xx audio standards fit. Recommended by the International Telecommunication Union (ITU), H.323 defines how audio/visual conferencing data is transmitted across networks.

In addition to the algorithms that compress/decompress sampled voice, some phoneware products include optimization techniques to deal with the inherent delay of the Internet. The packets may take different paths to their destination and may not all arrive in time to be reassembled in the proper sequence. If this were ordinary data, late or bad packets would simply be dropped and the host's error checking protocols would request a retransmission of those packets. But this concept cannot be applied to packets containing compressed audio without causing major disruption to voice conversations, which are supposed to be conducted in real time.

If only a small percentage of the packets are dropped, say 2–5%, the users at each end may not notice the gaps in their conversation. When packet loss approaches 20%, however, the quality of the conversation begins to deteriorate. Some products, such as VocalTec's Internet Phone, employ predictive analysis techniques to reconstruct lost packets, thereby minimizing this problem.

Occasionally, the Internet can become overloaded or congested, resulting in lost packets and "choppy" sound quality. A product called FreeTel has a feature called Booster that alleviates this problem and improves sound quality. Booster introduces an artificial delay into the signal, and uses this extra time to retransmit lost packets using a proprietary algorithm. The end result is better quality, at the expense of increased but predictable delay.

2.4 Operation

Once the call is placed, either by IP address or e-mail address, the users at each end speak into the microphones connected to the sound cards in their respective computers. The phone software samples the speech, digitizes and compresses the audio signal, and transmits the packets via TCP/IP over the Internet to the remote party. At the other end, the packets are received and pieced together in the right order. The audio is then decompressed and sent to the sound card's speaker for the other party to hear. The compression algorithm compensates for much of the Internet's inherent delay. As the packets are decompressed and the audio signals are being played, more compressed packets are arriving. This process approximates real-time conversation.

To improve overall sound quality early users of IP telephony software often found it necessary to fine-tune the sampling rate and compression level to suit their modem's speed. For example, to overcome the annoying problem of clipped speech, the user could reduce the sampling rate until smooth speech resumed.

Camelot Corp.'s DigiPhone,[2] for instance, allows users to adjust its recording and playback quality in response to the speed of the modem connection. The user can start this manual tuning process by connecting at the default sampling rate, then increase the sampling rate by increments of 500—the increments can be smaller or larger depending on preference. With the new setting, DigiPhone renegotiates the connection at the higher sampling rate. The user can continue to increment the sampling rate until the other party's speech begins to break up, then back it down until it is clear again.

The sampling rate can be set from 4,000 to 44,000 bps, depending on the capabilities of the sound card and the speed of the Internet connection. In general, the higher the speed of the modem connection, the higher the sampling rate can be set. In conjunction with the sampling rate, users can set the compression level. With a lower speed modem, a higher compression level can be selected to improve performance, but with some loss in sound quality. With a higher speed modem, users can select a lower compression level for better sound quality.

Some products, such as VocalTec's Internet Phone, provide real-time statistics that can help users determine the quality of the Internet connection at any given time (see Figure 2.2). The network statistics window provides a count of incoming and outgoing packets, the average round-trip delay of packets, and the number of lost packets in both directions—incoming and outgoing. This information helps the user pinpoint the source of the problem as originating from the network or locally (e.g., sound card, modem, or software) so corrective action can be taken.

Another problem phoneware users had to contend with is audio volume level. This problem is most often encountered with the installation of a new sound card; specifically, when the sound card ships with the microphone input disabled or the sound level turned off. Most sound cards come with a set of utilities, including a mixer program. With the mixer utility, the card's microphone input can be enabled and the volume

---

2. The latest version of DigiPhone is the Deluxe 1.0 version. Camelot Corp. has no plans to issue further upgrades of this product.

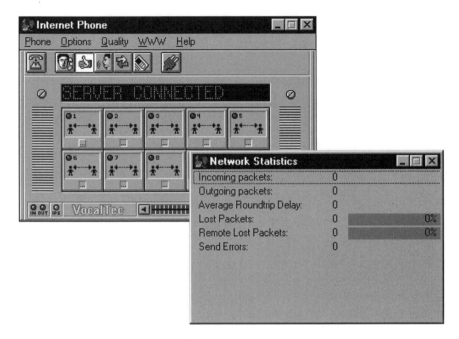

Figure 2.2   Statistics window of VocalTec's Internet Phone.

adjusted. Depending on the type of sound card, it is often possible to adjust microphone volume from the phoneware's interface. With Quarterdeck's WebTalk,[3] for example, there were three microphone settings on the main screen (see Figure 2.3):

> ▶ Sensitivity—controls microphone volume.

> ▶ Idle level—controls how loudly the user must speak to activate the microphone which helps prevent background noise from being picked up during conversation.

> ▶ Idle after—controls how soon after the user stops speaking before control of the microphone is turned over to the other user.

Some sound cards provide automatic gain control (AGC). This feature boosts the microphone level automatically when speaking into it and reduces the level during silent periods. This cuts down on ambient background noise that makes it difficult for both parties to hear each other.

---

3. Quarterdeck has discontinued offering WebTalk.

Figure 2.3 Microphone settings provided by Quarterdeck's WebTalk.

Taking advantage of this feature requires that it be enabled. Adjustments can be made by accessing the sound card's configuration utility, which offers a slider control for adjusting AGC, as well as bass and treble.

2.5 Features

Internet phone products offer many features and new ones are being added all the time. The benefit of using a computer for telephone calls—rather than an ordinary phone—is that the user can take advantage of integrated voice-data features. The following list provides the most common features found across a broad range of products:

▶ Adjustable volume control—allows the volume of the microphone and speakers to be adjusted during the conversation.

▶ Advanced caller ID—not only is the calling party identified by name, but some phoneware products offer a brief introductory message about what callers want to talk about, which is displayed as the call comes through. This information can help users decide whether or not to answer the call.

▶ Advanced phone book—not only holds contact information, but offers a search capability by name, e-mail, country, company, or any other parameter that can identify a particular person.

▶ Audio date/time stamp—notifies unavailable users of call attempts by date and time.

▶ Automatic notification—with this feature, the phoneware automatically looks for and provides notification of when specific users come online so they can be called.

▶ Busy notification—if a call is placed to someone who is busy with another call, an appropriate "busy" message is returned. Some products allow callers to send an e-mail message, voice mail, or other notification to the busy party indicating that they have tried to call.

▶ Call blocking—annoying or unwanted incoming calls can be blocked by fixed IP or e-mail address.

▶ Call conferencing—the ability to converse with three or more people at the same time.

▶ Call duration timer—provides an indication of the amount of time spent on each call.

▶ Call hold—allows an initial call to be put on hold while the user answers another incoming call. The user may continue the first conversation after holding or hanging up the second call.

▶ Call log—records information about incoming and outgoing calls, allowing the user to keep track of incoming and outgoing calls.

▶ Call queue—a place where incoming calls are held until they can be answered.

▶ Caller ID—identifies the caller by name, nickname, e-mail address, or phone number so a user can see who is calling before deciding whether to take the call.

▶ Configuration utility—scans the computer system to determine if the proper hardware is installed to use the phoneware and offers the user advice on configuring various system parameters to prevent conflicts with other communications hardware.

▶ Database repair utility—the phoneware maintains one or more databases to hold such things as private phone books and configuration data. If a database gets corrupted or destroyed, the user will be notified and have the option of running the database repair utility to restore it.

▶ Dedicated server—for those who receive a large volume of incoming calls, some phoneware vendors offer special servers to faciliate call handling.

▶ Directory assistance—a searchable directory of users currently online is automatically maintained. Users can initiate a phone call simply by mouse-clicking on a person's name, or by typing in the first few characters of a name.

▶ Dynamic, on-screen directory—provides the latest information on users who have registered with the server, indicating that they are online and ready to take or initiate calls. This display is periodically refreshed with new information.

▶ Encryption—to ensure secure voice communication over the Internet, a public-key encryption technology such as PGP (pretty good privacy) can be applied. Depending on vendor, PGP is integrated into the phoneware or may be licensed separately for use with the phoneware. Other products may not accommodate encryption at all.

▶ Event message system—allows users to view the ongoing status of the phoneware to determine what features and functions are active at any given moment.

▶ File transfer—allows the user to transmit a file to the other party during a conversation. The file transfer process takes place in the background and does not interfere with the conversation.

▶ Greeting message—when a user is not available or too busy to take a call, a recorded message can be played to callers.

▶ H.323 compliance—a worldwide standard for audiovisual communication over packet data networks, such as the Internet. Users of different H.323-compliant products are capable of conversing with each other over the Internet.

▶ Last-party redial—allows the user to redial the last party called without having to look up the address in a directory.

▶ Map—displays the connection of the call against the background of a U.S. or world map, showing the points of origination and destination.

▶ Multiple calling mechanisms—some phoneware products offer multiple methods of initiating calls, including by fixed IP address, domain name, e-mail address, saved address, and online directory of registered users.

▶ Multiple lines—some phoneware products allow users to carry on a conversation on one line and take an incoming call on another line, or put one call on hold while another call is initiated.

▶ Multiple user configurations—if several people share the same computer, some phoneware products allow each of them to have their own private configuration, including caller ID information and address books.

▶ Music on hold—plays music to a caller on hold, until the call can be answered.

▶ Mute—a mute button allows private, "offline" conversations.

▶ Online help—offers help on the proper use of various phoneware features without having to resort to a manual or opening a separate read-me file.

▶ Picture compression—some phoneware allows the user to call up a photo of the person he or she is talking to (if the remote user supplies one). Compression allows fast photo loading over the Internet of the commonly supported file formats.

▶ Programmable buttons—allow users to configure quick-dial buttons for the people they call most frequently. In some cases, buttons are added automatically and written over based on the most recent calls.

▶ Redial—if a person is not reachable on the first call attempt, the phoneware can be configured to automatically redial at designated time intervals until the connection is established.

▶ Remote information display—displays operating system and sound card information of the remote user.

▶ Remote time display—displays the remote time of the called party.

▶ Selectable codec—provides a choice of codecs, depending on the processing power of the computer. A high-compression codec can be used for Pentium-class machines and a lower-compression codec can be used for 486-based machines. The choice is made during phoneware installation.

▶ Silence detection—detects periods of silence during the conversation to avoid unnecessary transmission.

▶ Statistics window—allows the user to monitor system performance and the quality of the Internet link.

▶ Text chat—some phoneware products offer an interactive text or "chat" capability to augment voice conversations. The chat feature can be used before, during, or apart from voice mode.

▶ Toolbar—icons provide quick access to frequently performed tasks such as hang up, mute, chat, view settings, and help.

▶ Toolbox mode—the interface can be collapsed into a compact toolbox to save desktop space. This makes it easier to work in other applications until the phoneware is used for calls.

▶ User location service (ULS) compliance—ULS technology enables Internet phone users to find each other through existing Internet Directory Services such as Four11, Banyan's Switchboard, WhoWhere, DoubleClick, and BigFoot.

▶ User-defined groups—allows users to set up private "calling circles" for calls among members only or establish new topic groups for public access.

▶ Video—some phoneware products allow the calling and called parties to see each other as they converse. This requires a video camera connection to the computer.

▶ Voice mail—allows users to record and play back greeting messages as well as send voice mail messages for playback by recipients. Depending on vendor, this might include the ability to give specific messages to callers when they enter a personal code.

▶ Voice mail screening—allows users to delete voice messages before they can be downloaded to their computer.

▶ Web links—allows users to put links into their Web pages that, when activated, establish a call with the visitor.

▶ Whiteboard—some multifunction products allow participants to draw or annotate shared text and images while conversing.

While early phoneware products focused on PC-to-PC communication, they had the advantage of combining audio, video, and text capabilities. Such products are still available and are continually being enhanced with new capabilities and features.

The business version of NetSpeak's WebPhone, for example, offers four lines with call holding, muting, do not disturb, and call blocking options. It also offers a large video display area with self and remote views (see Figure 2.4).

Figure 2.4 NetSpeaks's WebPhone has a video capability that provides a remote view (shown) and self view.

2.6   Internet calls to conventional phones

In a demonstration of the feasibility of originating calls on the Internet and receiving them at conventional phones, the Free World Dialup (FWD) Global Server Network (GSN) project went online in March 1996. Organized by volunteers around the world, the noncommercial project was entirely coordinated in cyberspace via Internet telephony, e-mail, and chat software.

Using popular Internet telephony software, users were able to contact a remote server in the destination city of their call. This server "patches" the Internet phone call to any phone number in the local exchange area. This means, for example, that a user in Hong Kong can use an Internet-based server in Paris to effectively dial any local phone number and talk with a friend or family member in that city. A global server keeps a list of all servers and the real-time status of each.

The specific steps required to place a call through the FWD Global Server Network and answer it with a conventional phone are as follows:

1. Connect to the Internet as usual with a PPP or SLIP connection.

2. Start the phoneware and register with the vendor's server, if necessary.

3. Start the FWD client software and connect to a server in a select city.

4. Click on the connect button and enter the domain name of the server.

5. Once connected to the server, a message indicates that the connection is made, and a ring signal is sent.

6. Upon receiving the ring signal, the user enters the telephone number in the FWD client of the person in the local calling area, leaving out the area code.

7. The phoneware dials the number entered.

8. When the called person answers, both parties can start conversing.

This procedure is certainly more complicated than just picking up the phone and dialing the long-distance number, and its first implementation was limited to processing calls in only one direction—from the Internet to conventional phones. Critics cited these limitations as the reason why Internet telephony would not pose a threat to long-distance carriers for many years to come.

However, the technology demonstrated by the FWD Global Server Network has been continually improved to the point where it is now used in commercial products. Among the dozen or so companies that offer such products is VocalTec. The company's Telephony Gateway not only streamlines the calling process, it also provides many advanced features which make IP telephony services commercially viable.

To place a call, the user dials the nearest gateway from any phone or PC and then enters the destination number. After the gateways at each end establish the connection, the local gateway digitizes and compresses the incoming voice signals into packets, which travel over the Internet or intranet to the remote gateway where they are decompressed and reconstructed into their original form, making them suitable for transmission over the PSTN. From there, the call is then routed over the local loop connection to the destination telephone or PBX.

Calls also can be routed from the gateway to mobile phones on cellular networks and to computers attached to cable television networks. For the portion of the trip a voice call travels over the Internet/intranet, there are no costs incurred beyond that of the network connection. Per-call phone charges apply only for the portion of the trip the call travels over the PSTN or cellular network. On cable networks, operators charge a flat monthly fee for IP phone calls.

The IP/PSTN gateways offer many features that expedite administration. The VocalTec gateway, for example, provides features for call monitoring, security, and billing. It is even equipped with an Interactive Voice Response application, which acts as the interface between the PSTN/PBX and the IP network. The application includes an auto attendant, which guides users through the calling process.

2.7 Implementation perspectives

There are four implementation perspectives for commercial IP telephony services—traditional and next-generation carriers, ISPs, cable operators, and corporations. The choice often depends on the objectives of the organization.

Small businesses might rely on a carrier or national ISP to reduce their long-distance phone bill. Telemarketing firms with a Web presence might outsource their call center operations to a carrier that supports IP telephony, which is much cheaper than paying for an 800 number. Depending on its circumstances, a large company can choose a specific implementation or rely on a mix of all four.

For example, a company with many telecommuters and/or mobile professionals might rely on a national ISP to give remote users the ability to place a local call that connects them to the headquarters location. This can result in a substantial reduction in long-distance charges. A carrier-provided IP telephony service might be used to add value to an outsourced electronic commerce application. At the same time, the company might deploy an IP-PBX to route internal calls over its own intranet or VPN. A local or regional branch office might support telecommuting by tying in home users through the local CATV network.

2.7.1 Carriers

Many large carriers have built or are in the process of building extensive IP backbone networks capable of supporting data and voice traffic. AT&T is among the growing number of traditional carriers to embrace voice over IP. The telecom giant has announced a long-term plan to migrate all of its entire long-distance voice traffic to an IP platform. As part of AT&T's global joint venture with British Telecom, the two companies will build a new $10 billion global IP network over the next four years, which will deliver integrated voice and data services to multinational business customers.

The regional service providers are also planning extensive IP networks to position themselves for the emerging era of more vigorous competition from CLECs (competitive local exchange carriers). SBC (formerly Southwestern Bell), for example, will spend up to $600 million over the next few years to transform its circuit-switched data network to IP. Circuit switches will still exist, but the new SBC infrastructure will be capable of carrying multiple services, including voice, with a high degree of reliability.

Even traditional data-oriented value-added network (VAN) service providers are offering IP telephony service. Infonet, for example, offers the voice-over-IP (VoIP) service of NetWorks Telephony Corp. (NTC). Infonet users place telephone calls just as they would normally. NTC delivers the call to any phone, logs its duration, and generates a billing statement in real-time. Since Infonet operates a private, globally managed network, the quality of the voice calls is assured. Infonet's services are supported locally in 59 countries and its network is accessible from over 180 countries.

2.7.1.1 Gateway deployment

To support voice traffic, IP-PSTN gateways are deployed, which perform the translations between the two types of networks. When a standard voice call is received at a near-end gateway, the analog voice signal is digitized, compressed, and packetized for transmission over the IP network. At the far-end gateway the process is reversed, with the packets decompressed and returned to analog form before the call is delivered to its intended destination on the PSTN.

The gateways support the one or more of the internationally recognized G.7xx voice codec specifications for toll-quality voice. The most commonly supported codec specifications are:

▶ G.711—Describes the requirements for a codec using pulse code modulation (PCM) of voice frequencies at 64 Kbps, providing toll-quality voice on managed IP networks with sufficient available bandwidth.

▶ G.723.1—Describes the requirements for a dual-rate speech codec for multimedia communications (e.g., videoconferencing) transmitting at 5.3 Kbps and 6.3 Kbps. This codec provides near toll-quality voice on managed IP networks.[4]

▶ G.729(A)—Describes the requirements for a low-complexity codec that transmits digitized and compressed voice at 8 Kbps. This codec provides toll-quality voice on managed IP networks.

The specific codec to be used is negotiated on a call-by-call basis between the gateways using the H.245 control protocol. Among other things, the H.245 protocol provides for capability exchange, enabling

---

4. The mean opinion score (MOS) used to rate the quality of speech codecs measures toll-quality voice as having a top score of 4.0. With G.723.1, voice quality is rated at 3.98, which is only 2% less than that of standard analog telephony.

the gateways to implement the same codec at the time the call is placed. The gateways may be configured to implement a specific codec at the time the call is established, based on predefined criteria, such as:

- Use G.711 only—in which case, the G.711 codec will be used for all calls.

- Use G.729 (A) only—in which case, the G.729 (A) codec will be used for all calls.

- Use highest common bit rate codec—in which case, the codec that will provide the best voice quality is selected.

- Use lowest common bit rate codec—in which case, the codec that will provide the lowest packet bandwidth requirement is selected.

This capability exchange feature provides carriers and ISPs with the flexibility to offer different quality voice services at different price points. It also allows corporate customers to specify a preferred proprietary codec to support voice or a voice-enabled application through an intranet or VPN.

2.7.1.2 Call capacity

Many of the first-generation gateways have been limited in terms of the number of calls they can handle simultaneously—generally, they could not handle more than a few hundred calls at a time. As noted in Chapter 1, there are now carrier-class platforms available that let service providers add new services and features quickly and permit the processing of tens of thousands of simultaneous IP calls. The traditional telephone switch providers offer such platforms, including Ericsson, Lucent Technologies, Nortel, and Siemens. Nortel's DMS IP Gateway, for example, is an application of the Nortel DMS-100 central office switch that offers configurations supporting up to 100,000 IP ports and full access to all the end-user features available on the PSTN over an IP network.

A number of technologies may be employed to maintain consistent call quality over IP nets, including the use of ATM on high-capacity fiber backbones and routing protocols that give preference to real-time traffic. Although delay will not be entirely eliminated, neither will it remain an ongoing problem. As managed IP backbones branch out to reach major metropolitan areas, delay will be less of an issue to cost-conscious consumers and businesses.

2.7.1.3 Affordability

At the start of 1998, most IP telephony gateways were priced in excess of $500 per port. By year-end 1998, the price dropped to $200 per port. At this price, service providers and corporations can easily be dissuaded from making further investments in large-scale central office systems and enterprise communications systems. Instead, they will be more motivated to leverage the data backbones that are in place today by adding voice at only an incremental cost.

The increasing popularity of IP telephony has attracted new companies to this growing market. Sonus Networks and Salix Technologies, for example, are two smaller companies that offer carrier-class switches at breakthrough prices of less than $200 per port. Both the Sonus Gateway Switch (SGW) and the Salix ETX5000 are SS7-compatible[5] devices that can support both PSTN and packet data network (PDN) protocols.

The SGW can handle more than 70,000 simultaneous calls and is capable of moving voice traffic over managed IP networks with less than 100 ms of delay. It also features hot-swappable hardware to facilitate restoral and add new features. SS7 support enables traditional call services such as one-stage dialing as well as new features such as pre-call device negotiation.

2.7.1.4 Service creation

While it is generally accepted that the movement of voice traffic onto an IP network will result in substantial operational savings, the creation of new services and their associated revenue streams is the single most compelling driver for converged IP telephony. The key to realizing this potential is enabling the rapid introduction of new services. Sonus is addressing these needs with its Open Services Architecture (OSA).

OSA provides a framework upon which new services can be added. By tapping OSA's policy- and script-based service definitions and

---

5. In its simplest form, SS7 is a common channel out-of-band signaling system that evolved from and became the replacement for multi-frequency in-band telephony signaling systems. However, SS7 is much more powerful and flexible than the earlier signaling systems because it is a message-based system designed to operate on separate digital facilities. This capability has led to SS7 growing beyond the initial support of voice services to support for enhanced services for the mass market such as credit and debit card validation, follow-me services, 800 number routing, wireless applications, and automatic call distribution capabilities. The robust capability of the SS7 architecture will also serve the needs of the emerging personal communications services (PCS). Work continues within the ITU to evolve and improve SS7.

coupling them with open APIs and internetworked links to SS7, carriers can create enhanced services and make them available transparently to users on their networks. The architecture also allows carriers to offer services that their customers can configure and manage themselves via open interfaces such as Web browsers.

2.7.1.5 Class independence

In the conventional PSTN, there are different classes of switches. A Class 4 switch performs the toll/tandem switching functions for long-distance calls. A Class 5 switch performs switching functions for local calls and provides value-added features such as voice mail. With IP telephony, however, there is no need for differently equipped switches on the network.

Salix Technologies offers the ETX5000 switch, which is distinguished by its so-called class-independence. The switch interfaces between the PSTN and PDN and supports IP, TDM, or ATM. These features and capabilities ease as well as speed a service provider's ability to deploy next-generation enhanced telephony services. These same capabilities also facilitate the deployment of VPNs—all from a single switch. The platform scales up to 1.2 million ports, handling up to 600,000 calls simultaneously.

With its class-independence, the type of service offered is no longer tied to the location of a switch in the network, thereby making it easier to provide new services. This means the ETX5000 can be used for either Class 4 or Class 5 central office installations. Feature management software allows any combination of service features to be directed to any port, eliminating the rigidity of the traditional Class 4/5 hierarchy of network switching.

For example, for IP-based carriers, the ETX5000 can function as a Class 4 switch. This means that it will perform the toll/tandem switching functions needed to interconnect IP networks with local exchange telephone clients, acting as a distributed Class 4 switch. At the same time, however, the ETX5000 can support the incremental addition of new advanced service features that are not traditionally associated with Class 4 switches, such as call redirection on busy and voice mail—features normally associated with Class 5 switches. This architecture permits any telephone switch feature to be distributed to any switch interface on demand.

2.7.1.6   Interoperability

Although IP/PSTN gateways have been around for several years, they tended to be proprietary, making it difficult to use them on the same network. This interoperability problem is being addressed by the vendors themselves through the iNOW interoperability specification developed by Lucent Technologies, ITXC Corp., and VocalTec Communications Ltd. The iNOW specification describes how vendors' gateways should handle such things as call initiation, security, and other media-level steps needed to support IP calls among gateways.

2.7.1.7   Billing

A factor critical to the success of Internet telephony is a customer management and billing system. In the early years of IP telephony, there were no systems available for monitoring IP network usage and generating call records that could be used for customer billing. Several companies offer such systems today, including NetSpeak. The company's Event Management Server (EMS) provides a centralized, back-end, data repository for all transactions in carrier, service provider, electronic commerce, and enterprise network environments.

EMS securely logs event records from other NetSpeak servers into a SQL database engine. Data is made available to any ODBC-compliant third-party application for report generation and analysis. The EMS uses standard ODBC statements inside encrypted NetSpeak Communications Protocol (NSCP) packets to transfer transaction records from other servers. The ODBC statements can instruct the EMS to create, query, or remove tables; insert, update, or delete records; or call up stored procedures. Data returned from inquiries contains delimited columns with one record per packet.

Another company that provides usage-based pricing, real-time tracking and rating in support of IP telephony is Portal Software. The company's Infranet software unifies the management of all core Internet business operations including the ability to register, track, manage, and bill subscribers. Its real-time architecture and open development environment also provides a platform for Internet telephony service creation and management. Toward these ends, Infranet includes the following capabilities:

▶ Register customers in real time so that service can begin immediately; this can be done with a customer service representative or via the Web;

▶ Authenticate calls in real time to reduce fraudulent use of the service;

▶ Verify credit limits in real time to reduce expenses associated with bad debt;

▶ Track prepaid usage in real time and proactively notify customers to purchase additional usage time;

▶ Manage balance information in real time to determine how to optimize call routing based on costs from affiliate originating and terminating providers;

▶ Track usage levels to determine quantity and location of POPs;

▶ Generate invoices for subscriber usage;

▶ Reconcile and manage settlement among originating and terminating providers;

▶ Quickly and easily mix and match subscription and usage-based plans along with any type of discount-based promotion, including offerings bundled with traditional IP services.

In terms of billing, the Infranet software can split charges or credits between multiple billing entities during the rating process. For example, providers can enable corporate customers to charge standard usage to the company and premium services to employees. The software can manage billing activity of business subscribers via purchase order, which includes real-time customer notification when charges reach a specified threshold. A Java-based invoice design tool and API permit integration with third-party document production systems.

With regard to accounts receivable management, the Infranet software supports fine-grained adjustments of accounts receivable through the reversal of individual billable events, including automatic handling of tax and general ledger adjustments. An intuitive graphical view of accounts receivable across hierarchical accounts simplifies and facilitates the management for complex corporate customers.

Fast implementation of new or modified pricing plans is critical in the fast-changing Internet business. Accordingly, Portal's Infranet software comes with a graphical tool for creating pricing plans, which features product creation wizards, enhanced dialogs, and context-sensitive help. This tool allows providers to quickly and easily configure Infranet to implement custom business policies without programming.

2.7.1.8 Wireless local loops

A relatively new development is the support of IP telephony over wireless local loops (WLLs). BreezeCom Ltd. and Cisco Systems offer a wireless local loop solution called BreezePHONE, which also provides high-speed wireless Internet data access at rates of up to 3 Mbps. By using the BreezePHONE system, telecom operators can obtain significant savings in infrastructure costs. The inherent flexibility of using wireless radio connections provides the infrastructure required for toll quality services to be installed simply, cheaply, and rapidly—especially in areas where the physical infrastructure is limited. It also allows competitive carriers to provide voice and data services equivalent to ADSL without the need for wired infrastructure.

The solution is comprised of the BreezePHONE Remote Unit (BRU), BreezePHONE Access Point (BAP), and a gateway connection to the PSTN using Cisco routers equipped with the company's voice/fax network module and Internetwork Operation System (IOS). Each module supports two or four voice channels through a variety of voice interface cards. The modules support several toll-quality compression schemes, including G.711 for high bit-rate applications and G.729 for WAN applications.

The BreezePHONE uses frequency-hopping, spread spectrum radio technology and operates in the license-free 2.4 GHz ISM (industrial, scientific, and medical) band. The embedded DSP technology allows it to operate reliably, with data speeds of up to 3 Mbps within a single-cell radius of 5 km. BreezePHONE's Committed Information Rate (CIR) capability allows bandwidth availability to be specified on a per-user basis.

2.7.1.9 Market drivers

The reason long-distance carriers are so interested in Internet telephony is that there is great pent-up demand among consumers for an economy class of telephone service that breaks the dime-a-minute barrier.[6] Carriers recognize that they can easily break this barrier by taking advantage of efficient, reliable, economical IP networks. Such networks have a proven record of efficiency and reliability that spans almost 30 years. They are also cheap to build—a router-based network costs only 20% of a traditional circuit-switched network of comparable size. New carriers like Quest LCI view managed IP networks as a viable way to get into the long-distance service business quickly and economically, possibly in partnership with the former "Baby Bells."[7]

While cost savings is an important market driver for IP telephony services, it is not the only one. Another market driver for IP telephony is the opportunity to combine it with data services for building value-added applications. Examples of these applications include online help desks, telemarketing call centers, and customer care operations. In fact, some carriers view the provision of value-added services as a key ingredient in their IP telephony strategy.

2.7.2 Internet service providers

Most ISPs view Internet telephony as a value-added service that can attract new subscribers and retain existing subscribers, while providing a new source of revenue. As noted in Chapter 1, others see Internet telephony as a viable means of competing with established carriers. However, the lack of WAN infrastructure—or the resources to build one—has kept many smaller ISPs from participating in the IP telephony market.

This problem can be overcome through peering arrangements with other domestic ISPs or participation within global consortia that have IP telephony gateways in other countries. These arrangements give ISP subscribers local-call access to the phones in most major metropolitan areas in the United States and around the world.

---

6. Large corporations can negotiate much lower rates with carriers, based on call volume. Depending on the carrier, per-minute rates of between three and five cents can be negotiated for long-distance domestic calls and between seven and ten cents for international calls.

7. As noted in Chapter 1, a joint marketing agreement between Qwest LCI and U S West, which would have provided U S West customers with a discounted local and long-distance service package, ran into legal trouble in 1998 because U S West had not yet received regulatory approval to offer long-distance service within its 14-state region. Nevertheless, the arrangement illustrates the competitive impact IP-based next-generation service providers will have in the future. Such partnerships will enable them to compete in the long-distance market without having to invest in local loops. It also offers the Baby Bells a way into the long-distance market without having to invest in state-of-the-art long-haul transport facilities. When regulatory restrictions are lifted, such partnerships can mean more competition and new programs that provide customers with lower cost, value-added services based on IP.

IP telephony service provider Delta Three, for example, offers such a partnership arrangement. The company's Network Partners Program (NPP) is a turnkey solution jointly developed with Ericsson. The solution includes the Ericsson Internet Protocol Telephony Solution for Carriers (IPTC) platform operating over Delta Three's managed IP telephony network. Partners have the ability to generate revenue both by originating and terminating traffic over the Delta Three network. Delta Three makes it easy for partners to manage their business by providing them with online access to real-time data. Through Delta Three's Interactive Center, partners can view their customers call activity individually or in groups, access call detail records (CDRs), and instantly graph the information for management reporting needs.

Another company that offers this type of arrangement is Inter-Tel. The company offers ISPs the means to extend their infrastructures by allowing them to tap into its expanding network and also provides them with a package of services for account management, billing, and settlement.

Inter-Tel itself is part of a larger alliance that includes NTT International Corp. (NTTI), Glocalnet AB, and Telba Telecomunicacoes Ltd. Each alliance member is committed to maintaining the quality of their network connections, while expanding network coverage within their regions by either building additional infrastructure or forming relationships with smaller Internet telephony service providers (ITSPs).

Numerous other vendors are addressing the ISP market for IP telephony service. For example, a system offered by eFusion is the eStream Enhanced Internet Services (EIS) application gateway, which provides call-handling options for end users. Normally when a subscriber is browsing the Web, the person has no way to recognize and accept incoming calls because the line is being used for the Internet access connection. eFusion's Internet Call Waiting application allows the user to accept the incoming calls using Internet telephony, transfer the call to a traditional telephone line, or have the caller leave a voice mail message—all without leaving the Internet session.

While a subscriber is online, a screen will pop up to notify the subscriber that they have an incoming call. From this screen, a user can choose to:

▶ Accept the incoming call using any H.323 Internet phone and continue to browse the Web;

- Transfer the call to a telephone number specified in the subscriber's profile—such as a second phone line or cellular phone;

- Transfer the call to the subscriber's network-based voice mailbox;

- Take a message and save it on the eStream EIS application gateway.

When messages are stored, the subscriber can use the Internet call waiting application to see a list of messages on the system—who called, when, and length of each message. The subscriber can then play back the message using their Internet phone, e-mail the message to a predefined address for playback later, or delete the message from the system. The user only needs a standard multimedia PC equipped with an H.323-compliant Internet phone like Microsoft NetMeeting or Netscape Conference.

2.7.3 Cable operators

Cable operators offering Internet access now are beginning to offer IP telephony services as well. Although this has been possible for a long time, the new wrinkle is the ability to support data and voice through the same cable modem. The modem, in turn, is connected to the same coaxial cable that delivers TV service. Previously, to put voice and data on the cable network required a modem and a separate voice gateway, which is more complex and expensive.

The new type of cable modem is known as an integrated multimedia terminal adapter (MTA). Motorola's MTA, for example, has a telephone port and a four-port Ethernet hub built into it. Data speeds on the cable can be as high as 10 Mbps depending on how many other users are on the same shared cable subnet to exchange information. The integrated MTA uses proprietary traffic prioritization techniques to ensure that there is enough bandwidth to prevent voice packets from getting delayed.

The MTA sits on the customer premises (see Figure 2.5), where it converts phone and PC traffic into packets for transport over the cable network. The MTA sends the packets to a router or packet switch on the cable network, where voice and data are sorted out. Data is routed to the cable operator's packet network, while voice is routed to an IP/PSTN gateway. The gateway decompresses the packets and returns them to analog form so voice can be received by an ordinary phone on the public network or a corporate phone in back of a PBX.

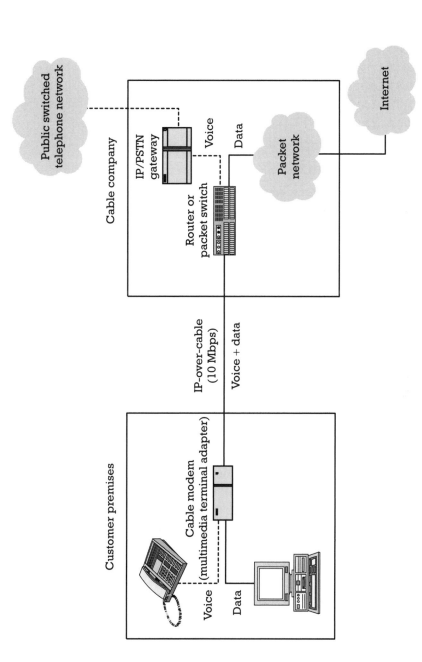

Figure 2.5 The route of a telephone call over a two-way cable network.

With many corporations devising ways of implementing VoIP internally, the cable option allows them to extend it out to the telecommuter. Cable modems already support data at the Ethernet speed of 10 Mbps, which gives telecommuters a viable way to retrieve information from the corporate LAN via the Internet access connection. Although ISDN is more flexible in terms of bandwidth allocation and call handling features, it does not provide enough capacity for frequent LAN access and it is typically more expensive than cable.

2.7.4 Corporations

Many corporations appreciate the benefits of being able to use their extensive IP-based intranets for telephony applications. In the closed environment of the intranet, congestion is more manageable and the levels of it are smaller and less frequent. Latency can be more easily controlled by upgrading routers with protocols that handle real-time traffic more effectively and by adding more bandwidth when baseline performance metrics reach critical thresholds. Companies can also increase router capacity to handle more traffic and reconfigure their networks to eliminate excessive router hops that cause delay.

2.7.4.1 Cost savings

Companies with as few as a dozen employees in a remote location who call the main office several times a week should be able to pay for an IP telephony gateway in three or four months. Once paid for, the system provides virtually free phone service between offices. A larger company, running T1 lines connecting 30–35 locations, would save about $475,000 over five years, assuming that the IP telephony gateway replaced PSTN calls billed at an average of 10 cents per minute.

From the perspective of the enterprise IT professional, the addition of telephony services would normally impose an undue management burden. Now there are tools available that automate traffic discovery and bandwidth control, such as Packeteer's PacketShaper.

2.7.4.2 Bandwidth management

A major obstacle to ensuring voice traffic the guaranteed availability and quality of service needed to satisfy users accustomed to smooth, jitter-free phone conversations has been that TCP data traffic has the habit of "bursting" up and interfering with voice transmissions that reside on the same network.

PacketShaper discovers and classifies voice and other multimedia traffic on the corporate intranet, and automatically suggests and enforces appropriate bandwidth policies that protect voice traffic traveling into the LAN or out to the WAN from TCP bursts caused by Web surfing, file transfers, or large e-mails with graphical attachments—something routers cannot do. And if a network manager knows that it is file transfers that need protection—from a bandwidth-hungry video application such as NetShow, for example—appropriate policies also can be set to facilitate file transfers.

PacketShaper supports H.323 but goes beyond basic compatibility to include full traffic discovery for all underlying protocols that constitute the H.323 standard: Q.931 (digital subscriber signaling), H.245 (capability exchange), T.120 (data collaboration and whiteboarding), RTP (Real-Time Transport Protocol), and RTCP (Real-Time Transport Control Protocol). NetMeeting, which is H.323-compliant, and NetShow, a traditional streaming video application, are specifically discovered and identified by name.

After discovering the traffic, PacketShaper suggests appropriate policies for constructing a voice application SLA (service level agreement). An organization that wants to connect its headquarters to a branch office with VoIP over an existing frame relay service, for example, could use PacketShaper to set rate-control policies for Web traffic, a low-bandwidth rate guarantee for RTCP traffic, an 8-Kbps rate guarantee for call session (RTP) flows, and complementary latency bounds for all RTP traffic to minimize jitter over the shared frame relay link. This would allow efficient access-link sharing without requiring new frame relay ports or permanent virtual circuits.

Determining such policies is particularly complex because VoIP traffic is not only bi-directional (e.g., a two-way dialog) but also involves multiple data (RTP) and control (RTCP) streams whose ports are dynamically negotiated at setup time, thus requiring more intelligence than would be needed for a simple TCP port- or socket-based classification scheme.

PacketShaper applies its rate-control technology to all TCP traffic on the network, dynamically guaranteeing availability for voice calls and protecting voice traffic from delay-causing TCP bursts. In addition, UDP latency controls can be used to set delay bounds at the millisecond level, so that non-rate-controllable UDP traffic can be scheduled and delivered within latencies appropriate for voice.

2.7.4.3   Virtual private networks

An alternative to a corporate intranet is an IP-based VPN. Although the links are shared among other subscribers of the carrier, the VPN is actually a closed environment in which congestion and delay can be minimized to facilitate voice traffic. With an IP-based VPN, business users essentially carve out their own WAN within a carrier's high-speed IP backbone using various tunneling protocols. In this way, VPNs running between corporate offices can provide the quality connections businesses demand for voice.

Several tunneling protocols are used to implement IP-based VPNs— the Point-to-Point Tunneling Protocol (PPTP), the Layer 2 Forwarding (L2F) Protocol, the Layer 2 Tunneling Protocol (L2TP), and the IP Security (IPSec) Protocol.

PPTP supports flow control and multiprotocol tunneling between two points (see Figure 2.6). L2F also supports multiple protocols, but can be used to create tunnels between multiple locations. Both protocols are vendor driven—PPTP is backed by Microsoft and 3Com, among others; L2F is backed by Cisco Systems, Nortel Networks, and Shiva. Since PPTP and L2F are not interoperable, L2TP is under consideration as an IETF standard. L2TP is a combination of PPTP and L2F, and offers the advantage of interoperability. The drawback to all three tunneling protocols, however, is their lack of integrated authentication, encryption, and integrity features.

As noted in the previous chapter, IETF's IPSec provides for packet-by-packet authentication, encryption, and integrity. These functions must be applied to every IP packet because Layer 3 protocols such as IP are stateless; that is, there is no way to be sure whether a packet is really associated with a particular connection. Higher-layer protocols such as TCP are stateful, but connection information can be easily duplicated or "spoofed" by knowledgeable hackers. The key limitation of IPSec is that it can only carry IP packets, whereas the other three protocols support IPX and AppleTalk, as well as IP.

The major carriers and ISPs provide service-level guarantees for their managed VPN services to overcome concerns about latency and other quality of service issues traditionally associated with the public Internet.

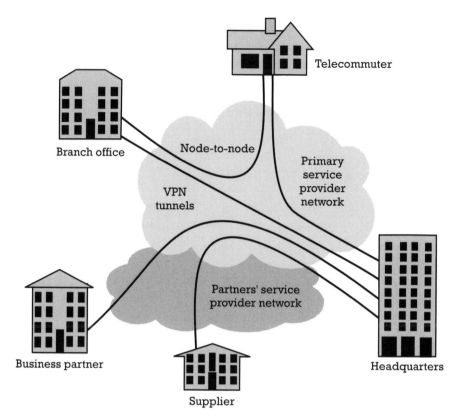

Figure 2.6 Tunneling offers the means of transporting multiprotocol traffic in a private session over the Internet.

Carriers and ISPs are able to guarantee network performance because they control of all the equipment, protocols, and bandwidth end-to-end. This is not possible on the public Internet because there is no central management authority—as a result, the performance of real-time applications, including voice, suffers.

Some industry experts have even made the case that well-managed IP networks are actually more reliable than the PSTN. For example, if a circuit-switched network experiences an outage, calls in progress will be disconnected. If an IP network experiences an outage, calls are not vulnerable to cut-off because if packets get blocked along one path, they simply find another path to the destination.

Another incentive to implement voice service over IP-based VPNs and corporate intranets is that the calls will probably be less likely to be subjected to the kind of access charges that may be slapped by regulators on carrier-based VoIP services as they come to more closely resemble the kind of PSTN long-distance service offered by traditional carriers.

2.7.5 Customer premises equipment

As noted in Chapter 1, some of the new customer premises equipment (CPE) becoming available routes voice traffic over corporate intranets, including IP-based PBXs and IP trunk modules for existing PBXs. There are also remote-access gateway devices that tie branch offices into the corporate IP network and larger multiprotocol switches for the enterprise.

2.7.5.1 Remote access units

Multi-Tech Systems, for example, offers an easy-to-install remote-access unit that offers two voice channels. The MVP200 connects a headquarters PBX to a remote office PBX or key telephone system via a company's IP network or the public Internet. For basic voice operation, one unit is installed between a PBX and an IP router located on a corporate Ethernet LAN. A second MVP200 unit is installed at a remote office between its phone equipment (key system or handset) and existing IP router. To make a voice call, a user at the corporate end dials a special extension, gets a dial tone, and then dials the number of the second unit at the remote location. When the link is made between the two units, the remote phone rings.

The MVP200 uses G.723 voice compression to enable voice conversations over an IP connection at only 6.3 Kbps. When more bandwidth is available, the compression ratio can be reduced for higher clarity.

A more versatile remote access system from 3Com—the PathBuilder S200 Voice Access Switch—supports voice and fax over both frame relay and IP, letting users choose between either protocol. The switch dynamically allocates bandwidth as needed to funnel voice and data traffic over IP or frame relay. This gives companies the option of using the more familiar frame relay platform before venturing into IP for the delivery of voice and data.

PathBuilder comes in two versions—a four-port unit for small and branch offices, and a 62-port device for PBX deployments at the corporate headquarters. Both devices handle basic voice transport options such as

G.711 pulse code modulation and compression down to 5.3 Kbps. In addition, the switches support enhancements such as echo cancellation, alternate destination routing, and Voice Activity Detection (VAD) to reduce bandwidth usage during silent periods of the call.

The dynamic voice and data delivery mechanism built within PathBuilder comes via dynamic cell packet transport (DCPT). In essence, when the switch detects voice traffic, DCPT fragments data packets to reduce the disparity between voice and data. In this way, small voice packets are not delayed as they wait for much larger data packets to be pumped through a network. When voice traffic is not detected, DCPT automatically returns the data packets to their non-fragmented state.

PathBuilder features network voice switching, where call switching takes place within the device itself instead of a PBX. It provides streamlined common channel signaling, in which commands can be read by PathBuilder and communicated directly to PBXs instead of having to rely on tunneling through network connections. This allows separate ISDN PBXs to be linked into a cohesive whole.

PathBuilder also offers simplified administration support. With the Voice Name Server (VNS), the administrator can centralize mapping information between phone numbers and the physical location of phones or PBXs. Acting much like the Domain Name Server (DNS) in conventional IP networks, dialed digits received on voice interfaces of remote nodes are sent to the VNS, which responds with the information necessary for the remote node to route the call through the network. Unlike traditional voice switching solutions that often require every node to be altered to add a new node or phone number, the administrator need only modify the VNS to provide all-site access to new users.

2.7.5.2 Premises gateways

Most current corporate VoIP product offerings are simple gateway solutions that route voice traffic over the Internet instead of the PSTN. This solution solves the cost problem, but adds two significant obstacles—unpredictable QoS and loss of PBX features, both of which companies have come to rely on as critical business tools. Businesses want a solution that saves them money while allowing them to retain or enhance the PBX feature set. To meet these requirements, a premises gateway is required which integrates traditional PBX features with

PSTN reliability over a corporate WAN. In the event of congestion or voice quality degradation on the IP network, the server routes calls onto the PSTN.

One vendor that offers a premises gateway is StarVox. The company's StarGate Server brings PSTN reliability and QoS to voice calls over the data network. To do this, the StarGate Server provides two key features—Failsafe and Fallback. Failsafe automatically detects the failure of any end-point in the data network for a specific call before completing the connection. If such a failure exists, Failsafe routes the call over the PSTN with no loss in service.

Fallback, in turn, constantly monitors the QoS during the call. If the data network service levels drop below preset packet loss, jitter, and latency thresholds, Fallback automatically reroutes the degrading WAN call over the PSTN without dropping the call. In the event of a catastrophic WAN failure, the StarGate Server will audibly inform the connected parties to stand by while it reroutes the call. Together, Failsafe and Fallback ensure that companies have the PSTN reliability and resiliency they need to safely deploy IP telephony throughout their organization.

As a network server that connects to a corporation's PBX through voice links and call control, the StarGate Server overlaps the existing PBX infrastructure in a non-intrusive way. A corporation's existing voice and data networks are connected only through StarGate; the organization's existing PBX and telephones as well as its network of PCs and data network remain untouched.

Users interact with the StarGate system through client software installed on their PCs. The StarCall Client is a browser-based Java applet that users activate by selecting a bookmarked URL (Uniform Resource Locator). Once the applet is loaded and a call comes through, the user receives a screen pop-up message showing the calling party's name. When placing a call, the applet provides access to a corporate "white pages." Users can point and click to dial any co-worker worldwide. If the called party is busy, the applet allows the caller to set a callback or send a message, instead of being sent to voice mail without any choice. The applet also provides a call log of all incoming calls to the desktop with the name of the caller and the result (i.e., hang up, voice mail, and so forth). Caller names are available for any call from a StarGate equipped site, as well as calls from the PSTN if caller ID is received with the call.

The Java applet implementation simplifies administration of the Star-Gate system. The VoIP application resides on the server, while the applet acts as the application front end. Keeping the VoIP application on the server means that installation happens only once—not hundreds or thousands of times at individual desktops. Since periodic updates to the application occur only at the server, all users are assured of having the same version of the application every time it is accessed through a freshly downloaded applet.

StarGate also integrates directory services, helping corporations to reap the full advantages of marrying voice and data. Not only can organizations save time and money by combining directory services for data networks with those of the voice network, but they can virtually configure their users rather than hardwire them. This adds a valuable level of flexibility when it comes to implementing moves, adds, and changes, which are continuous, resource-intensive tasks at most large companies.

The integration of StarGate with directory services such as Novell NetWare Directory Services (NDS), Microsoft Active Directory Service (ADS), and the Lightweight Directory Access Protocol (LDAP) provides uniform administration, eliminating the need for costly parallel administration or manual synchronization of user information and feature access rights across the integrated network. The StarGate Server also has a set of NetWare NWAdmin snap-in modules that enable a single point of control for the corporation's voice and data networks. StarGate supports more than 50 commercially available PBX systems through computer-telephony integration (CTI). This gives the StarGate Server both PBX status information as well as the ability to issue call control commands to the switch. In some instances, minor off-the-shelf hardware upgrades are required where the existing PBX is not configured with CTI or an ISDN PRI control channel.

The combination of directory services integration (DSI) and CTI allows the StarGate Servers to monitor when a user dials a call, determine the path through which the call should be sent, and instruct the PBX to route the call accordingly. This not only enables simple VoIP, but also gives the StarGate Servers the means to provide more sophisticated functionality that is required in today's corporate environment. These advanced features include routing calls around network failures; performing dynamic FallBack to the PSTN; presenting common, consistent point-and-click call control menus to all corporate users; and allowing users to set filter-based call forwarding and to set call-back on busy requests. Without DSI and CTI, VoIP products are merely "dumb pipes" which lack the intelligence to deliver the PBX features and QoS that corporate users require.

2.7.5.3 Routers

Implementing QoS for packetized voice will require changes to routers on the corporate intranet. Since the intranet will be carrying both voice and data, the routers must be equipped with protocols that can give priority to voice traffic. If voice is not given priority, then jitter, latency, and packet loss may degrade communication and negate whatever benefits the organization hoped to achieve with voice-data convergence over IP.

To provide the required voice quality, the QoS capability must be added to the traditional data-only network. This is accomplished in the router's operating system, such as Cisco's IOS. The QoS features implemented in the operating system give VoIP traffic the service it needs, while providing the traditional data traffic with the service it needs.

Cisco Systems, which supplies most of the routers used on the Internet and corporate intranets, provides several solutions for implementing QoS, including priority output queuing, weighted fair queuing, and the RSVP, which it developed and promoted as an IETF standard.

With priority output queuing, network managers can classify traffic into four priorities and provide the available bandwidth to the queues in the order of their priority. The highest priority queue gets as much bandwidth as it needs before lower priority queues get serviced.

Custom output queuing enables multiple queues to be defined, with each assigned a portion of the total bandwidth. For example, when used on a Novell IPX/SPX network, custom queuing gives network managers the ability to meet the QoS needs of several applications simultaneously. Mission-critical applications can be assigned to receive 40% of available bandwidth at all times, a videoconference session on TCP/IP can be assigned 30% of capacity to ensure smooth reception, while other network applications can share the remaining 30%. When the videoconference terminates, that amount of bandwidth goes back into the pool, ready to be reallocated to other network tasks.

Custom queuing is performed by the routers, but the processing delay is negligible because only the packet headers are examined. If there is no congestion on the outgoing link, queuing is not an issue and no further processing is done—the packets are put onto the outgoing interface as fast as they are received. If there is congestion on the outgoing link, then there will be some further processing to translate the queuing parameters. Since the outgoing link is congested anyway, overall performance is not affected one way or another by the extra processing.

Another technique of Cisco's is weighted fair queuing, which ensures that the queues do not starve for bandwidth and that the traffic gets guaranteed service. This is accomplished by classifying data into low-priority traffic and high-priority traffic and treating them differently. High-priority flows receive immediate handling, while low priority flows are interleaved and receive proportionate shares of the remaining bandwidth. Packets from low-priority flows may be discarded during periods of congestion.

2.7.5.4 Network management

With the growing popularity of VoIP services, network management vendors are offering tools that help IT administrators and network managers monitor voice and data traffic performance on IP nets. NetReality, for example, offers the WanTel module that integrates with its WiseWan product family to offer end-to-end QoS for integrated voice and data traffic. The module provides real-time traffic monitoring, prioritization, adaptive bandwidth allocation, and accounting for voice and data traffic over both IP and frame relay networks.

The WanTel module allows network managers to monitor information such as callers and destinations, call duration, time of day distribution, and associated costs. Voice and data traffic is dynamically prioritized in real time, based on actual bandwidth availability at the circuit level. Voice sessions can be accounted and billed for according to factors such as priority, user type, and time of day. Corporations can use these detailed accounting features to bill back departments, while service providers may use them to bill customers on a per-use and/or priority basis.

2.7.5.5 System administration

Through an IP telephony system administration application, an IP telephone system or commercial service could be configured to assign users a class of service, just as users are assigned a class of service through a conventional PBX or telephone service.

In the corporate environment, certain users, such as clerical staff, might be assigned a class of service that limits them to making calls over the corporate intranet, for example, while higher level staff might be assigned a class of service that allows them to use another type of service when the IP telephone system has no ports available. Managers and executives might be assigned a class of service that permits calls over any type of service. Different call features can also be assigned to a class of service. For example, if clerical staff have no need for voice mail or conferencing features, their class of service would not allow them to access these features.

2.8 Analysis

To be competitive, every business should have a strategy for implement- ing and leveraging IP networks with the addition of telephone service. Companies can evaluate IP telephony services in three ways. One way is to trial a commercial offering such as IDT's Net2Phone Direct on a call- by-call basis to get a sense of the service's quality before making further commitments. Another way is to add an IP module to an existing PBX or ACD and review the call detail reports for clues about performance. Alter- natively, it is possible with some vendors to obtain a small-scale IP-PSTN gateway for a free evaluation.

VoIP technology is ready to deploy today, but works best on private, managed IP networks where performance can be closely monitored and fine-tuned. According to various industry estimates, 50% of all compa- nies that have a private intranet are already running or experimenting with integrated voice-data applications and IP telephony.

Companies that have intranet links to the public Internet must be concerned about security. Firewalls and routers with firewall capabilities have amply demonstrated their effectiveness in stopping intruders from gaining access to corporate resources. Sensitive IP voice communications can be further protected by adding encryption, in the same way as sensi- tive data applications are protected.

The technology has progressed to the point where there is little or no reason for companies to delay implementation of VoIP solutions, especially for internal communications. Even traditional long-distance carriers have come to recognize that IP telephony is the wave of the future. Companies that are still skeptical of the technology should at least experiment with it to see how it might benefit them in the future.

The business decision to subscribe to the VoIP services of a carrier or ISP hinges on such factors as coverage, voice quality, features, cost, and the demonstrated commitment of the service provider to invest in the IP infrastructure.

In selecting a VoIP service provider, it is important to determine existing calling patterns. To minimize call costs, the service provider must be able to reach the most frequently called areas with as little "off-net" traffic delivery as possible. In cases where calls are frequently made between major cities across the country, a long-distance carrier or nationwide ISP would be best. If most calls are regional, a smaller ISP or RBOC might be the best choice, provided of course that they offer VoIP service. If a significant number of calls go to international locations, it would be important to choose a service provider that has IP-PSTN gateways in the countries where most of the calls go.

The service provider should be using the latest DSP technology in its IP-PSTN gateways to ensure the highest voice quality. It should be using routers that employ QoS mechanisms to give voice priority over ordinary data traffic that might also be included in the service, such as fax, e-mail, and file transfer. The IP backbone network itself should be dedicated (i.e., not part of the public Internet) and be continually monitored for peak performance.

Call features is another important criterion when choosing a VoIP service provider. Saving money with VoIP services need not mean giving up access to common call handling features. Some VoIP services are enhanced with such features such as single-stage dialing, voice mail, call forward, three-way calling, calling line ID, and pre-paid calling cards, among others.

The cost per call is another important selection criterion. AT&T, which offers WorldNet Voice, is experimenting with price models of 7.5 to 9 cents per minute. IDT Corp. offers Net2Phone Direct in 50 U.S. cities and charges about 5 cents per minute. Qwest LCI offers Q.talk in 25 U.S. cities and charges 7.5 cents per minute. For many companies these call charges represent a savings of up to 50% over long-distance calls placed over the PSTN. Of course, these rates are constantly changing.

Finally, it is a good idea to determine how committed the service provider is to the concept of VoIP service. Some ISPs just add VoIP service to help distinguish themselves from their competition and have no intention of investing in infrastructure to improve the service and expand the customer base. Others, like Qwest, are building high-speed fiber backbones and expect to offer a quality VoIP service that is capable of competing against the traditional long-distance carriers.

At this writing, no carrier or ISP offers an SLA specifically for IP telephony services. SLAs are contracts that specify the performance parameters within which a network service is provided. These contracts define the expected performance level in terms of delay, error rate, port availability, and network uptime. Response time to system repair and/or network restoral also can be incorporated into the SLA, as can penalties for noncompliance.

In mid-1998, a major step toward guaranteed performance for IP telephony was taken by ISP consortium Gric Communications. The consortium, which consists of over 350 ISPs that offer customers local dial-up access to the Internet from around the world using their standard ID and password, offers an IP telephony service that guarantees latency will not exceed 400 ms across the network. Any member of the consortium that cannot meet this guarantee will not be able to offer the IP telephony service, called GricPhone. Eventually, carriers and major ISPs will offer SLAs for VoIP services that include credits for poor performance.

The IP telephony market will experience tremendous growth over the next few years. The market will go through various phases, during which demand will increase, technology will improve, more service providers will appear, and IP networks will expand, until there is little to distinguish traditional phone calls from IP phone calls.

2.9 Conclusion

The poor voice quality offered by first-generation Internet telephony products condemned them to hobby status—but only briefly. Many of the issues that plagued users of first-generation Internet telephony products have been addressed by hardware and software vendors with the goal of facilitating the growth of commercial, carrier-grade VoIP services. The use of DSP technology has improved voice quality. Call processing has migrated from user desktops to Internet servers, which has improved performance even more. Speech encoding standards have been issued by the

ITU as part of the broader H.323 standard, facilitating interoperability between competing products.

There are now scalable IP switches that compete with traditional central office switches in terms of call-processing capacity and call-handling features. There are roaming agreements between service providers that help ensure the broadest possible coverage for commercial IP telephony service. There are gateways that enable ordinary phones to be used for placing and receiving calls between the PSTN and private IP networks. For businesses, there are even devices that route calls over intranets or extranets, and when service degrades on these nets, transparently route calls to the PSTN.

Mechanisms have been developed to accurately meter usage and charge for voice calls over IP networks. Administrative tools allow individual users or groups of users to be assigned a class of service. There are even tools that allow network managers to monitor the performance of the IP network in real time and check on the quality of service being delivered to each user at any given time. These and other developments have prompted carriers, ISPs, and corporations to take a serious look at this once-spurned technology.

The value of IP telephony is not confined to savings on long-distance calls. The ability to converge voice and data within the same packet stream over well-managed IP nets opens up the opportunity to create value-added applications that can have a tremendous impact on daily business operations. Combining voice and data within value-added applications can allow companies to better meet customer and employee needs, further reduce costs, and increase staff productivity in ways that have not been economically feasible with other technologies and carrier services. This is the subject of the next chapter.

More information

The following Web pages contain more information about the topics discussed in this chapter:

3Com: http://www.3com.com/
AT&T: http://www.att.com/
BreezeCom Ltd.: http://www.breezecom.com/
Cisco Systems: http://www.cisco.com/

Delta Three: http://www.deltathree.com/
eFusion: http://www.efusion.com/
Ericsson: http://www.ericsson.com/
Gric Communications: http://www.gric.com/
IDT: http://www.net2phone.com/
Inter-Tel: http://www.inter-tel.com/
International Telecommunication Union: http://www.itu.org/
ITXC: http://www.itxc.com/
Lucent Technologies: http://www.lucent.com/
Motorola: http://www.motorola.com/
Multi-Tech Systems: http://www.multitech.com/
NetReality: http://www.net-reality.com/
NetSpeak: http://www.netspeak.com/
Nokia: http://www.nokia.com/
Nortel Networks: http://www.nortel.com/
Packeteer: http://www.packeteer.com/
Portal Software: http://www.portal.com/
Qwest LCI: http://www.qwest.com/
Salix Technologies: http://www.salix.com/
Sonus Networks: http://www.sonusnet.com/
StarVox: http://www.starvox.com/
VocalTec: http://www.vocaltec.com/

CHAPTER

3

Contents

Value-added application: The Internet call center

Although the digitization of telecommunications has been occurring for decades—the long-distance telephone network in the United States is now almost entirely comprised of digital switches and fiber optic transmission links—the infrastructure has been optimized for the efficient transport of voice. In contrast, the Internet can transmit any form of data, including packetized voice. Internet protocols are sufficiently flexible to overcome the boundaries between voice and data services. This allows new services to be developed and immediately loaded onto the existing Internet infrastructure. This convergence creates new markets and new efficiencies because particular applications are no longer locked into specific lines, services, and technologies.

Telephone service on the Internet is software-driven. It works by sampling speech input and converting it into data, which can be compressed and assembled into packets for transmission over the Internet like any other packets and reassembled as audio output at the receiving end. Internet telephony is also technically different from conventional telephone service. A circuit-switched voice call uses an entire 56-Kbps channel; Internet telephony uses digital compression techniques that can encode packetized voice to as little as 4 Kbps. Internet telephony is also packet-switched, which means that it does not tie up a call path for the duration of the call.

The convergence of voice and data over IP networks opens new opportunities for the provision of value-added services, provides global reach, and lowers the cost of doing business. Among the value-added applications that this voice-data convergence makes possible are call centers (inbound and outbound), help desks, reservation centers, distance learning, telecommuting, and telemedicine. Although these services have been available for many years over the traditional circuit-switched PSTN or private leased lines, the ability to mix voice and data over IP networks enables users to take advantage of new features and changes the cost-benefits equation in a way that can stimulate much broader usage. Although this chapter focuses on Internet call centers (ICCs), the same principles can be applied to other applications, especially help desks and reservation systems, which are variations of the call center.

3.1   Concept of the call center

A call center is a facility that is configured, staffed, and managed to handle a high volume of incoming calls, with multiple agents responding to calls. A traditional call center is typically equipped with an automatic call distributor (ACD) connected to the company's PBX. The function of the ACD is to take certain calls coming into the PBX and route them to a given extension or any unused extension without operator intervention (see Figure 3.1).

Call centers typically are used to serve customers via toll-free 800 numbers or pay-per-minute 900 numbers. Mixed calls can also be fed into the PBX and ACD via ISDN PRI lines. About $900 billion of goods and services were sold through call centers in 1998 and that figure is expected to surpass $1.5 billion in 2001, according to Dataquest. In recent years, call centers have evolved from mere order-taking operations to "customer

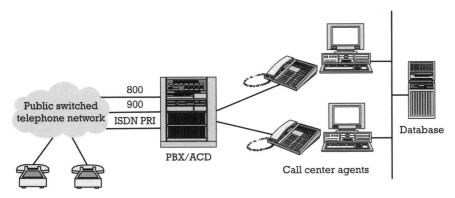

Figure 3.1 Traditional PBX and ACD call center configuration.

care" centers that build relationships and cement loyalty—in other words, they have evolved from cost centers to profit centers. As more value-added services and market differentiators become part of the product mix, the complexity of transactions passing through the call center is increasing. Customer contact with the call center is reaching far beyond the phone, to include faxes, e-mail, and direct two-way contact via the Web.

The integration of call centers with the Internet enables companies to personalize the electronic commerce experience for Web site visitors, and do so more economically than traditional 800-number service. As corporate intranets become more popular, large companies are leveraging their global IP networks with call centers that offer help desk and other types of support services aimed at trading partners as well as internal staff.

With more business operations migrating to the Internet and with voice quality continuing to improve, there will be a booming market for Internet call centers in the next three to five years. Large vendors such as Lucent Technologies and Nortel Networks are already aggressively pursuing the market for Internet call centers, along with smaller firms such as eFusion and PakNetX. These call center offerings take voice-data integration to a higher level by integrating multimedia messaging capabilities as well. These messaging capabilities might include e-mail, Web messaging, facsimile, real-time text chat, dynamic help, escorted Web browsing, shared page markup, and collaborative form completion. Some even offer the option for video communication, if the customer has a video camera and uses an H.323-compliant Web browser. For companies that prefer to

outsource this activity, AT&T, MCI WorldCom, and Sprint are among the
carriers that provide Internet call center services.

3.2 Benefits of ICCs

Internet call centers allow companies to personalize relationships with
Web site visitors by providing access to a customer service agent during a
critical moment—when the visitor has an important question, the answer
to which will influence the decision to buy. With the ability to intervene
in the online purchase decision and influence the outcome, a company
can realize several benefits:

> ▶ Competitive differentiation—by providing a convenient value-
> added service that improves customers' Web experiences via live
> agent assistance.

> ▶ Sales generation—by removing obstacles in the buying process
> with immediate interaction between customers and knowledge-
> able call center agents.

> ▶ Increased customer satisfaction—by delivering technical support
> and responding to customer needs quickly with personalized one-
> on-one service.

One company that understands these concepts is Micron Electronics,
which has a Web page (www.micronpc.com) that allows consumers
to configure and buy computer products online. Customers can buy
standard desktop PCs, notebooks, and servers or configure their own
system with desired features and options. Selected items are added to a
virtual shopping cart, which keeps a tally of the purchases. Changes can
be made to the configuration to stay within budget. Configuration
conflicts are even pointed out, giving the customer an opportunity to
resolve the problem from a list of possible choices.

As the customer makes changes, the new purchase price is displayed,
along with the monthly lease cost, in case the customer wants to consider
the finance option. When the customer is ready to buy, the shopping cart
adds shipping and handling charges and applicable state sales tax. The
customer completes the transaction by entering contact and payment
information and hitting the "submit" button to send the purchase order to

the company. Micron's secure server software encrypts credit card and personal information, ensuring that Internet transactions are private and protected. Customers can check the status of a recent purchase by entering their order number into an online search field.

Micron makes it easy for customers to request online assistance. At any time, customers can ask questions by selecting a preferred method of online communication (see Figure 3.2). A question can be asked by entering it into an online form, in which case a Micron sales representative will respond via e-mail, phone, or fax at a convenient time specified by the customer. Customers can also initiate an interactive text chat session with a sales representative or place a telephone call over the Internet to talk to a representative.

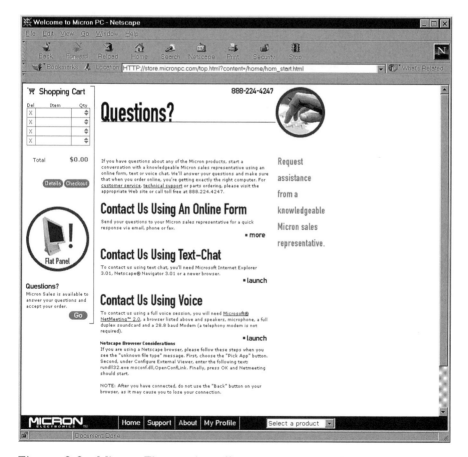

Figure 3.2 Micron Electronics offers customers a choice of communication methods: an online form, text chat, or voice call.

If the customer wants to initiate a voice call, but is not familiar with the procedure, a help window can be requested which provides information on system requirements and step-by-step instructions for placing the call (see Figure 3.3).

Regardless of what form of communication is selected, all go to Micron's Internet call center for handling by the next available agent or the agent who can most effectively respond to a customer's request. While having a real-time conversation with the Web site visitor, the call center agent can "push" Web pages to the customer's computer with appropriate text and images that help answer complex questions or illustrate key points.

In implementing an Internet call center—plus offering an online configuration and pricing tool, securing payment information with encryption, and providing an order status checker—Micron Electronics has not only provided online shoppers with a new level of convenience, it has also removed key potential barriers to online sales. These barriers include customer uncertainty due to lack of decision-making information and doubt about the safety of electronic commerce.

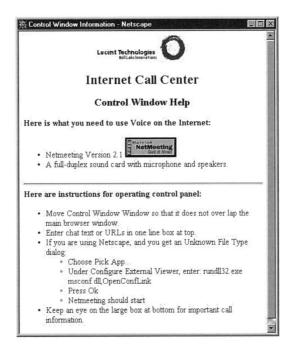

Figure 3.3   A help window provides customers with assistance in placing a voice call over the Internet.

3.3 ICC applications

One can infer from the Micron Electronics example that there are a number of practical applications for Internet call centers. Travel companies, for example, can provide a personalized travel planning service and increase sales by capturing potential travelers at their peak interest level. Web pages can be pushed to the Web site visitor to show a tropical island, a sample travel package, hotel accommodations, or current specials, all while having a real-time text or voice conversation with an agent.

Financial institutions can offer increased customer service and gain a competitive advantage by providing easily accessible, relevant information on complex products and services. A visitor to the company's Web site could be one click away from speaking to an agent to apply for a mortgage, get detailed information on financial services, or receive answers to individual-case questions not specifically addressed on the Web site.

Technical support and customer care departments can deliver help services by routing the call to a knowledgeable technical agent, who can answer specific customer questions over the phone, often with the assistance of Web-based information.

Virtually any type of company can benefit from an Internet-based call center, including those in the retail, travel, insurance, banking, and technology sectors. Internet call centers provide a convenient way for Web site visitors to get the information they need to make more informed buying decisions, while enabling companies to reduce their reliance on expensive toll-free lines.

Even companies that are not engaged in electronic commerce can leverage their global intranets to provide employees, customers, suppliers, and strategic partners with a variety of technical, administrative, and human resource services through an internal IP-based call center. Because the agents are connected to the Internet, they can be located anywhere, using any available infrastructure. This creates a virtual, round-the-clock, call center, without expensive networking packages or remote agent equipment.

The Internet call center can also be implemented on extranets. These are specialized VPNs that are similar to corporate IP-based intranets, except that they are designed for closed access by multiple companies and their constituents. As an extension of a corporate intranet, an extranet enables business partners to exchange private information and engage in financial transactions.

An Internet call center can augment the self-help capabilities of the extranet. For example, an extranet set up by a health care cooperative to serve the needs of community-owned health care organizations could implement an Internet call center as a help desk. This would enable IT managers of member organizations to call for technical assistance in troubleshooting an unruly application that was purchased through the cooperative. Or it could provide a referral service that helps local physicians obtain information on specialized clinics that can be recommended to their patients.

3.4 ICC components

A key component of the Internet call center is a gateway that provides interconnectivity between the packet-switched Internet and circuit-switched telephone network, enabling users to originate phone calls from their Internet-connected PCs (or new IP phone sets) to conventional telephones on the public network and vice versa.

The gateway is actually software that runs on a Windows NT or UNIX server, which is installed as a front-end to an ACD or PBX. Depending on the vendor, the gateway is scalable in terms of ports from as few as ten to as many as several hundred. Multiple gateways can be daisy-chained together to accommodate thousands of users.

Web site visitors need no more special equipment than a standard multimedia PC equipped with any H.323-compliant Internet phone software, such as Microsoft NetMeeting, Intel Internet VideoPhone, or Netscape Conference, which are included with virtually every multimedia PC shipping today. (Comparable software also is available for Macintosh and UNIX computers.)

As discussed throughout the previous chapter, H.323 is an umbrella recommendation from the ITU, which provides a foundation for audio, video, and data communications across IP-based networks, including the Internet. By complying with H.323, multimedia products and applications from multiple vendors can interoperate, allowing users to communicate without concern for compatibility at the other end of the network. The standard addresses call control, multimedia management, and bandwidth management for point-to-point and multipoint conferences. H.323 also addresses interfaces between LANs and other networks. H.323 also includes the G.7xx standards for compression and decompression of speech over IP nets.

Push technology is a key capability offered by many Internet call centers. This technology is used to automate the distribution of information over the Internet. There are several ways companies can use push technology: send information to all subscribers (broadcast), to select subscribers (narrowcast), or to an individual (personalcast). The personalcast method of information delivery is typically used with one-on-one agent-customer conversations. Depending on the particular interests of the Web page visitor, the call center agent can choose one of the following:

▶ Select a specific presentation for that individual from a list of canned presentations.

▶ Launch a process that will generate a custom presentation on the fly, which may interact with other databases (e.g., frequent flyer account).

▶ Escort the viewer through various Web pages, in which case the viewer's browser will synch to the page(s) selected by the agent.

3.5 Typical operation

The typical components of an Internet call center are depicted in Figure 3.4. There is nothing unique about the hardware and the connections. Basically, they are the same hardware and connections that support myriad other applications. The difference is the software that provides the convergence of different traffic types to a single point—the call center.

With a call from the Internet, the user places a call to the company's call center by clicking on a "call" button displayed on its Web page. This is actually a hypertext link in the familiar HTML format, which activates the Internet telephony program registered with the Web browser. This program can be a plug-in application of the Web browser, or it can be a separate application that is called by the browser from somewhere on the user's desktop.

There may be other features associated with the call button, depending on the specific purpose of the call center. For example, when a company offers free technical support (i.e., help desk service), the Web server may respond first by asking the user for a customer ID so it can pull up

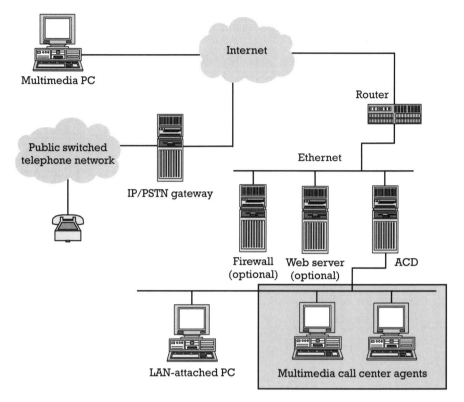

Figure 3.4 Topology of a simple Internet call center, capable of accepting calls from the Internet, PSTN, or LAN.

account records from a database and deliver that information to the next available agent, along with the call. On repeat calls, the account information may be retrieved automatically from a "cookie," which is a special memory area in the Web browser that holds information about the user for a specified period of time. If the caller uses the service again within that time frame, a customer ID would not be required because the cookie passes that information to the server automatically. However, after the cookie expires, the caller must again enter a customer ID to obtain service.

When the call button is clicked and the Internet telephony application is opened, a connection is established with the company's Web server. Some Internet call center products and services establish the call with the next available agent immediately, while others are capable only of notifying the agent to call back the user.

In some cases, the Web server only sends a call message, which triggers the scheduling of a call on the call center's outbound queue, specifying what agent or class of agent should handle the call. A response is sent back to the Web server with an estimated time of a callback, and the Web server relays that information back to the caller.

If the callers wait in queue for the next available agent, they might listen to music or advertising and get periodic barge-in messages informing them of their queue status. While in queue, callers can continue to browse the Web. If a call cannot be queued because all Internet call center resources are in use, the application pushes a Web page that apologizes for the delay and asks the customer to try again later or to call the company's 800 number as an alternative.

The call center can even push a Web page that tells the caller the time that he or she can expect a return call. If the call will be in 5 minutes, for example, a graphical indicator can display a clock that ticks down based on actual queue status. The indicator can even provide the caller with an opportunity to schedule the return call for a later time or cancel the call.

When the agent answers the call, several things can happen, depending on how the call center is configured to operate. In the simplest scenario, the agent addresses the caller's immediate needs and enters all relevant information into a computer database. This information can be referred to the next time the person calls. In addition, this information can be used for a variety of other purposes, including the dispatch of field service technicians, the generation of shipping labels for ordered merchandise, follow-up sales calls, direct mail advertising, and consumer surveys.

In a more sophisticated setup, the customer service representative can actually view the same Web page as the caller as well as any associated plug-in application the visitor happens to be using. This enables the company representative to be immediately helpful by answering questions, directing the viewer to appropriate information on the Web site, or taking a purchase order online.

To implement shared browsing, typically an applet is downloaded to the customer's PC, which captures a bit image of the screen about once every second. In this way, the Web server application that provides the browser sharing function "reads" the image on the customer's screen and then paints it on the agent's screen. The process also works in reverse, so the customer can see what the agent sees. This is especially useful when a page markup feature is available, which allows the agent to draw on the Web page with a mouse to emphasize specific content.

3.6 Integrated messaging

Many call center gateways also support interactive text chat in case there are users who still have low-speed modem connections to the Internet, do not have the right version of a browser, or do not have the right hardware configuration to enable voice conversation over the Internet.

Low-speed modem users can also be provided with a callback option. Typically, the user fills out a form, entering his or her phone number, the topic he or she wants to discuss, and the day and time he or she will be available to receive the company's call. The form is e-mailed to the call center where it is routed to the most appropriate agent for follow-up.

Sometimes Web site visitors will want information delivered to them by e-mail or fax instead of a follow-up call. For such cases, the call center application may include several built-in message-handling tools, including templates that allow agents to quickly create a personal e-mail response, place the message on hold to consult with other agents within the call center, or forward/transfer the message for handling by a more appropriate agent. Some call center platforms can also be integrated with automated fax response systems. The fax response system, in turn, can be configured to send faxes over the Internet as e-mail attachments.

Multimedia call transfer and conferencing improve responsiveness to customer needs for more information by allowing agents to transfer or conference complete Web and voice sessions. Multimedia queuing allows customers to continue browsing while in queue and/or provide infomercials that inform or entertain.

In addition, depending on company needs, the Internet call center's message-handling capabilities can be integrated with other application software, such as order entry, processing, and status monitoring. This would give users a convenient means to confirm an order, check on order fulfillment progress, and monitor delivery status—all without tying up corporate staff for these types of routine inquiries.

3.7 Cost-benefits analysis

The integration of call centers with the Internet allows businesses to save on staffing costs by offloading basic information requests to the Internet, while allowing customers easy access to live agents when required. By supporting calls from the Internet, a business gives its customers the freedom to browse any online material, such as a product catalog, until they have questions or comments. Adding this capability to Web pages could attract those more willing to provide credit card numbers via phone than type them into an online form.

Internet call centers can be easily cost-justified, especially if the company already conducts electronic commerce over the Web. By analyzing the Web server's log files, a company can quickly determine whether the traffic volume and Web browsing patterns make a call center worthwhile.

There are now Internet call center products that enable businesses to begin with as few as 10 ports. This enables smaller call centers to realize a faster return on investment, while concurrently implementing a highly productive and cost-efficient electronic-commerce program into their overall sales strategy. Such systems also give companies that are unsure of how the Internet fits into their marketing strategy a low-risk means of trialing an Internet call center, fine-tuning complementary applications, and testing reorganized business processes before making a long-term commitment.

The cost of an Internet call center system is more easily justified if a company already runs a traditional 800-number call center. An Internet call center can be easily integrated into a traditional ACD-based call center at incremental cost. Since Internet and PSTN calls are routed to agents without regard for where they originated, there is no need to add more staff just to handle Internet calls. In fact, a company can develop a migration plan to reduce its dependence on 800 lines and encourage more Internet calls to reap considerable cost savings.

For example, 800 number charges can be minimized by eliminating time-consuming navigation of audiotext menus that attempt to help customers get in touch with the most appropriate agent. There are no 800 number charges while the customer waits for an agent to become available over the Internet. There are no wasted minutes on 800 lines while customers explain the problem or get passed from one agent to another. There are no 800 number charges at all if customers find their own solutions to problems by searching the company's user-friendly Web pages.

Because the Internet call center application integrates seamlessly into a business's existing call center, there is no need for dedicated Internet call agents—agents can easily receive both 800 and Internet calls. Intelligent routing ensures that the call gets to the right agent the first time—customers are directly connected to the agent that has the answers they need. Call management reports can be applied to tracking calls on the Internet, just like any other call. For instance, the reports can show the average time Internet callers wait in queue, so the call center manager can take steps to improve response time. These call management reports can also be used to help companies evaluate the sales effectiveness of promotions, campaigns, and contests.

If the cost of system ownership is prohibitive, several pay-as-you-go alternatives are available from AT&T, MCI WorldCom, Sprint, and other carriers. AT&T's interactiveAnswers, for example, entails only a $295 installation charge, plus a $195 monthly service charge. For customers who do not use AT&T Web hosting for their sites, the monthly service charge is $295. In addition, there is a per-transaction charge of $1.95 for the first 10 minutes and an additional charge of $0.95 per 5-minute increment thereafter. MCI WorldCom's Click'nConnect entails only a $300 installation charge, and about $1.75 per call, less discounts for volume and/or term contract.

For companies that are relatively new to electronic commerce over the Web and are unsure of how much benefit they can derive from an Internet call center, an economical approach might be to go with a carrier-provided service to see how things pan out over the course of a year. If there is steady growth in the number of transactions that are facilitated by the call center service, and the service charges are approaching the purchase price of a system, then ownership is clearly justified by the anticipated return on investment.

3.8   Vendor approaches

In addition to the services offered by carriers, both large and small equipment vendors offer Internet call centers. The ways vendors implement Internet call centers can be categorized into the high-end solution that integrates multiple systems through a computer-telephony integration (CTI), proprietary solutions that attach to the vendor's own line of PBXs, and the open system that acts as a front-end to any vendor's ACD or PBX. There are also standalone products that address specific functions of the call center such as e-mail management.

3.8.1 Lucent Technologies

Lucent Technologies outlined its strategy for Internet telephony in early 1996. This strategy included its Internet Telephony Gateway, which is now the foundation of the most sophisticated call center product on the market today. The company's CentreVu Internet Solutions (formerly known by the generic-sounding moniker, Internet Call Center) is an example of the high-end, multisystem approach that relies on CTI to tie a company's data environment with its call center to accommodate mixed media communications, including voice, data, fax, e-mail, and eventually, video.

This level of mixed media integration enables companies to implement CentreVu for use in both internal business applications that allow employees to collaborate more effectively using an array of communications tools, and in external applications that support multiple access channels, allowing customers to communicate in the medium of their choice.

CentreVu Internet Solutions works with a company's existing PBX and ACD, specifically the DEFINITY Enterprise Communications Server (ECS) and DEFINITY G3 ACD, both of which are high-end systems offered by Lucent. The following key components are also required to implement CentreVu Internet Solutions:

- ▶ Call center software bundle, including Expert Agent Software (EAS);
- ▶ Internet Telephony Gateway (ITG);
- ▶ Adjunct Switch Application Interface (ASAI) software;
- ▶ Java call control application;
- ▶ CentreVu call management system;
- ▶ Web server, which is provided by the corporate customer.

At the customer premises, Lucent installs its ITG and loads the Java applications on the customer's Web server, both of which integrate with the DEFINITY ECS and DEFINITY G3 ACD to allow call center agents to receive voice and data calls from the Internet. Queuing and routing support for nonreal-time transactions such as e-mail, fax, and callback requests are provided through optional Message Care software and a POP3-compliant mail server.

In addition, high-speed LAN connectivity and ASAI software provide the links for the PBX, computer-telephony server, ITG, mail server, and agent PCs. EAS software works with the ACD to queue and route Internet calls, e-mail, and fax in the same way that ordinary 800-number calls are routed within the call center—by preprogrammed rules that select the most appropriate agent to handle the call. Two-way communication between the customer and agent occurs over the same communication link being used for the Internet connection.

The call center agent desktop is equipped with a PC and the appropriate 32-bit operating system—for example, Windows 95 or Windows NT—and a Java-enabled Web browser. No desktop speakers, microphone, or telephony application is needed because the voice portion of the call is delivered through the PBX or ACD to the agent's voice terminal just as if it was an ordinary voice call.

Because CentreVu Internet Solutions integrates seamlessly into a business's existing call center, there is also no need for dedicated Internet call agents—any agent can easily receive both 800 and Internet calls. Intelligent routing ensures that the call gets to the right agent the first time—customers are directly connected to the agent that has the answers they need, as determined by the rules set in the EAS software. Call management reports can track calls originating from the Internet, just like any other call. The reports are generated by the CentreVu call management system.

3.8.1.1   Call scenario

In a scenario where the end-user is equipped with a multimedia PC, a Java-enabled browser would be used to access the Web. While browsing a Web site, the customer simply clicks a button on the Web page to talk to a customer representative. There is no need for the customer to end the browsing session and wait for a call back. The customer simply uses the Web page interface to request the type of call he or she wants—an

Internet voice call, a text chat, or even a call back on an ordinary telephone line.

Clicking a button starts a call across the Internet to the call center ITG, using the same connection already established with the ISP. The button is actually a hypertext link written in HTML that is used to initiate this process. The link may look like this, with the IP address of the server and subdirectory where the application to find highlighted in boldface:

a<href="javascript:showWindow('**http://209.192.139.68/ITG/**

callerapps.pl?vdn_ext=55396&type=voice&browseWinURL=

http://store.XYX.com/icc/welcome.html');"><a>

In this case, a process written in Perl is used to find the appropriate caller application (the question mark indicates that this is a search process). The ITG downloads the Java call control applet to the customer's PC to start the Internet telephony application. The applet provides the interface through which the customer can also receive call status messages, engage in text chat, collaborate with a call center agent by sending a Web page, or drop the call (see Figure 3.5).

Figure 3.5   An example of Lucent's Java call control applet that is downloaded from a company's ITG to its customer's Web browser.

The use of Java to deliver this functionality to the customer has several advantages over other Common Gateway Interface (CGI) implementations. First, the applet is only downloaded when needed and does not take up permanent residence on the client machine. Second, all software maintenance tasks are performed at the server. This means the applet can be modified in any way—functionally or cosmetically—at the server. Since the applet does not stay on the client machine after use, customers would always get the latest version delivered to their Web browser when they want to initiate a call. Finally, since the Java applet extends server functions to the client, the amount of back-and-forth traffic between the client and server is minimized, allowing the connection to be used more efficiently.

CGIs written in Perl or Tcl, for example, require more communication between client and server, which increases the delay in implementing certain interactive processes.[1] If the call control applet requested a credit card number, for example, error correction (e.g., not enough or too many digits) and validation (e.g., not a valid account number) functions could be carried out by the applet itself, as would the issuance of appropriate error messages, if necessary. With a CGI built with Perl or Tcl, the credit card number would have to go to the server for error correction and

---

1. CGIs can be written in any language that produces an executable file, such as C and C++. Among the more popular languages for developing CGIs arre Perl (Practical Extraction and Reporting Language) and Tcl (Tool Command Language). Both are derivatives of the C language and offer the advantage of being able to speed the construction of applications to which new components can be added without requiring recompiling and restarting, as is required when the C language is used.

Increasingly, Java applets are being used to provide the interface between the Web server and back-end data bases and applications. The use of Java overcomes the limitations of Perl and Tcl. With a CGI implemented in Perl, for example, there are two sources of bottlenecks. First, a process in which to run the Perl interpreter must be launched, which in turn runs the Perl program. Creating such a process once a minute is not a problem, but creating it several times a second to support a busy e-commerce Web site is a major problem. Second, it takes time for the Perl program to send back the information it generates as output. Waiting for a small Perl script that does something simple like send the contents of an HTML form via e-mail is not a problem, but waiting for a large script that does something more complicated, like allowing customers to access account information stored on a mainframe and execute financial transactions, definitely is a problem.

With Java, application code is downloaded from server to client on demand. The applets do not need a Web server as an intermediary and, consequently, do not degrade performance. In most cases, the applications are stored in cache on a hard disk at the client location, and in others, they are stored in cache memory. Either way, the application does not take up permanent residence on the client machine. Since applications are delivered to the client only as needed, administration is done at the server, assuring that users have access to the latest application release level.

validation. Any error messages would have to be issued from the server as well.

Instead of Perl scripts, Java servlets can be used as a replacement for CGI scripts. While Java applets run on the client side, Java servlets run on the server side. Unlike CGI scripts, servlets can be persistent, meaning that one servlet can stay running over many Web accesses, allowing it to remember things about previous accesses. Since the servlet does not have to be restarted for each access, overall performance is greatly improved.

As the call control applet is downloaded to the customer's browser, the ITG hands the call off to the DEFINITY ECS over an ISDN PRI link. At the PBX, the call is queued according to its Vector Directory Number (VDN), which is specified in the initiating Web page. As all this is happening, the customer may continue to peruse the Web site. Content of interest can be pushed to the customer to hold his or her attention. This content can be an advertisement, supplementary information, or a combination of the two—an "infomercial."

The VDN is written into the same Web page that gives visitors the communications options they can select. The entire hypertext link that specifies a full voice session might look like this, with the VDN highlighted in boldface:

<ahref="javascript('http://209.192.139.68/ITG/callerapps.pl?**vdn_ext=55396**&type=voice&browseWinURL=http://store.XYX.com/icc/welcome.html');"><a>

In this case, a JavaScript is used to call up a window from which the user can select a preferred method of communication. This is followed by the IP address of the server and subdirectory containing the caller applications. A script written in Perl locates the application that will establish a voice call. The VDN specifies the 5-digit extension the call will be routed to at the call center, and tells the agent how the user wants to communicate (i.e., voice) and provides the URL of the Web page the call originated from.

The VDN can specify an individual agent, an agent group, or the ACD itself in which case the EAS software will help decide the most appropriate agent to get the call. An agent can handle ordinary voice calls as well as Internet voice calls using the same equipment—both call delivery processes are transparent to the agent. When the call is delivered to the agent's voice terminal, the CTI server simultaneously sends a "call answered" message to the ITG, which in turn delivers the URL to the agent's PC to display a screen "pop" of the Web page from which the customer initiated the call. If the customer is looking at another Web page when the agent receives the pop, he or she is returned to the original Web page from which the call was placed.

With both the customer and agent on the same page, they can converse via voice or text chat, and Web page collaboration can take place. The agent can lead the customer to related pages to provide more information, show a product photo, or steer the sale. Along the way, the agent can answer the customer's questions. While the customer fills out a form to place an order, the agent can view the form and offer assistance in filling it out.

The agent also uses information passed via the CTI link to review account status and other customer-specific information that is already present in the business database, which helps the agent understand the customer's needs so a more personal level of service can be provided. If the customer has no record on file, one is created from the order information sent by the customer, or a record can be created by the agent from information conveyed in a voice or text conversation. The record can be accessed by different call agents and updated with new information generated by future calls from the customer.

3.8.1.2   E-mail and FAX

The optional Message Care software available from Lucent supports free-formatted e-mail in addition to form-generated e-mail, allowing the call center to support general inquiries from customers who contact the business. This software helps manage:

▶ E-mail generated directly from the Web site—for example, customer responses to "write to us now" links created for specific Web pages and routed to skilled agents;

▶ E-mail sent to a specific address—such as direct to the call center and routed based on a corresponding VDN;

▶ Faxes forwarded to an "electronic mailbox" and stored in a POP3-compliant mail server.

Many business Web sites allow customers to send e-mail, but it takes a lot of time to route it to the most appropriate person, and usually that person is deluged with so much e-mail that days or weeks may go by before the customer receives a reply. By that time, the chance to influence a purchase decision is long gone. With e-mail integrated into the call center as if it were another call, response time is greatly improved.

In a typical scenario, a customer sends an e-mail message to the service center, either through a typical e-mail channel or by filling out a form on the Web site. The message travels over the Internet to the premises mail server. A "phantom call" through the DEFINITY ECS creates a virtual message call that is treated like an ordinary voice call, and is queued and routed according to the VDN and skills set specified by the application.

When an agent becomes available, the e-mail "call" is delivered to the agent's voice terminal and a screen pop to the agent's browser provides notification of its presence. A user interface is also delivered to the agent's desktop (see Figure 3.6), which includes several message-handling tools. With these tools, the agent has the ability to create an e-mail response, place the message on hold to consult with other agents within the call center, or forward/transfer the message for handling by another agent who is considered the subject matter expert. While the agent handles the message call, the agent's busy status is noted as such by the DEFINITY ECS, so subsequent calls will be routed to another agent.

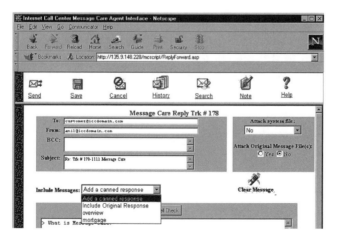

Figure 3.6 The Message Care user interface, as delivered to the agent's desktop via the Netscape browser.

The Message Care software can also route customer faxes to an electronic Internet call center mailbox via an e-mail server's fax interface, where the fax messages can be accessed and routed just like Internet calls or e-mail.

With this application, agents do not have to stop what they are doing to periodically check their e-mail box. They can handle both phone calls and electronic messages as they arrive at the call center. Since the Message Care software automates this process, agents spend less time handling each message, which significantly cuts the labor cost per message.

3.8.1.3 Text chat

Text chat is implemented as part of the call control applet that is downloaded to the customer's PC when a call is initiated. It supports the voice session between the customer and the agent, or it replaces the voice session for customers who do not have a multimedia-equipped PC. For example, the applet provides a convenient means of accurately transmitting strings of digits, such as account codes or serial numbers, verifying spoken communications—"Did you say 'D' or 'G'?"—and providing a method of access for those who are hearing impaired.

3.8.1.4 Callback requests

In addition, for those customers who have a second telephone line, CentreVu Internet Solutions allows businesses to accommodate callback requests. The customer simply indicates this preference and provides a telephone number. The request is queued and routed just like any other call coming into the call center. The first available appropriately skilled agent will then automatically return the customer's call using the PSTN. The agent is provided with the customer's telephone number within a pop-up window, and then selects "Call" to signal the DEFINITY ECS to automatically dial the number and connect the call.

For the customer who has a firewall that will not permit Internet telephony access to the desktop, CentreVu Internet Solutions can push a page that prompts the customer to enter callback information. The customer submission is routed to the first available appropriately skilled agent for follow-up.

In the future, Lucent will support additional access choices for CentreVu Internet Solutions—such as videoconferencing—further enhancing a business's ability to provide immediate, responsive, and personal support services. In addition, CentreVu Internet Solutions will be enhanced with the capability to route customer requests to the best available agent anywhere in a business's global call center network.

3.8.1.5 Computer-telephony integration

CentreVu Internet Solutions also provides an effective way for agents to collaborate with customers, by combining real-time Internet interactivity with existing CTI applications and standard call center functionality. In turn, this allows agents to access all the technologies and resources available to the call center to enhance the customer experience.

For example, in addition to providing a screen pop of the specific Web page that a customer was viewing when he or she initiated an Internet call, the Web application is configured to present the agent with enhanced information—such as account information drawn from the company database, a special offer related to the page being viewed, or another page that's linked to the original page. These links are based on data that the customer provides when initiating the call—such as a name, telephone number, account code, or registration number.

CentreVu Internet Solutions links are also configured to automatically present the customer with additional pages (such as a "thank you for calling" page) or information that is linked to the original page he or she was viewing, to help further enhance the online shopping experience.

CTI also permits escorted browsing. The agent and customer can be surfing the Web site independently and then, with the click of a "send page" button on the call control applet, synchronize their browsers to view the same page simultaneously. Security functions embedded into CentreVu Internet Solutions allow the agent and customer to browse in synch and control what information is sent to each other—ensuring that neither party will view inappropriate, confidential, or inaccessible pages.

3.8.1.6 Expert agent software

Calls that are received by the CentreVu Internet Solutions are converted into circuit switched calls, sent across an ISDN PRI facility, and routed using the same queuing and call vectoring capabilities already present in the DEFINITY ECS call center. The call center simply needs to modify the EAS criteria with any new skills that are necessary to handle calls generated from the Web site. Agents continue to handle ordinary voice calls in addition to Internet calls, based on EAS.

3.8.1.7   Queue messages

CentreVu Internet Solutions can be configured to provide customers with Web pages that present reassuring messages in addition to the applet messages seen while calls are in queue. For example, a business links a URL to a call status indicator to present a Web page that tells a customer his or her place in queue. If a call cannot be queued because all ITG resources are in use, the application presents a Web page that apologizes for the delay and asks the customer to try again later or to call the company's 800 number as an alternative.

3.8.1.8   Call statistics

Internet call statistics such as average talk time, queue time, the numbers of calls, and so on, are collected by the CentreVu Call Management System (CMS) and reported the same way that other call statistics are gathered. CMS tracks e-mail message handling, starting when the phantom call is launched within the DEFINITY ECS and including the entire time the agent takes to process the message.

CMS also gathers data on how many times a communication-enabled Web page is hit, as a basis for analyzing how often page hits result in customers actually accessing the call center. In addition, the CentreVu Internet Solutions detects and sends to CMS information on calls that could not be delivered to the ACD due to insufficient ITG resources or because no agents were staffed or logged into the call center.

These reporting capabilities consolidate data from various media—gathered via the standard call center as well as the Internet—into a single call report tool that reveals information pertinent to market analysis and sales and service strategies, as well as to the effective management of the call center and Web site.

3.8.2   PakNetX

Another vendor, PakNetX, offers the PNX ACD, a software-only telephone switch and ACD application that provides traditional telephone features for the Internet call center, including call queuing and distribution, information systems, agent and group management, CTI, and all the other functions typically available from an ACD. It uses the IP network for transport, but implements a virtual switch matrix, which does all the protocol, routing information, and application level processing in software.

Running on a Microsoft Windows NT-based server, the PNX ACD integrates with existing call center components, using CTI technologies. An integrated H.323/T.120 firewall, inherent in the software switch, provides security. All communications between callers and agents go through the PNX switch, which provides the firewall function as well as agent, supervisor, and management features.

A firewall is a packet filtering mechanism that guards the corporate network from unauthorized access. Packets from suspicious sources are blocked, while packets from familiar sources are allowed to pass. Another key function of the PakNetX integrated firewall is to re-map real IP addresses into virtual IP addresses.

When the customer initiates the call, the PNX ACD is called. H.323 call setup occurs between the customer's H.323 phone and the ACD. When an agent is available, an outbound call is placed to the agent, and the caller and agent are conferenced together. All of the H.323/T.l20 packets go through the ACD. Upon call arrival, the agent interface (see Figure 3.7) shows call context information, which can include the customer's name, account number, currently viewed URL, and subscriber service level.

As each call is connected through the ACD, packets pass through the switch where each packet address is remapped for routing to the appropriate agent. This process allows agents to be hidden from direct public access, simplifying LAN topology management issues and resolving potential security concerns.

The PNX ACD's supervisor features include group creation and management, agent management, hours-of-operation controls, call detail records (CDRs), and real-time agent status. Supervisors can assign one agent to multiple groups so that an agent can connect to customers from either the company's technical support or customer service organization.

Like other Internet call center systems, PakNetX's PNX ACD can be used to speed the adoption of Internet commerce by helping consumers

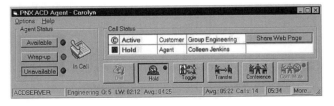

Figure 3.7   The agent interface of PakNetX's PNX ACD.

obtain more immediate answers to their questions and become more comfortable with using the Internet for online purchases. And like other products that integrate the Internet with conventional call centers, PakNetX offers the Web site visitor a choice of multimedia options to communicate with the call center agent. The Web customer may choose an IP telephony application installed on the desktop, a callback using a telephone on the PSTN, or text chat on theNWeb. The PNX ACD also offers video capabilities that can be tapped with the addition of a desktop camera (see Figure 3.8).

In addition, PakNetX provides a page markup feature that allows both the agent and the customer to draw on the screen, underlining and circling Web page elements. PakNetX supports Microsoft's NetMeeting T.120-based features, such as whiteboarding. What the agent marks, the customer sees, and vice versa. The page markup feature enables both the customer and the agent (see Figure 3.9) to use the mouse like a marker to circle, underline, and otherwise call attention to information on the screen. This markup capability greatly enhances collaboration, and

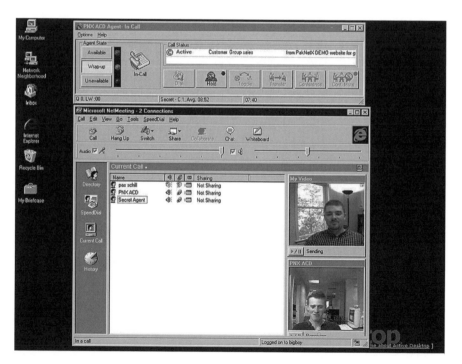

Figure 3.8 PNX ACD agent screen with a video session.

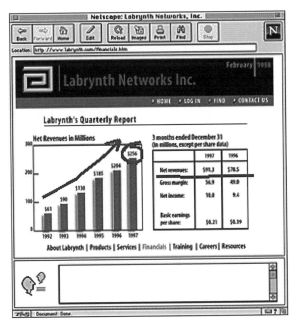

Figure 3.9 PakNetX's PNX ACD supports Microsoft's NetMeeting, which allows the customer and agent to view the same Web page and use their mouse to mark it up.

especially makes the online shopping experience more personal for customers.

PNX ACD also offers collaborative form completion. This gives the call center agent the ability to assist the customer in filling out online forms by answering questions and even filling out portions of the form. This can reduce customer frustration and ensure the accuracy of form content. Whatever the agent types into the form shows up on the customer's browser and vice versa.

Another interactive capability of the PNX ACD is dynamically enabled help. This feature can be used to cause a Help button appear on the Web page after a customer has engaged in a predetermined set of Web interactions. When the customer clicks on the Help button, it opens a communication channel for a voice call or text chat with an appropriate call center agent. Alternatively, a callback can be requested, in which case the agent will call the customer at the specified time.

These collaboration features are available to both the customer and agent with tools that conform to ITU H.323 and T.120 standards, such as

Microsoft NetMeeting—available with Microsoft Internet Explorer 4.0 and Windows 98—and Intel Business Video Conferencing with Pro Share technology. As noted in Chapter 2, the H.323 standards are important because they facilitate peer-to-peer communications and peer-server-peer communications in a nonproprietary fashion, enabling audio (and optionally video and/or data collaboration) communications over the Internet.

H.323 is important for another reason—it includes a capabilities exchange dialog, which is used to match the operating parameters between end-points before establishing the call. Capabilities information includes protocol identifiers (e.g., G.723.l audio encoding), video capabilities (none, one-way, two-way), and data session establishment. Once the call is established, customers and agents can speak to each other and potentially see each other, as well as push Web pages at each other. In addition, the customer and agent can engage in data collaboration.

Data collaboration includes whiteboarding, file transfer, text chat, and remote application control. Some of these capabilities lend themselves to help desk services. For example, remote application control enables a help desk operator to perform actions on a customer's PC, such as initiate a diagnostic routine to solve a problem with a software driver. Or, using file transfer, a software patch can be downloaded and installed to the customer's PC. The customer must agree to these actions, but can easily stop them in progress by clicking on a "panic button."

Since the PNX ACD is a self-contained, software-only solution, it can co-exist with a conventional call center. This enables the new Internet call center to be placed into operation quickly, with no risk or impact to the existing call center. For PSTN callers, an IP/PSTN gateway can be used in conjunction with the PakNetX solution to bridge the callers to the PNX ACD. This arrangement allows calls to be migrated to the Internet call center and expensive toll facilities to be phased out over a timeframe that is both convenient and consistent with the company's business objectives.

3.8.3 eFusion

eFusion, Inc. (formerly Telepresence, Inc.) offers an add-on call center solution in the form of its eBridge Interactive Web Response (IWR) system. This application gateway converts Internet calls into conventional telephone calls, which are passed on to a company's PBX/ACD within its call center. Like other products, the eBridge allows Internet-enabled consumers to browse the Web, talk directly with a customer service or sales representative, and conduct business, all over a single phone line.

A separate eFusion client application called Internet Call Assistant runs on the consumer's PC and works in conjunction with an Internet phone client. This application performs a variety of coordination tasks, such as registering the subscriber with the application gateway and executing Push-to-Talk buttons on a company's Web page.

Customers use the Push-to-Talk button to establish voice communication with an agent while they browse the Web, exchanging information and completing transactions, without the need for a callback. The technology works over the single telephone line installed in most residences, and users need only a standard multimedia PC equipped with any H.323-compliant Internet phone software and Microsoft Windows 95.

While the prospect of allowing Internet users to talk directly with call center agents is promising, inherent network congestion of the Internet today prohibits consistent delivery of voice quality consumers expect. eFusion addresses this issue with its DirectQuality technology, which enables a voice and data connection over the public telephone network to the call center. With this technology, call center managers have the option of selecting either a low-cost connection through the Internet with variable voice quality, or a reliable and secure connection with DirectQuality service.

Once callers are connected to the call center, they may be put on hold while waiting to speak with an agent. Through eBridge's MultiHold feature, they can receive a variety of multimedia content, such as a series of Web pages or audio announcements, streamed through their browser as they wait for the next available agent. Through intelligent call handling scripts—written with tools such as Visual Basic or Java—this content can be tailored to each caller's profile or interests and even provide the caller with answers before consuming an agent's time, thereby deflecting calls and increasing workforce productivity.

As they talk, the caller and the agent can simultaneously share Web pages using the TeamBrowse feature. The agent can see what the consumer is viewing, and either party can push visual Web-based content to the other. Clearer communication with the caller can reduce errors and increase opportunities to sell products, which saves time and increases the revenue potential of each transaction.

eFusion has targeted the rapidly expanding online banking, financial, service/support, travel, and retail markets for its eBridge. Based on the Windows NT platform, the eBridge is installed as a front-end to any ACD or PBX and integrates with existing management information systems and agent processes, allowing for quick, low-risk implementations. A 10-port system offers smaller call centers with the means to economically integrate electronic commerce into their overall sales strategy. For large firms unsure about the Internet's merits, the 10-port device offers the means to bring Web messaging capabilities into a conventional call center economically and scale up as demand increases.

An eFusion application gateway is comprised of a Pentium-class, multiprocessor server that provides direct connection to the PSTN using a variety of digital trunk interfaces, including T1/E1 and primary rate ISDN. The eBridge system is also available in 24-, 48-, 96-, and 120-port versions. Through the interconnection of multiple servers, an eFusion system can effectively support thousands of lines. Digital voice encoding/decoding between the PSTN and IP networks is performed by DSPs, which relieve the server's CPUs of this processing-intensive task. The DSPs are linked to the PSTN lines via a standards-based pulse code modulation (PCM), time division multiplexed (TDM) internal bus.

The eBridge system provides detailed management information, which can be generated automatically, manually, or on a user-defined schedule. Call detail record information can be read, exported to other systems, consolidated, or removed. The system uses Open Database Connectivity (ODBC) and industry-standard report writers to create reports and billing systems.

3.8.4 Genesys Telecommunications Laboratories

CTI vendors are blurring the line between the Internet and the traditional call center. Genesys Suite, the third generation of the vendor's CTI Framework product, treats customer e-mail queries like any other incoming call, matching names to the customer database and routing them to the appropriate agents for a response. The Suite also Web-enables call centers by letting companies place callback buttons on their Web sites. By clicking on the button, customers are immediately placed in the outbound call queue, and are provided with a countdown indicating their place in line and the time when they should expect a return call. The application is customizable—for example, the call center can create a form that asks for additional information when the customer clicks the callback, allowing the call center agent to respond with appropriate information.

The Genesys Suite provides companies with the tools to build customer interaction networks. This type of network extends beyond traditional telephony-oriented applications by combining multiple customer interaction components such as IVRs, ACDs, CIS applications, databases, and even call centers, making every customer interaction strategic—regardless of time, place, or medium.

A configuration manager enables call center configurations and agent groups to be set up just once and then used by all applications in the suite. This results in significant time and cost savings over traditional call center software where each component, often coming from a different vendor, can require a different configuration setup.

3.8.4.1 E-mail

Genesys's e-mail option allows enterprises to integrate e-mail handling into their existing telephony infrastructure. With this capability—which includes content analysis, response library, and customer history features—call centers can reduce the number of e-mails requiring a manual reply. For example, as with voice interactions, customer information is automatically delivered to an agent's desktop so that before an e-mail is answered, the agent has the right information to address the customer's issue.

3.8.4.2 Interaction router

Named for its ability to route multiple media, the interaction router gives an enterprise the capability to control calls at the network level from the customer premises. A call can be held at the network level and then routed to the appropriate site and agent in a matter of seconds directly from the network. This capability also allows an enterprise to route a call in the network based upon business data rather than traditional load balancing—ensuring that customers are routed to the best agent armed with the right information.

3.8.4.3 Remote option

Genesys's remote option moves the call center beyond site and switch boundaries, central or remote agents, on or off premise, across the enterprise and around the world. The remote option extends the call center to branch offices without PBXs or to agents working from home or in other locations outside the formal call center. As a result, companies can experience an increase in agent productivity, retention, and effectiveness in the same way they have been doing for years with telecommuting programs.

3.8.4.4 Outbound scripting

This option allows the dynamic creation, distribution, and modification of agent scripts. These scripts optimize customer interaction and increase revenue by stepping agents through a consistent sales dialog, giving them guidance for handling objections, and helping them recognize and take advantage of cross-selling opportunities. This tool also brings consistency and responsiveness to the messages delivered in an outbound telemarketing campaign.

3.8.5 Mustang Software

Companies that have invested heavily in electronic commerce are often unprepared for the endless avalanche of e-mail—purchase orders, questions, comments, complaints, and requests for product updates—to name a few. The resulting backup often leaves businesses with unattractive alternatives, including adding more workers to the payroll or using depersonalized automated response programs that turn off potential customers. Even worse is to simply ignore vast quantities of e-mail. These scenarios not only can undermine the company's e-commerce effort, but damage the corporate image as well.

Most corporate Web sites use aliases to help direct e-mail to the right individuals within the company. This is mail addressed to corporate functions instead of particular individuals. Examples include mail addressed to: sales@domain.com, support@domain.com, webmaster@domain.com, or info@domain.com. The trouble is that even mail addressed to an alias can be voluminous and difficult to handle. Although mail can be routed to support@domain.com, for example, it must eventually get to the individual with the right expertise so that an authoritative reply can be sent off in a timely manner.

While client software alone cannot deal with this kind of mail on an enterprisewide level, there are products that are designed to deal effectively with inbound Internet e-mail addressed to aliases. They simplify internal routing to improve responses to requests for customer service, technical support, and marketing information. This approach is similar to the way telemarketing call centers effectively handle inbound telephone calls by connecting callers to the right individuals. When applied to e-mail, this helps managers avoid having messages pile up unanswered and enables them to monitor message response times.

One such product, Mustang Software's Internet Message Center (IMC), works with Microsoft Exchange, Lotus Notes, Novell Groupwise, Netscape Mail, Qualcomm Eudora, and other leading messaging packages, enabling companies to add structure, management, and accountability to incoming Web and e-mail inquiries.

IMC consists of four components: IMC Service, IMC Monitor, IMC Reports, and the IMC Agent. Companies simply give the core application, IMC Service, an account on the existing messaging server so it can monitor incoming messages for specific aliases. When messages are received, they are retrieved and stored in an ODBC-compliant database, where they are assigned a tracking number. Once a tracking number is assigned, IMC Service sends an auto response message to the sender, acknowledging receipt of the message with the expected response time. The tracking number is included in the reply, in case the recipient wants to follow up later. This preliminary auto response might include the following helpful information:

> You should be receiving a personal response by e-mail from one of our staff shortly. In the event you need to contact us regarding your original message, please refer to the tracking number at the top of this message. This will help our staff locate and review your correspondence with us.

3.8.5.1 IMC operation

As in call centers, where groups of agents handle incoming telephone calls, personnel who can respond to company e-mail are assigned to pools, based on areas of expertise or responsibility. Messages are routed to the appropriate individual from the IMC Agent. The agent, available in both Java and Windows 95 versions, lets assigned personnel access the messages assigned to them from anywhere on the corporate intranet. They can see how many messages are waiting, and simply select "READY" for the next message.

IMC gives managers a good deal of flexibility in determining how to route incoming messages. The system can be instructed to look for particular words in the subject line of a message header, for example, which could be useful for filtering requests for a particular product so they can be routed to a specific group of marketing or support personnel.

The IMC Agent informs the IMC Service to take the next message from the pool and e-mail it to the next available agent's Internet e-mail address. This allows the appropriate person to read and answer the message using their existing Internet e-mail client. Once they respond, the message goes back to the IMC Service, which looks up the tracking number and replies to the original author of the message.

3.8.5.2 Tracking messages

The tracking number also lets senders check the status of their inquiries and lets managers see all correspondence related to a specific message to follow the history of a particular inquiry (see Figure 3.10). Managers can use IMC Monitor to view pool messages from customers and responses from agents. A search function allows managers to find messages corresponding with a tracking number, which is useful when the tracking number is known, but not the agent or pool it is assigned to. The tracking capability is a useful supervisory tool for monitoring quality control and investigating complaints.

The tracking numbers are created according to this format: Pool Initials + Date (YYYYMMDD) + Number of the message in a sequence

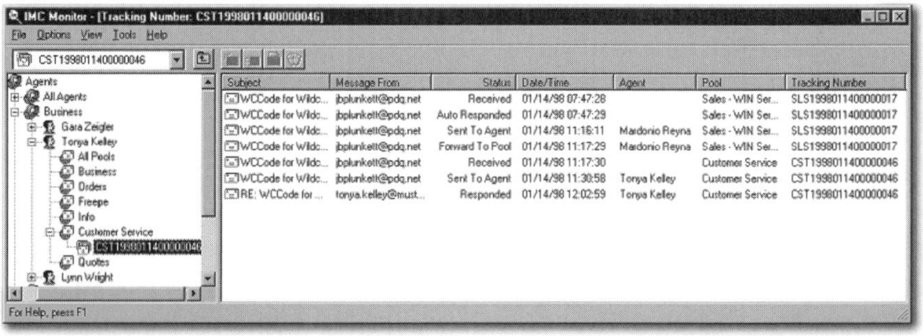

Figure 3.10 Mustang Software's IMC allows managers to view a list of all correspondence related to a specific message through tracking numbers assigned to each inquiry (see right-hand column).

that started at midnight. A typical tracking number looks like this when it is first assigned to an incoming message: SPS1999021200000088.

IMC adds a unique message ID to the tracking number for each additional message associated with the original message. This allows agents to maintain an ongoing e-mail discussion with a particular user, while keeping track of every message that has been sent and received. A typical tracking number for a message in reply to the original message might be as follows: SPS1999021200000884359.

IMC's monitoring and reporting tools let managers see how many transactions each agent has handled. IMC Monitor allows the status of each agent to be viewed as: Online, Offline, Idle, Ready, or Responding. New information on pool and agent status can be displayed automatically at a specified time interval, or requested manually at any time. This information can be used to make changes in the distribution of work to ensure timely responses.

Several reports are available, including an agent summary report that shows how many messages were received by each agent pool and the average time it took to answer incoming messages between their arrival and response by an agent (see Figure 3.11). This information can also be used to baseline the performance of the e-mail response center and identify the need for more staff, for example, when response time slows to an unacceptable level.

3.8.5.3 Installation and setup

IMC does not entail equipping every computer in the e-mail response center with new client software. A POP3-compatible e-mail client is all that is needed to send and receive messages, and operators probably already have this if they have been using e-mail all along. Easily installable agent software also goes on the client machines. The setup wizard guides operators through the procedure of selecting directories, installing the agent, and connecting to the IMC Service. No IS involvement is required.

Once IMC Service software is installed, it is easy for the e-mail response center administrator to set it up, again without IS involvement. The procedure consists of filling in various fields in the IMC setup window with appropriate information concerning such aspects of the operation as the names and addresses of pools and agents. Administrators can also use standard templates that call certain types of information into e-mail replies. In the IMC setup window, administrators can select message

Agent Summary Report
Weekly - Sunday, January 04, 1998 through Saturday, January 10, 1998
Generated - Sunday, January 11, 1998, 12:30:35 AM
All Agents

Agent	Messages Taken	Messages Answered	Messages Forwarded	Messages Open	Messages Resolved	Time Online	Time Ready	Time Responding	Time Idle
Bob Allman	7	4	4	0	0	07:59:11	00:00:03	06:07:04	01:52:04
Bob Jones	0	0	0	0	0	00:00:00	00:00:00	00:00:00	00:00:00
Brent Rosales	0	0	0	0	0	00:00:00	00:00:00	00:00:00	00:00:00
Brooke Graham	0	0	0	0	0	00:00:00	00:00:00	00:00:00	00:00:00
Charlene Brandon	0	0	0	0	0	00:00:00	00:00:00	00:00:00	00:00:00
Dan Cooper	0	0	0	0	0	02:11:22	00:00:00	00:00:00	02:11:22
Dan Hom	0	0	0	0	0	00:00:06	00:00:00	00:00:00	00:00:06
Debi Fryatt	84	63	21	1	3	01:19:27:01	00:00:40	20:03:25	23:22:56
Don Leonard	1	1	0	0	0	01:04:11:40	00:13:23	00:08:24	01:03:49:53
Ernie Leyva	0	0	0	0	0	00:00:00	00:00:00	00:00:00	00:00:00
Gara Zeigler	17	19	4	1	0	01:12:31:05	08:49:36	23:53:48	03:47:41
Greg Hewgill	31	26	4	0	1	01:12:03:15	00:00:12	10:10:05	01:01:52:58
Gwen Barnes	4	3	1	0	0	01:17:11:59	00:36:38	00:38:04	01:15:57:17
Jim Harrer	0	0	0	0	0	00:00:00	00:00:00	00:00:00	00:00:00
Jim Opperman	0	0	0	0	0	00:00:00	00:00:00	00:00:00	00:00:00
Kim Cooper	0	0	0	4	0	00:00:00	00:00:00	00:00:00	00:00:00
Lance Cooper	4	0	4	0	0	04:36:25	00:00:03	04:32:18	00:04:04
Larry Frieson	94	75	4	4	16	01:15:53:09	00:00:36	05:44:06	01:10:08:27
Lynn Wright	40	37	7	3	0	01:16:03:25	00:00:21	09:04:15	01:06:58:49
Mardonio Reyna	42	31	17	0	0	18:12:10	00:02:29	03:00:09	15:09:32
Ozzie Yero	0	0	0	0	0	00:00:00	00:00:00	00:00:00	00:00:00
Paul Davis	16	12	8	0	0	03:09:13:13	00:00:28	02:25:35	03:06:47:10

Figure 3.11   Agent summary report generated by Mustang Software's Internet Message Center (IMC).

templates for standard responses IMC uses to automatically answer e-mail, and to provide additional information on agent replies. Using the IMC Agent client, the agents can quickly resolve messages via a personalized reply using the standard response library. Agents can also view and copy the text of the standard responses to the clipboard in order to create a customized response in their e-mail client.

Templates can also be used to add a header or footer to each outgoing message, or to inform users automatically of the status of their messages. Templates can contain macros that are automatically filled in with system information when the message is created. There are macros for such items as tracking number, agent name, date, time, and message status and pool name.

Mustang's IMC can be run as a standalone application or integrated into an existing call center. It works with firewalls such as Microsoft Proxy Server, and any others that use NT Domain Security for security validation. IMC must be able to log onto the NT domain under a user account that has the appropriate access privileges. In addition, the user account under which IMC Service logs on must have rights to go outside the firewall, or it will not be able to send and receive messages from a company's SMTP/POP3 servers.

3.9 Carrier approaches

Companies can buy the individual call center components and integrate them with the Internet, or subscribe to a carrier-provided service. MCI WorldCom's Click'nConnect, for example, enables Web users to click a button to speak to a call center agent. The service is implemented with NetSpeak's ITEL Call Center system, which consists of the following components:

- ▶ ACD Server—performs dynamic call routing and handling functions;

- ▶ Connection Server—provides real-time IP address resolution;

- ▶ Information Server—serves as a centralized listing directory;

- ▶ Event Management Server—acts as a data repository for all agent call statistics;

- ▶ Control Center—provides GUI-based configuration and supervisory monitoring;

- ▶ Mini WebPhone Kit—telephony client that can be incorporated into a Web site for fast customer download;

- ▶ Agent WebPhone (available in 5-seat increments)—specialized WebPhone client for call center agent desktops.

Users of MCI WorldCom's Click'nConnect service place a button on their Web site that, when clicked, launches an IP telephony call via an Internet gateway that terminates at a call center agent's telephone. The gateway supports Web browsing and voice conversation concurrently over a single telephone line.

MCI WorldCom's approach differs from other services, such as AT&T's interactiveAnswers (see Figure 3.12) and Sprint Corp.'s Give Me a Call service, both of which require the user to have two phone lines—one for Internet access and the other to receive a callback from a

Figure 3.12   Outrigger Hotels and Resorts in Honolulu, Hawaii, allows users to make reservations via the company's Web page through AT&T's interactiveAnswers technology. The agent will call back the user at no charge.

customer service representative. If the user has only one phone line, the return call will be made when the user logs off the Internet.

AT&T's interactiveAnswers service sets up a call between Web visitors and a live, personal agent who can instantly answer product or service questions. When the user clicks on the Call Me Toll-Free button, a notification is sent through the Web, indicating to an agent that a callback has been requested (see Figure 3.13). Since the agent knows the Web site from which the visitor entered, he or she will usually have a good idea of the kinds of questions the user has.

Despite the potential to extend call centers globally and offer 24-hr customer service, many companies will choose to keep normal business hours. In this case, Internet calls will have to be answered with an

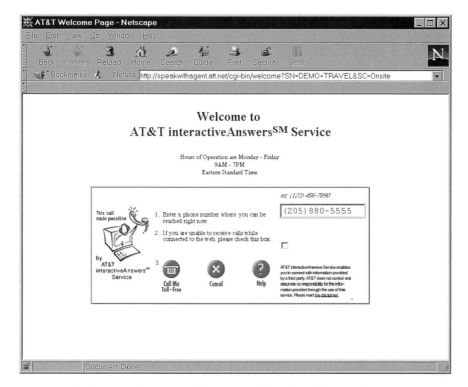

Figure 3.13 AT&T offers a demonstration of its interactiveAnswers on its own Web page. The user fills in the phone number for an AT&T representative to call in the data field located at the top right side of the form. Clicking on the Call Me Toll-Free button sends a notification through the Web, indicating to an agent that a callback has been requested.

appropriate message (see Figure 3.14). The message can be customized to include an apology, the business hours during which the customer can call, and perhaps a hyperlink to an e-mail form that allows the caller to send a callback request.

When the user clicks on the Call Me Toll-Free button, his or her phone will ring instantly, assuming the phone number of a second line was entered in the form's data field. Otherwise, the user must log off the Internet to get the ring on the same line. When the user answers the phone, a recorded message is played, which requests verification that the user wishes to talk with an agent. The agent will then be called, and after a short wait the user can actually speak with the agent.

AT&T, MCI WorldCom, and Sprint are targeting the retail, travel, insurance, banking, and technology industries. The three carriers use push technology in conjunction with their services, which enables them to do things like provide pictures of hotel rooms or vacation destinations to users booking reservations.

Companies subscribing to carriers' call center services usually pay an installation charge, a monthly fixed charge, a per-call transaction fee, and the regular negotiated rate for toll-free 800/888 numbers. Some carriers

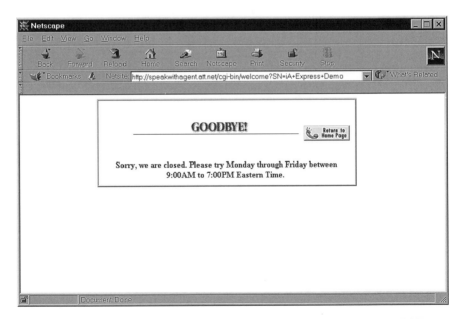

Figure 3.14 If a call is received during nonbusiness hours, AT&T's interactiveAnswers returns an appropriate message to the caller's Web browser.

offer call center services that can be used with 900 numbers, enabling companies to offer fee-paid technical support hot lines. Value-added services provided by the carrier may include application integration, Web site development, and infomercial development.

3.10 Management issues

The management issues related to running an Internet call center are fairly similar to those of a traditional call center, especially if the Internet call center is seamlessly integrated with an existing operation. Basically, the issues boil down to the three main areas of data management, call management, transaction management, and staff management.

3.10.1 Data management

Data management refers to the ability of the system to gather, sort, and store information about Web site visitors and customers for use in routing calls to the most appropriate call center agent.

Whether the inquiry comes in the form of a voice call or an e-mail message, relevant information about the Web site visitor is gathered and stored in a database. Information can be gathered in a number of ways: cookies can collect preference data; Common Gateway Interface (CGI) scripts can gather such information as the visitor's Web browser and version, remote host and IP address, and the referring Web page; a Java applet can be used to request a user ID and password; and Web forms can be used to request specific information about the user.

There are a number of third-party data collection and management tools available. For companies interested in increasing the return on their online marketing campaigns, improving customer satisfaction, and increasing revenues, there are tools that are capable of capturing a complete picture of the online visitor experience.

Accrue Software, for example, offers an analysis tool called Insight, which offers reports that detail user behavior at the Web site. Unlike log file analysis tools that measure activity at the server level, Accrue Insight measures Web site traffic at the network level. This information allows Web site managers and marketing professionals to understand how visitors move throughout their site, what content is valuable to them, and how effectively content is being delivered to them.

For example, Insight provides reports that highlight individual URLs and identify where visitors enter and exit a site. Tracking entry and exit points provides important information about how a user navigates through a site, including what content attracts them to a site and whether they leave a site prior to viewing important content, such as registration or purchase pages. The information provided in these reports helps companies to define their audience, lower online user acquisition costs, and improve user retention rates.

The product's report filtering capabilities consolidate site data into customizable reports that can include information relevant to specific projects, departments, or budgets. Insight reports can be generated and analyzed through any Java-enabled browser. Users can export data to Microsoft Excel or HTML format, enabling easy customization and distribution of site analysis data throughout the enterprise.

Another traffic analysis application that observes the browsing patterns of site visitors is Whirl, which is offered by Interlogue Communications. Among its capabilities, the product reports on the actions taken after a visitor has clicked on a banner ad. The product's support for Oracle databases allows companies to associate collected data with sales and marketing information. Whirl also produces reports that can be exported to Microsoft Excel for extended analysis.

Whatever the means of data collection used, all of this information can be sent to a database where it can be retrieved the next time the visitor makes an inquiry. This helps agents address the needs of the customer promptly and accurately.

Some call center systems, such as Nortel Networks' Symposium Call Center Server, come with visual development tools that eliminate the need for a hodge-podge mix of data collection mechanisms. Call center supervisors can use the tool's cut and paste capabilities, pull-down menus, context-sensitive help, and advanced scripting language to configure the system to collect particular types of information and set up call routing based on that information.

As in traditional call center environments, routing options and customer service can be dramatically improved by using the caller information together with the contents of the database. In being able to use what is already known about each customer, the system can route customers with special needs to an appropriate agent, thereby increasing customer satisfaction.

3.10.2 Call management

Call management entails the ability to route inquiries—both voice and data—to the most appropriate agent, as determined by information sent with the call request or matched against a database of previously gathered information, before the conversation begins (or the return call is placed) or an e-mail reply goes out.

In a typical call center operation, there will be agents who are knowledgeable about specific products and services offered by the company. It is therefore imperative that callers get connected to the agent who can best answer their questions or render the proper technical assistance. In the case of a telephone call from the Internet, the call center system can identify what Web page the caller is viewing, so the call can be routed to an appropriate agent. The call center supervisor can configure the system to identify which agents should respond to calls originating from certain Web pages.

In the case of e-mail inquiries, agents can be assigned to "pools" based on areas of expertise or responsibility. Managers generally have a lot of flexibility in configuring the system to route e-mail appropriately. When a message comes in addressed to laptop@domain.com, for example, it can be routed to a queue assigned to agents trained to answer questions about the company's line of laptop computers. In addition, the system can be instructed to look for particular words in the subject line of a message header, for example, which could be useful for filtering requests for a particular product so they can be routed to a specific group of marketing or support personnel.

The next available agent accesses this incoming mail queue to see how many messages are waiting, and simply selects the next message. The reply can be tailored to the individual message or a canned reply can be selected from a database and then edited appropriately by the agent before it is sent.

The correspondence can be tagged with a tracking number so it can be accessed or referenced in the future. The tracking number lets managers, supervisors, and agents view all correspondence related to a specific message to follow the history of an inquiry. In some cases, the customer can be furnished with the tracking number to follow such things as problem resolution process, to check the status of an order, or to monitor delivery of an item.

3.10.3 Transaction management

Transaction management provides a flexible, near real-time view of dynamic call center activity and alerts the supervisor when a potential problem exists. For example, when a call threshold is reached, appropriate warnings are issued so there is enough time to react before caller satisfaction is compromised. The alerting feature can provide visual or audible alarms or warnings. Among the actions that can be taken automatically are the routing of calls to supervisors and managers to handle the extra load and the returning of messages to callers informing them that their calls will be returned as soon as possible.

During this automated decision-making, the system's transaction manager application considers such factors as the number of active agents, the skills of those agents, the volume of calls in queue, and the number of calls being served. Not only do these transaction management features let the Internet call center handle more calls without additional staffing, they can improve customer satisfaction since calls are processed faster.

3.10.4 Staff management

As in traditional call centers, the objective of an Internet call center is to route each incoming call to an agent trained to handle a particular type of inquiry and to automate as many functions as possible to ease the workload burden of each agent. The management process begins by matching the system's objectives with the skills available within the resource pool. Efficient management relies on a skills matrix that provides a quick way to assess agent skills. Individual strengths are examined, and agents are placed where they will be most productive. Skills assessment also allows call center managers to establish job descriptions, performance standards, salary surveys, and personnel-testing procedures.

In an integrated call center environment, where calls from the PSTN and Internet are often handled by the same agents, everyone must be familiar with the content of the company's Web page, know how and when to apply push technology, understand how to retrieve and respond to e-mail inquiries, and handle callbacks when required.

Agents are most likely to be successful in this demanding environment if they receive a combination of classroom and on-the-job training before full productivity is expected. Classroom training usually includes responding to practice calls using basic scripts. As training moves to the live environment, a supervisor can listen in as new agents handle calls. Periodic critiques of their performance, including positive reinforcement, are used to sharpen their skills. Within 30 to 60 days, trained agents should be able to achieve productivity levels that meet organizational standards that are quantified by the number of calls answered per hour, average call length, and average time between calls.

An efficient call center operation should connect callers to an agent within 20 seconds. Within this time, the agent should have all the available information about a customer displayed on the desktop. Since this pace can be a source of stress and fatigue when done over time, call center managers and supervisors should make arrangements for several breaks during the course of the day. The trick is to allow all agents to have breaks and still handle all the incoming calls. By consulting call reports, managers can identify the peak traffic hours when all agents must be on hand. Breaks can then be arranged to avoid the peak hours. Other ways to sustain high performance include flexible work hours, bonuses, contests, and awards.

3.11   Interactive support services

Companies that do not have call centers, but want to offer interactive support services over the Internet, can combine different media in other ways to help customers find the information they need quickly and easily. Instead of receiving telephone calls over the Internet, these companies can offer live chat, audio and video streaming, push technology, and online trouble ticket processing. Ascend Communications, acquired by Lucent Technologies in mid-1999, offers a glimpse of how interactive support services can be delivered to customers.

The company, a leading provider of wide area network (WAN) and intelligent network (IN) products, has enhanced its Web site to provide content-rich interactive self-service technical support. Using third party tools and applications from such companies as Acuity (for interactive chat) and Documentum (for managing Web content development), the company incorporates a Web-based service delivery model that enables customers to manage and maintain Ascend hardware and software products through assisted self-help. At the same time, the service delivery model enables Ascend to make the most efficient use of its service and support resources.

Through the application of real-time text conferencing, push, and streaming audio and video technologies, Ascend Online Services (AOS) allow customers to proactively manage their networks and work directly with an Ascend network support engineer without leaving the company's Web site. AOS also enables customers to obtain information that provides them with a clearer understanding of the company's new products and technologies.

Central to AOS is an extensive self-help library that includes software downloads and thousands of pages of continuously updated technical information such as FAQs, configuration help, and troubleshooting tips. AOS customers can quickly and easily search for the information they need or click on an icon to begin a live session with an Ascend technician. Users are queued and handed off to the most appropriate operator. During the session, the engineer can push the appropriate pages and links directly to the customer's browser, speeding service by eliminating the typing errors and tedious site or file searches that often bog down telephone-based support.

Only those customers who purchase standard service contracts get interactive online access to Ascend staff. Support options include Advantage On.Line, a service package for customers who prefer to handle most support issues on their own, and which includes interactive Web-based support. There is also Advantage On.Guard, which offers an assigned network service engineer, up to 2-hour on-site response, unlimited telephone support, as well as full access to AOS. Both support offerings are available on a 24 × 7 basis, with unlimited access to one or more of the AOS interactive features. These features include:

▶ *AOS Live:* This feature creates a private text chat session between a customer and an Ascend network support engineer (see Figure 3.15). The engineer assists the customer in finding relevant information by pushing answers and instructions to his or her screen, or by providing step-by-step troubleshooting and configuration assistance. The solution is automatically entered into Ascend's online knowledge base and can be e-mailed to the customer for future reference.

▶ *AOS Casts:* This feature uses audio and video streaming to provide seminars and education on new voice and data services opportunities, configuration guidelines, troubleshooting tips, and

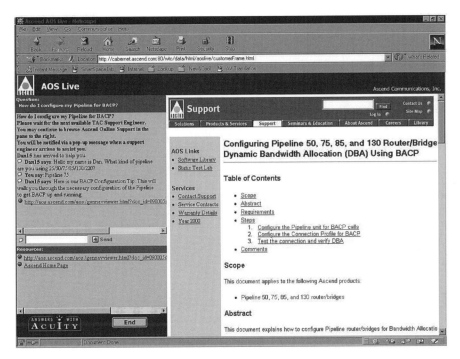

Figure 3.15 With AOS Live, a Java applet is sent to the user, providing a window with three views—two for the private chat session, and a third for displaying Web content pushed to the customer.

other topics. These sessions are scheduled throughout the day and week, allowing customers to sign up at their convenience. At the scheduled time, customers receive the multicast sessions on their computers. The AOS Casts are interactive using multicast chat technology, and the sessions may be moderated or unmoderated.

▶ *AOS Alerts:* This proactive network management tool uses push technology to automatically deliver information vital to the performance of customers' networks via e-mail. Customers are notified of such things as hardware issues and software bugs, and receive the information and patches they need to resolve problems before they impact their systems. Customers also receive troubleshooting and configuration tips as well as notification of new white papers and technical documentation.

▶ *CaseView:* This management system allows customers to open and close trouble tickets and track the status of open tickets online. A collection of technical tips is also provided to help customers resolve issues quickly.

Ascend's approach is user-friendly and customer specific, a combination that can reduce the average problem identification-to-resolution time by as much as 20%. This represents a substantial improvement over previous online support delivery methods, especially when customers are faced with significant network downtime if they cannot get systems up and running.

3.12 Conclusion

The extraordinary growth of the Internet coupled with the emergence of standards-based technologies has the potential to revolutionize business communications, allowing companies to cost-effectively utilize the Internet for telephony, fax, and conferencing in addition to data transfer, file sharing, and e-mail. These and other capabilities can be integrated into the traditional call center.

The convergence of a variety of infrastructures, technologies, and communications media is becoming essential to the effectiveness of call center operations and self-help customer service systems. A key element of this convergence is leveraging the Internet to offload calls from more expensive facilities and collect as much information as quickly as possible to provide customers with a high quality of service. The blurring of the border between call centers and Web-based sales also works to the benefit of customers by giving users more communications options that make the online shopping experience more personal as well as convenient.

With videoconferencing being added to the communications mix, businesses have still another tool with which to provide immediate, responsive, and personal service. In addition, Internet call centers can route customer requests to the best available agent anywhere in the business's global call center network, providing 24-hr customer service. These and other capabilities help differentiate companies in competitive markets and facilitate their entry into new markets. Companies that do not wish to build and run Internet call centers of their own can outsource the operation and management to the major long-distance carriers, which are aggressively pursuing this market.

Through carriers and Internet service providers, consumers and small businesses can take advantage of call management capabilities once reserved for large companies. Lucent offers an Internet call management software suite for service providers, called Online Communications Center (OCC), which enables carriers and ISPs to offer their customers a broad range of easy-to-use options for answering and forwarding telephone calls received while connected to the Internet. Lucent's Internet Call Waiting service enables someone using a phone line connected to the Internet to be notified of incoming calls by an on-screen, pop-up message without disrupting the connection. Lucent's Online Communications Center adds VoIP capabilities, allowing users to remain online while answering phone calls at their computer. OCC also provides intelligent call management features, enabling users to decide how to handle incoming calls before they are received.

Using Online Communications Center's IN features provided by a local carrier, ISP, or enhanced services provider, a consumer or business user can:

▶ Answer a call and speak to the caller while maintaining the Internet session on the same line, using voice-over-Internet service provided by the network operator;

▶ Forward calls to another number;

▶ Play a pre-recorded message for the caller;

▶ Set up caller profiles to treat incoming calls according to priorities set by the consumer;

▶ Automatically forward all calls during priority Internet sessions using a "do not disturb" feature;

▶ Selectively screen or refuse certain calls, such as those that have blocked Caller ID, via an "auto handle call block" option;

▶ Sort or export incoming call records.

In the future, this type of call management capability will provide hooks to personal information management (PIM) applications that reside on the desktop. With an incoming call, the PIM will be opened automatically and provide the record of the person calling, showing his or her name, address, and contact information, as well as notes about the previous conversation.

More information

The following Web pages contain more information about the topics discussed in this chapter:

Accrue Software: http://www.accrue.com/
Acuity: http://www.acuity.com/
Ascend Communications: http://www.ascend.com/
AT&T: http://www.att.com/
Documentum: http://www.documentum.com/
eFusion: http://www.efusion.com/
Genesys Telecommunications Laboratories:
   http://www.genesyslabs.com/
Interlogue Communications: http://www.interlogue.com/
International Telecommunication Union: http://www.itu.org/
Lucent Technologies: http://www.lucent.com/
MCI WorldCom: http://www.mciworldcom.com/
Micron Electronics: http://www.micronpc.com/
Mustang Software: http://www.mustang.com/
NetSpeak: http://www.netspeak.com/
Nortel Networks: http://www.nortel.com/
Oracle: http://www.oracle.com/
Outrigger Hotels & Resorts: http://www.outrigger.com/
PakNetX: http://www.paknetx.com/
Sprint: http://www.sprint.com/

CHAPTER

4

Contents

Infrastructure for voice-data convergence

The convergence of voice and data will benefit consumers and businesses in several ways. In bringing voice and data to homes and offices over a single connection, the installation and ongoing costs of multimedia communication can be greatly reduced. There is more opportunity to put new interactive applications into practical use and access them in a variety of ways, including television, computer, and mobile appliances. Finally, there is the convenience of consolidated billing for a variety of services, including high-speed Internet access.

Yet, the simple fact is that there are three separate wires going into the home—a cable wire, a telephone wire, and an electrical wire. Fortunately, 95% of homes in the United States are wired for telephone service, cable passes 70% of homes, and virtually all are equipped with electricity. Each type of connection holds great potential for voice-data

143

convergence. Businesses have more connection options, but often find themselves having to set up parallel networks—one for voice and one for data—which greatly increases operating costs and management complexity. Of course, wireless technologies have made major inroads in delivering converged voice-data communications as well. The impact of wireless will continue to grow as next generation satellite networks come online with their promise of uninterrupted global coverage for voice calls and, in some cases, broadband data services.

For both home and office environments, several technologies are in the running that not only unify voice and data communications, but do so with much more bandwidth, efficiency, and configuration flexibility than traditional lines and services are capable of providing. The following discussion identifies the weaknesses of the present network infrastructure, starting with the local loop, and highlights progress being made towards overcoming these obstacles to achieve voice-data convergence over a totally revamped "next-generation" infrastructure.

4.1 The local loop

The local loop refers to the equipment and lines of a telephone company that provide access to the public switched telephone network (PSTN) for local and long-distance calls. Decades ago, mostly telephones and PBXs were connected to the local switch, which delivered plain old telephone service (POTS) via copper wires. In recent years, much of the technological innovation associated with the local loop has been data-oriented. The increasing popularity of the Internet, for example, has rekindled interest in ISDN and sparked the development of new broadband digital subscriber line (DSL) technologies to improve the performance of multimedia applications and facilitate access to multimedia content on the Web. In addition, consumer interest in more cable channels and value-added services such as video-on-demand and online shopping are driving the deployment of broadband distribution technologies such as fiber-in-the-loop (FITL) and hybrid fiber/coax (HFC). There is also interest among telephone companies for wireless technologies and IP switches to relieve central offices of the need to handle the growing number of data calls, which tend to tie up circuits for much longer periods than voice calls.

4.1.1 Central office switch

The local loop is serviced by a Class 5 switch, which is modular in design to accommodate growth, new services, and features with the simple addition of subsystems, cards, and/or software. Remote line termination units extend the capabilities of the central office to sparsely populated outlying locations.

The central office switch accommodates many types of hardware modules that provide the following key functions:

▶ Tone dialing and call setup as well as call handling and value-added messaging features;

▶ Signal conversion from analog subscriber lines to digital interoffice trunks and back to analog at the other end;

▶ Connections of subscriber lines and interoffice trunks with the common channel signaling network (i.e., Signaling System 7) for long-distance call setup and call routing;

▶ Traffic concentration over T-carrier (e.g., T1 or T3) or optical carrier (OC) trunks between the central office and smaller switches serving less populated remote areas;

▶ Operations support system (OSS) comprised of databases and tools that enable a telephone company to manage service ordering and provisioning, maintenance and repair, and billing.

There are also a number of service modules and subsystems that can interface with the switch to provide ISDN, ATM, and frame relay capabilities, as well as such emerging services as DSL. There are even subsystems that provide connections between the central office switch and IP-based adjunct switches to intercept data calls intended for the Internet, which frees up badly needed switch capacity for telephone calls and emergency services. Central office switches also have connections to local cellular networks. However, not every Class 5 switch (or even higher order Class 4 switches) is equipped to support these services. Only those serving a significant population of users who have such needs are equipped to provide such services. For example, not every Class 4 switch is equipped to provide ISDN. Even if a switch is equipped to provide ISDN, it does not mean every subscriber can order the service. ISDN lines are only specified for local loops up to 15,000 feet. Any location beyond this distance requires special conditioning of the circuit.

To maximize performance, the switch is usually engineered to provide one circuit for every eight subscribers. This assumes that a voice user typically places one or two six- to eight-minute calls each hour. Now that data traffic has increased to the point of overtaking voice traffic, these assumptions are no longer valid and the problem threatens to get worse. According to Nortel, there are about 60 million Internet users in the United States. Telephone companies cannot afford to upgrade their voice switches to keep pace with the continuing onslaught of data calls. As noted in Chapter 2, vendors like Lucent and Nortel now offer front-end IP switches that can help the traditional carriers meet the growing demand for data traffic and IP telephony without tying up more expensive voice circuits at the central office switch. Such switches route data to the carrier's IP or ATM network instead.

4.1.2 Operating environment

There are about 13,000 central offices in the United States, owned by some 1,400 telephone companies, that differ according to the equipment and lines they have and the services they offer. Since passage of the Telecommunications Act of 1996, there is even more diversity because competitors are allowed to collocate their equipment in the central office. This was deemed necessary in order to stimulate competition in local services and encourage the development of new services.

The central office typically serves subscribers via copper pairs within a 3-mile (15,000 feet) radius. To reach subscribers farther away, several strategies are employed. One is to set up a serving wire center (SWC) with trunks linking it to the local central office. Another is to set up a remote terminal (RT) to link a fairly limited number of subscribers to the network. For example, an SWC might be used to give tenants of a distant office park access to the telephone network, while an RT might be used to reach a gas station and a few fast food restaurants along an otherwise isolated stretch of interstate highway.

Through SWCs and RTs, the telco can extend the local loop to accommodate incremental growth, providing service to an increasing number of subscribers in any number of remote locations. Subscribers served by an SWC or RT have access to all the features of the central office to which they are ultimately connected (see Figure 4.1). If the link between the central office and the smaller systems is cut, basic telephone service is still available among the subscribers at the remote location. If redundant paths are available, remote subscribers will still have access to the rest of the world if the primary link becomes inoperable.

The overriding limitation of the present local loop is that it was designed for voice calls. It works very well for that purpose, but not for data. To support data, a variety of modulation schemes have been employed to convert the digital signals generated by a computer into analog signals suitable for transmission over dial-up telephone lines. The

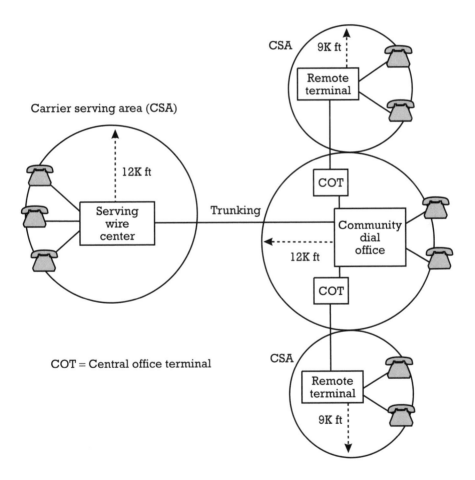

Figure 4.1 A typical local loop with serving wire center and remote terminals.

signals are carried in digital form over the PSTN, but at the remote local loop, the digital signals are converted back to analog. A demodulator connected to the remote computer must convert the analog signals back into digital form. Although long-distance lines are digital, most local lines are not, which explains why modulator/demodulators (modems) are required to access the Internet, send electronic mail, and connect to host computers from remote sites.

Today's modems are capable of sending data at only 56 Kbps, but rarely attain their maximum rate due to various line impairments, such as noise, that prevent data from flowing at consistently high rates in both directions (see Figure 4.2). Through modem bonding, however, two or more modems can be made to work in unison to achieve much higher data rates in the downstream (network to computer) direction.[1] This bonding is achieved with the Multi-Link Point-to-Point Protocol (MLPPP), an extension of the earlier PPP, an IETF WAN link protocol for connecting clients and servers and for interconnecting routers to form enterprise networks. As its name implies, PPP is intended for simple point-to-point connections.

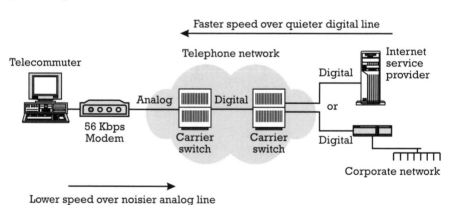

Figure 4.2 A 56-Kbps modem sends data at near top speed, but only from a digital source. Since it is already in digital form, the traffic is free of impairments from noise introduced when an analog modem signal is made digital within the carrier network. From an analog source, the top speed is quite a bit lower than 56 Kbps, since the traffic is subject to impairment from noise.

---

1. Vendors offer devices that bundle multiple modems in a single unit. A separate phone line connects to each port. MLPPP will work on the digital ports as well, enabling two ISDN BRI channels of 56/64 Kbps each to be used as a single channel.

Among the many advantages of MLPPP is that it permits multiple links to be combined and used as a single logical data pipe (see Figure 4.3). This is very useful on private networks, where network managers have control of equipment and software at each node. However, the drawback to this approach when used for accessing the public Internet is that the user must be willing to pay for multiple phone lines. Another potential problem is that the user must find an ISP that supports MLPPP without requiring multiple accounts. And as the Internet continues to offer increasingly sophisticated applications and denser multimedia Web content, even multiple low-speed logical connections may not be enough to overcome bottlenecks that impede the flow of information.

4.2 ISDN

ISDN made its commercial debut in 1980 with the promise of providing a high-quality, ubiquitous all-digital switched service that would support voice and data simultaneously. Although ISDN was intended to become a worldwide standard to facilitate global communications, this was not to

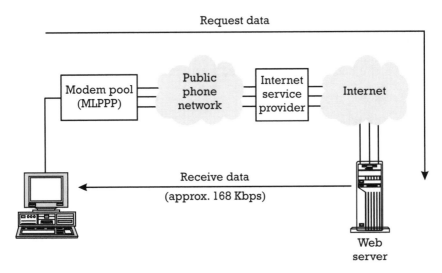

Figure 4.3 The modem pool device initiates two or three simultaneous calls, divvying up the separate TCP/IP sessions, making Web page downloads go much faster. The Web server sends the requested page to the modem pool device, and the page is assembled on the user's screen.

be the case until 15 years later. In the United States, nonstandard carrier implementations, incompatibilities between customer and carrier equipment, the initial high cost of special adapters and telephones, spotty coverage, and configuration complexity hampered user acceptance of ISDN through the mid-1990s.

By 1996, many of the problems with ISDN were finally resolved. In adopting national ordering codes, for example, the carriers not only have made it easier for customers to order ISDN, they have also simplified the process in which ISDN is installed. When customers call to order ISDN, they are asked about the application for the service and about their equipment to determine which type of ISDN will best suit their needs. The customer's requirements are matched against a database of about 70 different ISDN order codes. Installation wizards downloaded from the carrier simplify configuration of the user's equipment.

The increasing popularity of the Internet has spurred consumer demand for ISDN as a means of accessing the Web and improving the response time for navigation and viewing multimedia content. Today, ISDN is available throughout 90% of the United States.

Subscribers can assign channels to either voice or data and dynamically change these assignments to suit continuously changing requirements. For example, if 128 Kbps of bandwidth is being used for accessing the Internet, 64 Kbps can be immediately reassigned to support an incoming or outgoing phone call or fax. Upon completion of the voice or data call, the bandwidth is returned to supporting the Internet session.

4.2.1 ISDN channels

ISDN comes in several varieties, the most common of which are the basic rate interface (BRI) and the primary rate interface (PRI). The BRI provides two bearer channels of 56/64 Kbps each, plus a 16-Kbps signaling channel. The PRI provides 23 bearer channels of 56/64 Kbps each, plus a 64-Kbps signaling channel. Any combination of voice and data can be carried over the B channels. On private networks, additional voice or data channels can be created within the B channels through the use of compression.

ISDN's D channel has access to the control functions of the various digital switches on the network. It provides message exchange between the user's equipment and the network to set up, modify, and clear the

B channels. The D channel also gathers information about other devices on the network, such as whether they are idle, busy, or off. In being able to check ahead to see if calls can be completed, network bandwidth can be conserved. If the called party is busy, for example, the network can be notified before network resources are committed.

4.2.2 Always On/Dynamic ISDN

Normally, the D channel goes unused most of the time—it only takes a fraction of a second to tell the nearest switch to set up or tear down B channels. Whenever the D channel is not being used for signaling, it can be used as a bearer channel for the transport of X.25 data packets. With a feature called Always On/Dynamic ISDN (AO/DI), the D channel can be used for maintaining continuous access to the Internet, regardless of how and when the B channels are used. The D channel provides ample capacity for such routine tasks as checking for new mail, security authentication, and receiving text-based news feeds. Use of the D channel for these purposes keeps the B channels open for voice calls and faxing, which is especially beneficial to telecommuters. If the D channel becomes overloaded, the use of B channels can be triggered if the customer has subscribed to a related service called On-Demand B Channel Packet. Because the packets are accepted at the central office by the X.25 packet handler, it is possible to route these packets—from either the D channel or B channels—without the involvement of the time-division circuit-switched fabric, which reduces the load on the PSTN.

4.2.3 Terminating equipment

In addition to the line from the carrier, users will need terminating equipment (see Figure 4.4) that interfaces with the ISDN service and with other voice, data, and video equipment that may already be in use. Such equipment is typically available for under $200.

The Network Terminating Unit, or NT1, connects the line and the terminal equipment. The NT1 makes the conversion from 2-wire to 4-wire and synchronizes and tests the line. This unit comes in standalone versions or may be integrated into other terminal equipment. The NT1 includes an AC power supply. NT2 devices—usually integrated with PBX and key systems—include all NT1 functions, plus protocol handling, multiplexing, and switching.

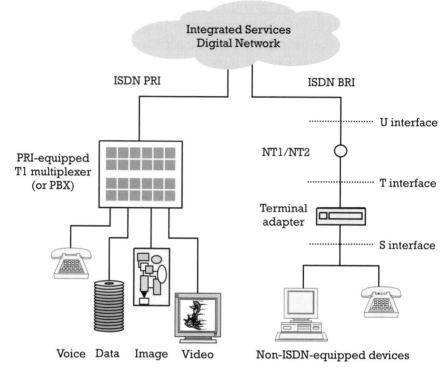

Figure 4.4 Summary of ISDN architectural elements and reference points.

The terminal equipment (TE) can take on various configurations and functions. Basically, the equipment falls into three categories:

▶ *Terminal adapter (TA):* converts the ISDN signal to a signal that is recognizable and usable by non-ISDN devices, like PCs, fax machines, and analog telephones. Like the NT1, a TA requires a power source.

▶ *ISDN telephone:* works specifically with an ISDN BRI line to perform voice telephony. Many ISDN telephones can be equipped with a TA to connect to an additional non-ISDN device, such as a PC.

▶ *PC or in-slot terminal adapter:* enables a PC to be connected to an ISDN BRI connection through an expansion card inside the computer.

Terminal adapters, both standalone and in-slot, can come configured in a variety of ways when used to interface with data communications equipment. The adapters range from a simple TA providing 19.2-Kbps serial connection to a PC communications port to ISDN BRI bridge/routers for connection to LAN adapters and other higher speed interfaces.

4.2.4 MLPPP

Some ISDN LAN modems allow two 64-Kbps ISDN B channels to be aggregated to achieve a single higher-speed 128-Kbps connection. This aggregation is handled by MLPPP. It allows the user to engage in a voice conversation on one B channel, while simultaneously sending or receiving data over the Internet on the other B channel. When the voice call is completed, the two B channels are re-aggregated to restore the higher speed Internet connection. Such modems are flexible enough to connect to two different ISPs over the same ISDN line. The modem automatically connects to an ISP whenever a Web browser is started and it automatically cancels the call when the browser is closed.

4.2.5 Pricing

While ISDN combines voice and data very effectively and offers unparalleled bandwidth efficiency and channel flexibility, it may not necessarily be the best option, especially if the preponderance of calls is data oriented. Typically, customers pay a one-time installation fee, a flat monthly service fee, and per-minute charges based on standard rates for voice calls. Alternatively, there are some ISDN services that are not metered, but which entail a higher monthly fee.

In Texas, for example, Southwestern Bell offers 2B+D ISDN service for about $55 per month, including all port and surcharges; this is an unmetered service, meaning that there are no per-minute charges. X.25 packet switched service over the D-channel has no added installation charge, but adds $2 per month to the user's bill. In addition, there is a set-up charge of $0.005 and a character transmission charge of $0.20 for each kilosegment (one thousand segments of 64 bytes each). The set-up charge is incurred each time a stream of packets is initiated. The set-up charge can be eliminated by a monthly virtual circuit charge of $5, which applies to each destination. On-Demand B Channel Packet costs another $10.35 per month.

Given this type of pricing scheme, ISDN for Internet/intranet access is more expensive than other technologies such as CATV cable and DSL. These alternatives become even more attractive if they include a voice channel. Currently, the local loop is being upgraded by the telephone companies to accommodate the demand for more advanced services.

4.3 Local loop enhancement

Driven by the seemingly insatiable demand for high-performance Internet access, and the growing desire to converge voice and data over the same medium, the local loop itself is being enhanced with a variety of technologies, including DSL, FITL, HFC, and wireless local loop (WLL).

With broadband becoming more and more of an issue, and competition changing the face of the entire telecommunications industry, the local loop is in a state of transition. In particular, the RBOCs are looking for ways to economically serve the needs of Internet users (and position themselves to offer value-added services) without having to tear up massive amounts of copper wire in favor of higher-capacity coaxial cable or optical fiber. Instead, they are seeking to leverage their local loops by increasing the bandwidth capacity of existing twisted pairs so that high-speed data can pass reliably.

One of the most economical ways for telephone companies to do this is by adding digital subsystems at the central office, which in effect turn noisy analog lines into clean digital lines capable of handling data at multimegabit speeds. These DSL systems not only increase the data rate, in some cases, they support voice as well, and extend the distance range beyond that of ISDN.

4.3.1 Digital subscriber line

Of the dozen or so variants of DSL technologies, the most popular among telephone companies is ADSL, which supplies up to 8 Mbps of bandwidth to support multimedia—video, audio, graphics, and text—to the customer premises. According to the Universal ADSL Working Group (UAWG), between 70% and 90% of the world's nearly 750 million telephone lines are upgradable ADSLs based on the current technology.

ADSL carves up the local loop bandwidth into several independent channels suitable for any combination of services, including POTS, ISDN, and video-on-demand. The electronics at both ends compensate for line impairments, increasing the reliability of high-speed data transmission. A significant benefit of ADSL is that it is "always on," providing a continuously available connection that can be shared by multiple users on a network.[2] This avoids the time-consuming connection procedures and busy signals of dial-up modems, and allows users to benefit from new services available from a permanent connection.

The UAWG developed a standard for a "lite" version of full-rate ADSL, also known as Universal ADSL or G.Lite, which is interoperable with full rate ADSL, but with fewer complexities and fewer overall requirements at a tradeoff for speed. The speeds being targeted for G.Lite are up to 1.5 Mbps downstream (from the network to the end user's PC) and 512 Kbps upstream (from the end-user's PC to the network). G.Lite also includes the voice channel of full rate ADSL, but eliminates the need for a technician installed "splitter" box installed outside the home.

The ITU refers to G.Lite as Recommendation G.992.2, which took effect in mid-1999. Anticipating this standard, many DSL modem vendors need only offer firmware upgrades to bring their existing products into compliance. In addition to the G.992.2 specifications, a number of complementary technical specifications have also been agreed upon by the ITU, addressing test procedures, system management, and handshaking procedures.

Other standards such as Home Phoneline (discussed in the next chapter), which enables multiple PCs in the home to be networked together over existing phone extension wiring, are interoperable with G.Lite. The Home Phoneline standard allows networked computers to share any Internet access connection, whether the connection is a dial-up line, CATV cable, ISDN, or G.Lite.

4.3.1.1 Other DSL technologies

There are other DSL technologies available that meet different connectivity needs. Symmetric DSL (SDSL), for example, offers 1.5–2 Mbps in

---

2. Although 56-Kbps dial-up modems can also be shared by multiple users through special modem sharing software, performance may be severely constrained when each user attempts to download information simultaneously. With its multi-megabit capacity, this is not a problem with DSL.

each direction. This is suitable for businesses as a low cost replacement for local T1 (1.544 Mbps) lines that are used exclusively for data. Table 4.1 summarizes the major DSL technologies.

Other DSL technologies include:

▶ Consumer DSL—CDSL is a Rockwell proposal that can be installed without extra hardware to filter the data stream from the voice channel.

▶ EtherLoop—A proposal from Nortel, this DSL variant offers 10 Mbps over very short wires.

▶ HDSL 2—This technology has the same characteristics as HDSL, except that it requires only one pair of wires instead of HDSL's two pairs of wires.

▶ ISDN DSL—IDSL is a dedicated, data-only service at 128 Kbps for which users pay a flat fee rather than the per-minute charges typical of switched ISDN, something that would be attractive to users who spend a lot of time on the Internet, for example. Whereas ISDN passes through the phone company's central office voice switch, IDSL bypasses the voice network by plugging into a special router in the central office.

▶ Multiple Virtual Lines (MVL)—A technology from Paradyne in which one MVL line supports one modem at every phone extension. The devices can talk modem-to-modem within the same building as if on a LAN.

▶ Total reach multi-bit-rate DSL (MDSL)—A technology from Adtran that balances speed and distance, allowing service providers to offer

Table 4.1
Summary of Major DSL Technologies

Technology	Downstream Speeds	Upstream Speeds	Max. Dist.
ADSL (asymmetric)	1.5 Mbps to 8 Mbps	64 Kbps to 640 Kbps	18,000 (ft.)
HDSL (high-bit-rate)	512 Kbps and 2 Mbps	512 Kbps and 2 Mbps	12,000
RADSL (rate adaptive)	128 Kbps to 7 Mbps	64 Kbps to 640 Kbps	25,000
SDSL (symmetric)	1.5 Mbps to 2 Mbps	1.5 Mbps to 2 Mbps	10,000
VDSL (very high-speed)	13 Mbps to 52 Mbps	1.6 Mbps to 2.3 Mbps	4,500

128 Kbps symmetric DSL services to customers located up to 30,000 feet away from their carrier's central office, 384 Kbps at 24,000 feet, 760 Kbps at 18,000 feet, and 1 Mbps at distances of up to 15,000 feet.

▸ Universal DSL—Similar to CDSL, this proposed international standard allows any DSL modem to be used with any DSL service provider, as long as the provider's equipment also meets the standard.

4.3.1.2 DSL service management

Service providers have had a difficult time provisioning DSL and tailoring the service to individual users. This has slowed the growth of DSL despite high interest among consumers and businesses. Now there are products available that make it easier for carriers to implement and manage DSL service.

Cisco Systems, for example, offers a product called the Service Selection Gateway (SSG), a dedicated system that sits in a carrier's central office or a service provider's point of presence. The system lets service providers offer users a way to instantly select additional services beyond the flat-rate DSL access service they may already have. The system even advertises services to customers so that they can instantly know what services are available and easily select a service. Once a user is authenticated—usually by username and password—access to the various network resources and services can be granted by the carrier.

Another problem with DSL has been the lack of integration between its element management system and the carrier's OSS. An OSS is a database that is used by the carrier to support telecommunications services to its customers. Among the functions of OSS are preordering, ordering, provisioning, maintenance and repair, and billing. To address this integration problem, Cisco and other vendors offer carrier-class service policy administration systems that interface with existing operations support systems.

4.3.1.3 DSL and USB

Typically, DSL modems connect to PCs through an Ethernet cable, which requires the user to install an Ethernet adapter card into the PC to use a DSL service. Because the installation of an Ethernet adapter card is beyond the expertise of many users, service providers have had to send out a technician to get the user up and running. This raises the support costs of the carrier, while putting users through considerable inconvenience.

Now there are DSL modems that connect to a PC using the Universal Serial Bus (USB). This is an easy to use and flexible interconnect specification that enables instant peripheral connectivity external to the PC. It allows users to add peripheral devices without expensive add-in cards or configuration challenges such as DIP switches and interrupt request (IRQ) settings. A single connector type simplifies connection of all USB-compliant devices, including telephony and broadband adapters, digital cameras, scanners, monitors, joysticks, keyboards, and other I/O peripherals.

Once the connections are made, a wizard-driven graphical user interface allows the user to easily get the modem ready for proper operation without technical support. Setup is as simple clicking a mouse to install a pre-configured service profile software module created by the carrier and bundled with the modem. By using service profiles, no other configuration tasks are required during installation.

4.3.2 Fiber-in-the-loop

FITL is a system in which services to contiguous groupings of residential and business customers are delivered using fiber optic media in either all or a portion of the local loop. The latest FITL systems offer a dedicated rate of 52 Mbps on the downstream path and 1.6 Mbps on the upstream path. FITL is actually an umbrella term that encompasses various approaches, including:

- Fiber-to-the-neighborhood (FTTN) brings fiber directly into the neighborhood and distributes the signals to businesses and residences via the existing copper wiring.

- Fiber-to-the-curb (FTTC) technology involves bringing fiber into the neighborhood and up to the curb where signals would be carried to businesses and residences via the existing copper wiring.

- Fiber-to-the-building (FTTB) technology involves bringing fiber all the way to the building (e.g., offices, apartments, condominiums).

- Fiber-to-the-home (FTTH) technology involves bringing fiber all the way to the home.

In each approach, a host digital terminal (HDT) at the central office performs optical/electrical conversion and signal processing and control functions (see Figure 4.5). The optical distribution network (ODN) provides a fiber optic splitter function from the HDT to one or more optical network units (ONU) in a point-to-point or point-to-multipoint architecture. On the subscriber side, an ONU performs optical/electrical conversion and signal processing. Usually, in FTTN, FTTB, or FTTC, the ONU is shared by multiple subscribers, while in FTTH the ONU is dedicated to one subscriber.

A key disadvantage of fiber is that the optical components required to send and receive data are still too expensive to deploy to each subscriber location. Therefore, some cable operators have adopted the intermediate approach of FTTN, in which fiber reaches into the neighborhood and coaxial cable branches out to each subscriber. This HFC arrangement increases the bandwidth that the plant is capable of carrying, while reducing both the total number of amplifiers needed and the number of amplifiers in cascade between the headend and each subscriber.

The total number of amplifiers is an important economic consideration because each amplifier must be upgraded or, more typically, replaced to pass the larger bandwidth that the fiber and shorter coaxial cable runs allow. The number of amplifiers in cascade is important for ensuring signal quality. Since each amplifier is an active component that can fail, the fewer amplifiers in cascade, the lower the chance of failure. Fewer amplifiers and shorter trees also introduce less noise into the cable signal. These improvements translate into higher bandwidth, better quality

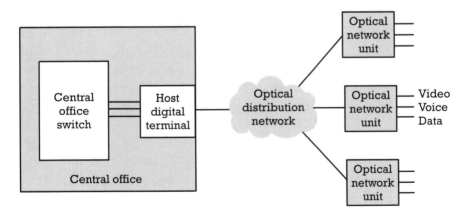

Figure 4.5   Fiber-in-the-loop architecture.

service, and reduced maintenance and operating expense for the cable operator.

4.3.3 Hybrid fiber/coax

As the term implies, HFC is the combination of optical fiber and coaxial cable on the same network. Some local telephone companies have partnered with cable companies to offer value-added services to consumers and businesses. With the advent of DSL technologies, there is no longer intense pressure on the telephone companies to partner with cable companies, since they now have a broadband capability of their own. However, some cable companies intend to upgrade their networks for bi-directional communications, enabling them to compete with local telcos for the provision of telephone service, as well as for Internet access and entertainment services. Such networks provide the foundation for future high-speed digital services to the home. A full HFC system could deliver:

- ▶ POTS;

- ▶ 25–40 broadcast analog TV channels;

- ▶ 200 broadcast digital TV channels;

- ▶ 275–475 digital pointcast channels that deliver programming at a time selected by the customer;

- ▶ High-speed two-way digital link for Internet and corporate LAN access.

HFC divides the total bandwidth into a downstream band and an upstream band. The downstream band typically occupies 50–750 MHz, while the upstream band typically occupies from 5–40 MHz.

Originally, cable operators envisioned a coaxial tree-and-branch architecture to bring advanced services to the home. However, the increasing capacity of fiber optic transmission technology has led many cable operators to shift to an approach that combines fiber and coax networks for optimal advantage.

In HFC networks, fiber is run from a services distribution hub (i.e., headend) to an optical feeder node in the neighborhood, with tree-and-branch coax distribution in the local loop (see Figure 4.6). Two overriding

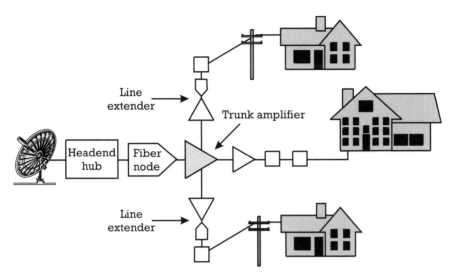

Figure 4.6 Typical topology of an HFC network.

goals in an HFC architecture are to minimize the fiber investment by distributing it over the maximum number of subscribers, and to use the upstream bandwidth efficiently for the highest subscriber fan-out.

There are approximately 200 million TV sets and 100 million VCRs deployed in the United States—and few of them are digital. This casts doubt about the immediate viability of all-digital FTTH networks because any serious attempt to provide video services must deal with this installed base. Services that provide digital television signals require a separate digital decoder for each TV tuner, including the VCR, making FTTH an expensive proposition for subscribers, especially since the digital set-top converter would be required even to receive basic television channels. The advantage of HFC networks, however, is that they carry RF signals, delivering video signals directly to the home in the exact format that the vast installed base of television sets and VCRs were designed to receive.

HFC networks also have the ability to evolve over time from a basic broadcast plant to a two-way network with interactive bandwidth equal to or even exceeding the stated goals of digital FTTH networks. This evolution is achieved incrementally by activating unused dark fiber and subdividing existing nodes to serve fewer homes per node.

During the initial stages of interactive services development, it is anticipated that consumer demand will be low with early subscribers scattered throughout the service area. To provide services, network operators would normally adhere to standard network design concepts, which entail bringing fiber to every curb whether or not the household wants to subscribe to the service. With HFC networks, scattered demand can be accommodated by providing modems as needed. Using modems, digital bandwidth can be allocated flexibly on demand, with virtually arbitrary bit rates. This allows the service provider to inexpensively add bandwidth capacity as demand builds, spreading out capital expenditures to meet subscriber growth in any given service area.

For example, if the HFC network operator wants to offer subscribers high definition television (HDTV) service, today's HFC network designs can already accommodate individual users of digital HDTV at 20 Mbps—even if the demand is scattered. Using quadrature amplitude modulation (QAM), a single HDTV channel could be supplied in a standard 6-MHz channel slot within the digital band between 550 MHz and 750 MHz. HFC network equipment operates equally well with National Television Standards Committee (NTSC) signals, which occupy 6-MHz channels, or Phase Alternating by Line (PAL) standard broadcast analog signals, all on the same network. Channel assignments are completely arbitrary within the 750-MHz spectrum.

One of the major difficulties with HFC is that the cable system was never intended for reliable high-speed data transmission. Cable was installed with the aim of providing a low-cost conventional (analog) broadcast TV service. As a result of the original limited performance requirements, CATV networks are very noisy and communications channels are subject to degradation. In particular they can suffer badly from very strong narrowband interference. This "ingress" noise plays havoc with conventional single-carrier modulation schemes such as QAM and quadrature phase shift keying (QPSK), which cannot avoid the noisy region.

A newer modulation technique—discrete wavelet multitone (DWMT)—is a multicarrier modulation scheme, which provides better channel isolation, thereby increasing bandwidth efficiency and noise immunity. DWMT divides the channel bandwidth into a large number of narrowband subchannels, and adaptively optimizes the number of bits per second that can be transmitted over each subchannel. DWMT provides throughput of 32 digital signal-level 0s (DS0s—64 Kbps each) over each megahertz of bandwidth on the coaxial cable. When a particular subchannel is too noisy, DWMT does not use it, thereby avoiding channel impairments and maintaining a reliable high-bit-rate throughput.

Using DWMT, higher data rates can be achieved—which translates to more channels—over longer distances. It uses the cable bandwidth and communications infrastructure more efficiently, and that allows cable operators to offer more services.

The growth of video and interactive communications services coupled with developments in digital compression have driven both CATV and telephone operators to seek effective ways to integrate interactive video and data services with traditional communications networks. With two-way communications, HFC enables cable operators to offer two-way communications, which will provide subscribers with telephone service over the cable as well as full interactive access to broadband signals, including hundreds of channels of interactive TV, digital services, and more. By using fiber links from the central site to a neighborhood hub, and coax cable from there to a few hundred homes, HFC provides an efficient and economical way to deliver the next generation of communications services, while supporting current services.

The HFC infrastructure offers interexchange carriers the means to bypass the local telephone company, eliminating the need to pay them access charges. AT&T's $48 billion merger with cable operator TCI and $62 billion acquisition of MediaOne illustrate how serious the interexchange carriers are about bypassing the local telephone companies' local loops. With access to an extensive cable TV infrastructure, which it can use for the last mile connections to the home, AT&T is expected to run IP over these connections, including telephone service.

4.3.4 Wireless local loop

A WLL is a generic term for an access system that uses wireless links, rather than conventional copper wires, to connect subscribers to the local telephone company's switch. Wireless local loop—also known as fixed wireless access (FWA), or simply fixed radio—entails the use of analog or digital radio technology to provide telephone, facsimile, and data services to business and residential subscribers.

Chapter 2 mentioned the WLL solution offered by BreezeCom Ltd. and Cisco Systems, which provides high-speed wireless Internet data access at rates of up to 3 Mbps, as well as IP telephony service. WLL solutions that combine voice and data have actually been around for many years, providing rapid deployment of basic phone service in areas where the terrain or telecommunications development makes installation of traditional wireline service too expensive. WLL systems are easily integrated into the wireline PSTN to quickly meet the growing demand for communication services.

WLL solutions include analog systems for medium-to-low-density and rural applications. For high-density, high-growth urban and suburban locations, there are WLL solutions based on code division multiple access (CDMA). Time division multiple access (TDMA) and GSM telecommunications systems are also offered. In addition to being able to provide higher voice quality than analog systems, digital WLL systems are able to support higher-speed fax and data services, including Internet access and IP telephony.

WLL subscribers receive phone service through a radio unit linked to the PSTN via a local base station.[3] The radio unit consists of a transceiver, power supply, and antenna. On the customer side, the radio unit connects to the premises wiring, enabling the customer to use existing phones, modems, fax machines, and answering devices (see Figure 4.7).

The WLL subscriber has access to all the usual voice and data features, such as caller ID, call forwarding, call waiting, 3-way calling, and distinctive ringing. Some radio units provide multiple channels, which are equivalent to having multiple lines. The radio unit offers service operators the advantage of over-the-air programming and activation to minimize service calls and network management costs.

The radio unit contains a coding and decoding unit that converts conventional speech into a digital format during voice transmission and back into a nondigital format for reception. Many TDMA-based WLL systems use the industry standard (IS-127) 8-Kbps enhanced variable rate coder (EVRC), which provides benefits to both network operators and subscribers.

For operators, the high-quality voice reproduction of the EVRC does not sacrifice the capacity of a network or the coverage area of a cell site. An 8-Kbps EVRC system, using the same number of cell sites, provides

---

3. The International Telecommunications Union (ITU) V5.2 open standard interface enables network operators to mix and match local exchange equipment and local access equipment, irrespective of competing suppliers. Enhancements to V5.2 enable the interconnection of WLLs with existing PSTN switching platforms.

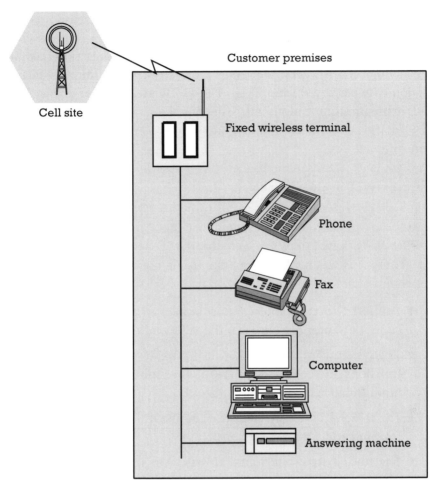

Figure 4.7 The fixed wireless terminal is installed at the customer location. It connects several standard terminal devices (telephone, answering machine, fax, computer) to the nearest cell site base transceiver station (BTS).

network operators with greater than 100% additional capacity than the 13-Kbps voice coders that are deployed in CDMA-based WLL systems. In fact, an 8-Kbps EVRC system requires at least 50% fewer cell sites than a comparable 13-Kbps system to provide similar coverage and in-building penetration.

For subscribers, the 8-Kbps EVRC uses a background noise suppression algorithm to improve the quality of speech in urban environments.

Voice privacy is enhanced through the use of a DSP-based speech encoder, an echo canceller, a data encryption algorithm, and an error detection/correction mechanism. To prevent eavesdropping, low bit rate encoded speech is encrypted using a private key algorithm, which is randomly generated during a call. The key is used by the DSPs at both ends of the communications link to decrypt the received signal. The use of a DSP in the radio units of analog WLL systems also improves fax and data transmission.

4.3.4.1 WLL architectures

WLL systems come in several architectures: a PSTN-based Direct Connect network, a Mobile Telephone Switching Office/Mobile Switching Center (MTSO/MSC)-based network, and proprietary networks.

There are several key components of the PSTN Direct Connect network:

▶ The PSTN-to-radio interconnect system, which provides the concentration interface between the WLL and the wireline network;

▶ The system controller (SC), which provides radio channel control functions and serves as a performance monitoring concentration point for all cell sites;

▶ The BTS, which is the cell site equipment that performs the radio transmit and receive functions;

▶ The fixed wireless terminal (FWT), which is a fixed radio telephone unit that interfaces to a standard telephone set acting as the transmitter and receiver between the telephone and the base station;

▶ The operations and maintenance center (OMC), which is responsible for the daily management of the radio network and provides the database and statistics for network management and planning.

An MTSO/MSC-based network contains virtually the same components of the PSTN Direct Connect network, except that the MTSO/MSC replaces the PSTN-to-radio interconnect system. The key components of an MTSO/MSC-based network are:

▶ MTSO/MSC, which performs the billing and database functions and provides a T1/E1 interface to the PSTN;

‣ Cell site equipment, including the BTS;

‣ FWT;

‣ OMC.

For digital systems such as GSM and CDMA, the radio control function is performed at the base station controller (BSC) for GSM or the centralized base site controller (CBSC) for CDMA.

In GSM systems, there is a base station system controller (BSSC), which includes the BSC and the transcoder. The BSC manages a group of BTSs, acts as the digital processing interface between the BTSs and the MTSO/MSC, and performs GSM-defined call processing.

In CDMA systems, there is a CBSC, which consists of the mobility management (MM) and transcoder subsystems. The MM provides both mobile and fixed call processing control and performance monitoring for all cell sites as well as subscriber data to the switch.

As in PSTN-based networks, the FWT in MTSO/MSC-based networks is a radio telephone unit that interfaces to a standard telephone set acting as the transmitter and receiver between the telephone and the base station.

Operations and maintenance functions are performed at the OMC. As in PSTN-based networks, the OMC in MTSO/MSC-based networks is responsible for the day-to-day management of the radio network and provides the database and statistics for network management and planning.

The PSTN Direct Connect network is appropriate when there is capacity on the existing local or central office switch. In this case, the switch continues to provide the billing and database functions, the numbering plan, and progress tones. The MTSO/MSC architecture is appropriate for adding a fixed subscriber capability to an already existing cellular mobile network or for offering both fixed and mobile services over the same network.

While MTSO/MSC-based and PSTN Direct Connect networks are implemented using existing cellular technologies, proprietary WLL solutions are designed specifically as replacements for wireline-based local loops. One of these proprietary solutions is Nortel's Proximity I, which is used in the United Kingdom to provide wireline-equivalent services in the 3.5-GHz band. The TDMA-based system was designed in conjunction with U.K. public operator Ionica, which is the source of the "I" designation. The system provides telecommunications service from any host network switch, providing toll quality voice, data, and fax services. The system is switch independent and is transparent to dual tone multifrequency (DTMF) tones and switch features. The Proximity I system architecture consists of the following main elements:

- Residential service system (RSS), which is installed at the customer premises and provides a wireless link to the base station;

- Base station, which provides the connection between the customer's RSS and the PSTN;

- Operations, administration, and maintenance system (OA&M), which provides such functions as radio link performance management and billing.

The RSS offers two lines, which can be assigned for both residential and home office use, or for two customers in the same 2-km area. Once an RSS is installed, the performance of the wireless link is virtually indistinguishable from a traditional wired link. The wireless link is able to handle high-speed fax and data via standard modems, as well as voice. The system supports subscriber features such as call transfer, intercom, conference call, and call pick-up.

The RSS has several components: a transceiver unit, residential junction unit (RJU), network interface unit, and power supply. The transceiver unit consists of an integral 30-cm octogonal array antenna with a radio transceiver encased within a weatherproof enclosure. The enclosure is mounted on the customer premises and points toward the local base station.

The RJU goes inside the house where it interfaces with existing wiring and telephone equipment. The Proximity I system supports two 32-Kbps links for every house, enabling subscribers to have a voice conversation and data connection for fax or Internet access at the same time. At this writing, work is under way to develop systems that can handle ISDN speeds of 64 Kbps and beyond. Further developments will result in RSSs that can handle more lines per unit for medium-sized businesses or apartment blocks.

The network interface unit, mounted internally or externally, is a cable junction box that accepts connections from customer premises wiring. The unit also provides access for service provider diagnostics and contains lightening protection circuitry.

The power unit is usually mounted internally and connects to the local power supply (110/220V AC). The power unit provides the DC supply to the transceiver unit. A rechargeable battery takes over in the event of a power failure and is capable of providing 12 hours of standby and 30 minutes of talk time.

The base station contains the radio frequency equipment for the microwave link between the customer's RSS and the PSTN, along with subsystems for call-signal processing, frequency reference, and network management. This connection is via radio to the RSS and by either microwave radio, optical fiber, or wireline to the local exchange. The base station—consisting of a transceiver microwave unit, cabinet, power supply, and network management module—can be configured to meet a range of subscriber densities and traffic requirements.

The connection from the base stations to the local exchange on the PSTN is via the V5.2 open standard interface. In addition to facilitating interconnections between multivendor systems, this interface enables operators to take full advantage of Proximity I's ability to maximize spectrum utilization through allocation of finite spectrum on a dynamic per-call basis, rather than on a per-customer basis. Concentration allows the same finite spectrum to be shared across a much larger number of customers, producing large savings in infrastructure, installation, and operations costs for the network operator.

4.4   Interexchange facilities

Interexchange facilities are the trunks between central offices. The term *interexchange* is typically used in reference to the long-distance portion of the PSTN; specifically, the networks of carriers like AT&T, MCI World-Com, and Sprint—all of which carry telephone calls between the local loops at each end of the circuit. The distinction between local and interexchange carrier, however, is becoming blurred. The regional Bell holding companies, which own the local loops, want to get into the long-distance business, while the interexchange carriers want to get into the local access business so they do not have to share their revenues with the local telephone companies. A number of strategies are being pursued by both camps to achieve these objectives, but their plans must pass muster with the FCC. The topic of regulation in reference to these and other matters is covered in Chapter 9.

4.4.1 T-carrier

Among the most prevalent technology used for interexchange facilities is T-carrier, a circuit-switching technology, which had its origins in the 1960s. It was first used by telephone companies as the means of aggregating multiple voice channels onto a single high-speed digital backbone facility between central office switches. The most widely deployed T-carrier facility is T1, which has been commercially available since 1983. It provides 24 individual 64-Kbps (DS0) channels over 1.544 Mbps of bandwidth. This is often called digital signal level 1, or DS1. T1 is also used in the local loop, mostly to carry business traffic, and is the facility over which ISDN PRI and other digital services are provided in support of voice and data.

T1 lines are commonly leased by large organizations for their private networks, which typically support a mix of voice and data. The permanent end-to-end circuit consists of the local loop segment at each end and the interexchange facility in between. The charges include a fixed element (generally called a channel termination) at each end and a mileage-sensitive element for the long-haul portion of the link. Mileage is calculated for the shortest airline distance between serving wire centers using vertical and horizontal coordinates. These rate elements consist of monthly recurring charges only; usage is not metered.

4.4.1.1 Digital signal hierarchy

To achieve the DS1 transmission rate, selected cable pairs with digital signal regenerators (repeaters) are spaced approximately 6,000 ft apart. This combination yields a transmission rate of 1.544 Mbps. By halving the distance between the span line repeaters, the transmission rate can be doubled to 3.152 Mbps, which is called DS1C. Adding more sophisticated electronics and/or multiplexing steps makes higher transmission rates possible, creating a range of digital signal levels, as summarized in Tables 4.2 and 4.3.

For example, a DS3 signal is achieved in a two-step multiplexing process whereby DS2 signals are created from multiple DS1 signals in an

Table 4.2
North American Digital Signal Hierarchy

Signal Level	Bit Rate	Channels	Carrier System	Typical Medium
DS0	64 Kbps	1	—	Copper wire
DS1	1.544 Mbps	24	T1	Copper wire
DS1C	3.152 Mbps	48	T1C	Copper wire
DS2	6.312 Mbps	96	T2	Copper wire
DS3	44.736 Mbps	672	T3	Microwave/fiber
DS4	274.176 Mbps	4,032	T4	Microwave/fiber

Table 4.3
International (ITU) Digital Signal Hierarchy

Signal Level	Bit Rate	Channels	Carrier System	Typical Medium
0	64 Kbps	1	—	Copper wire
1	2.048 Mbps	30	E1	Copper wire
2	8.448 Mbps	120	E2	Copper wire
3	34.368 Mbps	480	E3	Microwave/fiber
4	44.736 Mbps	672	E4	Microwave/fiber
5	565.148 Mbps	7,680	E5	Microwave/fiber

intermediary step. DS1C is not commonly used, except in highly custom-ized private networks where the distances between repeaters is very short, such as between floors of an office building or between buildings in a campus environment. Some channel banks and multiplexers support DS2 to provide 96 voice channels over a single T-carrier facility. DS4 is used mostly by carriers for high-traffic trunking between central offices.

4.4.1.2   Channel utilization

T1 lines are used for economical and efficient voice and data transport over the wide area network. Economy is achieved by consolidating multi-ple lower-speed voice and data channels over the higher-speed T1 line through a process called time division multiplexing. With TDM, each input device is assigned its own time slot, or channel, into which data or digitized voice is placed for transport over the high-speed link (see Figure 4.8). The link carries the channels from the transmitting

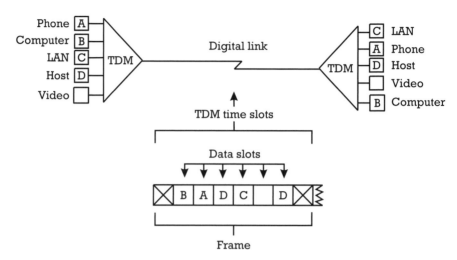

Figure 4.8 Data from multiple input sources is interleaved by the time division multiplexer for transmission over the high-speed link. Note the empty time slot. If an input device has nothing to send, this amount of bandwidth goes unused.

multiplexer to the receiving multiplexer, where they are separated out and sent on to assigned output devices. This is more cost-effective than dedicating a separate lower-speed line to each terminal device. The economics are such that only 5 to 8 analog lines are needed to cost-justify the move to T1.

Further efficiencies can be obtained by compressing voice and data to make room for even more channels over the available bandwidth. Individual channels also can be dropped or inserted at various destinations along the line's route through the use of an add-drop multiplexer (ADM) or digital cross-connect system (DCS). An 8-Kbps channel can be embedded in each 64-Kbps channel for basic supervision and control, leaving 56 Kbps available for the application.

Usually a T1 multiplexer provides the means for companies to realize the full benefits of T1 lines, but channel banks offer a low-cost alternative. The difference between the two devices is that T1 multiplexers offer higher port capacity, offer more bandwidth configuration options, support more types of interfaces, and provide more network management features than channel banks.

The disadvantage of TDM is that when an input device has nothing to send, its assigned channel goes empty and that amount of bandwidth is wasted. A better approach is statistical time division multiplexing (STDM)—if a device has nothing to send, the channel it would have used is taken by a device that does have something to send. If all channels are busy, input devices wait in queue until a channel becomes available. STDMs are used in situations where efficient bandwidth usage is valued and the applications are not bothered by delay.

Now that voice can be carried effectively over IP nets along with data, more businesses are using T1 for building intranets because they are more economical and efficient than TDM and STDM. Instead of multiplexers at each end of the T1 link, lower cost routers are used. And instead of putting voice and data into circuit-switched time slots for delivery, voice and data are put into individually addressed packets that are sent out to a number of locations on and off the net. Since packet switching dispenses with the need for "time slots," hundreds more voice calls and data sessions can be handled over the same amount of bandwidth.

4.4.1.3 Virtual private networks

Many companies do not want to set up intranets of their own. An economical alternative to leased lines is a virtual private network (VPN), which lets businesses set up their own IP-based WANs within the carrier's high-speed IP backbone. As noted in Chapter 2, the traffic of various organizations that share the network is kept separated through the use of tunneling protocols, such as the Layer 2 Tunneling Protocol (LT2P) and IP Security (IPSec). Corporate locations usually access the VPN via a frame relay service, but remote offices and telecommuters can access the VPN through alternative means such as modem dial-up, ISDN, DSL, cable, or wireless. The VPN can even be extended beyond employees to include customers, suppliers, and business partners. Such VPNs are often called "extranets."

4.4.2 Optical fiber

Fiber optic transmission systems have been in commercial use for almost 25 years, and carriers have been eagerly deploying it because it offers far more bandwidth than T-carrier transmission systems. Fiber optic cable is also more efficient for data transport because the media does not have to contend with the physical limitations of copper wire, such as its high attenuation and susceptibility to various forms of interference. With rare exception, all long-distance traffic is now carried over fiber optic cable. In addition, most metropolitan areas are served by fiber optic rings that provide high-bandwidth, fault-tolerant data communications for businesses.

Although newer technologies are becoming available, most fiber optic transmission systems in use today rely on light-emitting diodes (LEDs) or injection laser diodes (ILDs) as their light source. The transmitter circuitry translates electronic signals into optical signals by modulating the drive current to the emitter. Both LEDs and ILDs emit an intense beam of monochromatic light in the form of visible red light or invisible infrared light. Monochromatic light ensures a uniform frequency whereby emitted photons travel at a constant speed; ordinary light is unacceptable for fiber transmission because it contains a range of frequencies that creates "chromatic dispersion"—a data-corrupting condition caused when photons travel at slightly different speeds and cause the energy bursts to disperse at varying distances.

At the receiving end of a fiber link, equipment with a light-detecting device is used to create an electrical current from the transmitted light. Commonly used light-detecting devices include:

▶ Positive-intrinsic-negative-channel (PIN) photodiodes, which produce an electrical current in proportion to the amount of light energy projected onto them;

▶ Avalanche photodiodes (APDs), which operate in much the same way as PIN photodiodes, but require more complex receiver circuitry and provide faster response times;

▶ Phototransistors, which function as amplifying detectors, which are useful for fiber optic systems with long spans between amplifying stages.

4.4.2.1 Types of fiber

There are two types of optical fibers: single-mode and multimode. The former transmits only one light wave along the core, while the latter transmits many light waves. Single-mode fibers entail lower signal loss and support higher transmission rates than multimode fibers and, therefore, are the type most often selected by carriers for use on the public network. Over 90% of fiber optic cable installed by carriers is the single-mode type.

Multimode fibers have relatively large cores. Light pulses that simultaneously enter a multimode fiber can take many paths and may exit at slightly different times. This phenomenon, called intermodal pulse dispersion, creates minor signal distortion and thereby limits both the data rate of the optical signal and the distance that the optical signal can be sent without repeaters. For this reason, multimode fiber is most often used for short distances and for applications in which slower data rates are acceptable.

Multimode fiber can be further categorized as step-index or graded-index. Step-index fiber has a silica core encased with plastic cladding. The silica is denser than the plastic cladding; the result is a sharp, step-like difference in the refractive index between the two substances. This difference prevents light pulses from escaping as they pass through the optical fiber. Graded-index fiber contains multiple layers of silica at its core, with lower refractive indices toward the outer layers. The graded core increases the speed of the light pulses in the outer layers to match the rate of the pulses that traverse the shorter path directly down the center of the fiber.

The individual fibers in most of today's fiber optic cable have an outside (or cladding) diameter of 125 micron (m). (A micron is one-millionth of a meter.) The core diameter depends on the type of cable. The cores of multimode fibers comprise many concentric cylinders of glass; each cylinder has a different index of refraction. The layers are arranged so that light introduced to the fiber at an angle will bend back toward the center. The bending results in light, which travels in a sine-wave pattern down the fiber core, and allows an inexpensive noncoherent light source, such as an LED, to be used. Almost all multimode fibers have a core diameter of 62.5m. Bandwidth restrictions of 200 to 300 MHz/km limit the maximum length of multimode segments to a few kilometers. Wavelengths of 850 and 1,300 nm are used with multimode fiber optic cable. Single-mode fiber consists of a single 8- to 10m core. This means that a carefully focused coherent light source, such as a laser, must be used to ensure that light is sent directly down the small aperture. Single-mode fiber is normally operated with light at a wavelength of 1,300 nm. Because possible light paths through the fiber are restricted, there is essentially no bandwidth limit for single-mode fiber. Frequencies of many gigahertz can be carried tens of kilometers without fiber optic repeaters.

4.4.2.2 Performance advantages

Optical fiber provides many performance advantages over conventional metallic cabling, starting with high bandwidth. The laser components at each end of the optical-fiber link allow for high encoding and decoding frequencies. For this reason, optical fiber offers much more bandwidth capacity than copper-pair wires. Data can be transmitted over optical fiber at multigigabit speeds, as opposed to a maximum transmission speed of only 44.736 Mbps over T3 copper-based digital facilities, and 56 Kbps over unconditioned analog facilities.

The speed of transmission over fiber facilities is further increased by SONET, which permits speeds of up to 13 Gbps. Experimental techniques have transmitted data at one trillion bits per second—the equivalent of sending the contents of 300 years' worth of daily newspapers in a single second, or conveying 12 million telephone conversations simultaneously.

When a photonic technology called dense wave division multiplexing (DWDM) is applied to SONET links or rings, transport capacity can be increased several times, without the carrier having to lay any additional fiber optic cable. However, DWDM only works on the newer grade of optical fiber called nondispersion-shifted fiber, which can support up to 80 channels on a single hair-thin strand. Dispersion-shifted fiber suffers from the problem of signal bleeding. This problem limits the older fiber to eight channels per strand, which are derived from the older WDM technology.

Fiber also exhibits lower signal attenuation than copper. Measured in decibels (dB), signal attenuation refers to signal "loss" during transmission (i.e., the signal received is not as strong as the signal transmitted). Signal attenuation is attributed to the inherent resistance of the transmission medium. For transmissions over metallic cable, loss increases with frequency, and limits the amount of available bandwidth. The characteristics of optical fiber, however, are such that little or no inherent resistance exists. This low resistance allows the use of higher frequencies to derive enough bandwidth to accommodate thousands of voice channels. Whereas an analog line on the local loop has a frequency range of up to 4 KHz for voice transmission, a single optical fiber has a range of up to 3 GHz.

Fiber also surpasses copper in terms of data integrity, which is expressed as the bit error rate (BER). A typical fiber optic transmission system produces a bit error rate of less than 10^{-9}, while metallic cabling typically produces a bit error rate of 10^{-6}. Because of their high data integrity, fiber optic systems do not require extensive use of the error-checking protocols common in metallic cable systems. Because the error-checking overhead is eliminated, the data transmission rates are enhanced. In addition, because the required number of retransmissions is reduced with fiber, overall system performance is greatly improved.

Electromagnetic and radio frequency interference (EMI and RFI) are the primary sources of data errors in transmissions over metallic cable systems. Fiber's immunity to interference not only results in far fewer data errors, but makes fiber more economical to install since it does not have to be rerouted around sources of EMI/RFI and there is no need to build special conduits to shield fiber from the external environment. In addition, optical fibers do not radiate signals—another source of electromagnetic interference—that often cause cross-talk on metallic cables.

Since fiber does not radiate signals, it is a more secure transmission medium than metallic cable. In fact, to intercept an optical transmission the fiber's core must be physically broken and a connection fused to it (see Figure 4.9). Although this splicing procedure is routinely used to add nodes to the fiber cable, it momentarily blocks the transmission of light beyond the break point, and therefore makes unauthorized access easily detectable.

If there is any disadvantage of fiber optic transmission systems it has been the lack of equipment that can make routing decisions at speeds above the DS3 rate of 44.736 Mbps. But even this barrier has fallen by the

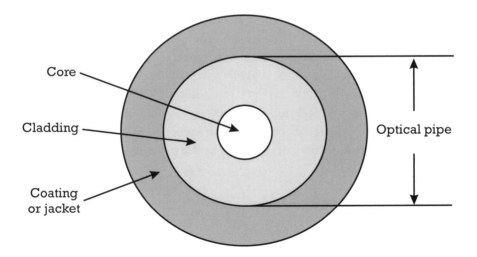

Figure 4.9 Cross-sectional view of an optical fiber strand.

wayside. Tellium, a company spun off from Bellcore (now known as Telcordia Technologies), offers an optical cross-connect switch called Aurora that eliminates the need to demultiplex optical signals into an electrical cross-connect that can only operate at DS3 speeds. Before Aurora, the only way service providers could receive and transmit such high-speed data signals was to upgrade their networks to SONET-based OC-192, which entails a multimillion-dollar forklift upgrade.[4] The new switch can also make existing SONET networks more efficient by eliminating the expensive demux/remux process that SONET multiplexers and cross-connects require for every signal. Half to three-fourths of the traffic on public networks does not really need processing at every node because it is just passing through. Aurora off-loads this burden by handling the pass-through traffic optically. What this is leading to is the concept of "all optical networks."

4.4.2.3 All optical networks

To handle the increasing demand for multimedia applications—and enable carriers to position themselves for more demanding ones in the future—requires a new paradigm for optical-network service deployment in the next century. This new paradigm will give carriers the means to provide wavelength-allocated bandwidth, wavelength routing, and wavelength-translation capabilities.

Wavelength-allocated bandwidth will provide the capacity needed for large users, who will need ubiquitous access throughout the country via optical virtual private networks. Wavelength routing will enable diverse point-to-point and point-to-multipoint optical payload transport for users in regional networks. Wavelength translation will enable optical services to traverse multiple carriers, regardless of vendor technologies. It enhances optical-network topology designs by enabling a wavelength frequency to be selected at one port and switched over to another frequency for acceptance by another port in the network.

---

4. SONET is a transmission standard originally developed for circuit-oriented voice networks and today provides the transmission foundation for public networks. However, SONET networks mandate a rigid architecture that is limited in its ability to scale to support high-speed services. Forklift upgrades are required to increase capacity—a task that is expensive and takes months to implement. With this infrastructure, service providers are constrained in the ability to increase bandwidth and provide new services quickly.

The key elements of the all-optical network architecture consist of DWDMs, optical ADMs, and optical cross-connects. DWDM systems are already in operation today, particularly on long-distance and so-called next-generation networks. These systems multiplex channels on a single fiber to vastly increase its capacity for point-to-point applications. Current DWDM systems multiplex up to 16 SONET 2.5-Gbps signals into one 40-Gbps composite multiwavelength signal. Carriers are optimizing the best attributes of SONET ring protection and survivability with DWDM virtual capacity. This allows them to maintain existing SONET rings by deploying DWDM as fiber expansion, instead of using DWDM for network replacement.

Optical ADMs are being integrated into DWDM-based networks. This type of multiplexer enables carriers to access wavelength-based services and create route diversity for network topologies, as well as provide a migration path to optical ADM rings. In conjunction with ADMs, rings provide a fail-safe means of surviving the most severe disasters. The ADMs work together to take traffic off the affected ring segment and place it onto another ring with spare capacity.

Optical cross-connect systems are being deployed that give carriers more options in building network topologies consistent with today's telecom infrastructure, including mesh, ring, and star. Mesh designs, for example, make it possible to build network topologies with enhanced survivability. Essentially, optical cross-connects let carriers establish mesh designs at the optical layer, regardless of payload or service application. Optical cross-connects will complement optical ADM by centralizing all network topologies.

The new paradigm for optical network services also includes the concept of transparency—the use of the light path itself as the transmission medium, which eliminates the need for optical-to-electrical conversions in the network. Existing transmission networks—which consist of fiber, DWDM multiplication products, and SONET transmission equipment—are congested because they require optical signals to be converted to electrical signals and then to be converted back into optical. Much of the congestion can be relieved by eliminating the need for these conversions and making the light path become the transport medium rather than the fiber.

Sycamore Networks, for example, envisions a network that will initially coexist with SONET and DWDM equipment, before phasing into an all-optical network, where services are mapped to light paths, optical/electrical conversions are eliminated, and WAN services are delivered at LAN speeds over distances of up to 10,000 km without the need for signal regeneration.

SONET is valued for its intelligence and restoration capabilities and DWDM for its bandwidth multiplication capabilities. Accordingly, the company's products combine optical networking software with SONET and DWDM to provision and deliver services over light paths, eliminating the expensive optical/electrical conversions associated with a SONET infrastructure.

Sycamore's first phase focused on optical-networking add-drop nodes aimed at increasing fiber capacity by providing up to 44 wavelengths to relieve congestion on links spanning distances of up to 500 km without regeneration for such applications as corporate Internet access and virtual private networks.

All-optical networks are essential to providing the bandwidths that will be required in the future. Improved technology in optical components—such as lasers, amplifiers, and filters—will enable information to be sent over longer distances and at lower cost. Being able to use light waves that can flow freely without having to be converted into electrical energy and then back again will one day enable carriers to offer OC-3 service (155 Mbps) at the price of T3 (45 Mbps) service. This transparency also provides additional advantages such as reduced delay through the elimination of optical-electrical conversion, and reduced cost due to a need for fewer system components and the added flexibility carriers will have in service provisioning and management.

4.4.3 Wireless IP

Wireless IP achieves voice-data convergence over cellular networks and at the same time bridges PSTN and Internet environments. Also known as cellular digital packet data (CDPD), wireless IP provides a way of passing data packets over cellular voice networks at speeds of up to 19.2 Kbps. Although CDPD employs digital modulation and signal processing techniques, the underlying service is still analog. The media used to transport data consists of the idle radio channels typically used for AMPS cellular service. Wireless IP is available in 3,000 cities in the United States through

AT&T and through other carriers that have interconnection agreements with AT&T.

The CDPD infrastructure employs existing cellular systems to access a backbone router network that uses IP to transport user data. Personal digital assistants, palmtops, and laptops running applications that use IP can connect to the CDPD service and gain access to other mobile users or corporate computing resources that rely on wireline connections.

To take full advantage of wireless IP networks, users should have an integrated mobile device that operates as a fully functional cellular phone and Internet appliance. For example, the AT&T PocketNet Phone contains both a circuit-switched cellular modem and a CDPD modem to provide users with fast and convenient access to two-way wireless messaging services and Internet information. GTE provides a similar service called SuperPhone for its users. In addition to handling data, both enable users to make ordinary phone calls as well.

The PocketNet Phone and SuperPhone enable both end users and corporations to capitalize on the content and messaging power of the Internet. PocketNet Phone users, for example, have access to two-way messaging, airline flight information, and financial information. In addition, users can access Web content and legacy data. Although CDPD services might prove too expensive for heavy database access, the use of intelligent agents can cut costs by minimizing connection time. Intelligent agents gather requested information and report back the results the next time the user logs on to the network.

Unlike voice cellular charges, which are based on call duration, CDPD fees are based on the amount of data transferred. In addition to monthly service plans, which vary from $8.00 for low usage to $49.00 for high usage, it costs from $.05 per kilobyte to use the CDPD network for usage above the number of kilobytes included in the plan. There are also monthly service plans for unlimited usage, which range from $55 for unlimited local usage to $65 for unlimited nationwide usage. Service activation fees range from $35 to $50.

Although CDPD piggybacks on top of the cellular voice infrastructure, it does not suffer from the 3-KHz limit on voice transmissions. Instead, it uses the entire 30-KHz radio frequency (RF) channel during idle times between voice calls. Using the entire channel contributes to CDPD's faster data transmission rate. Forward error correction ensures a high level of wireless communications accuracy. With encryption and authentication procedures built into the specification, CDPD also offers more robust security than any other native wireless data transmission method. As with wireline networks, users can also customize their own end-to-end security.

4.4.3.1 Underlying technologies

CDPD is in fact a blend of digital data transmission, radio technology, packetization, channel hopping, and packet switching.

Digital transmission technology is reliable and more resistant to radio interference than analog transmission technology. The digital signals are broken down into a finite set of bits, rather than transmitted in a continuous waveform. When signal corruption occurs, error-detection logic at the receiving end can reconstruct the corrupted digital signal using error-correction algorithms. Digital technology also enables the use of processing techniques that compensate for signal fades without requiring any increase in power.

In accordance with the Internet Protocol (IP), the data is packaged into discrete packets of information for transmission over the CDPD network. In addition to addressing information, each packet includes information that allows the data to be reassembled in the proper order at the receiving end and corrected if necessary.

Channel hopping automatically searches out idle channel times between cellular voice calls. Packets of data select available cellular channels and go out in short bursts without interfering with voice communications. Alternatively, cellular carriers may also dedicate voice channels for CDPD traffic to meet high traffic demand. This situation is common in dense urban environments where cellular traffic is heaviest and the channel-hopping mode is difficult to use.

4.4.3.2 Network architecture

The CDPD specification defines all the components and communications protocols necessary to support mobile communications. Figure 4.10 shows the main elements of a wireless IP network.

The backbone router, also known as the mobile data-intermediate system (MD-IS), uses the location information derived from the mobile network location protocol to route data to the mobile units, which are referred to as mobile-end systems (M-ES). Information on the link

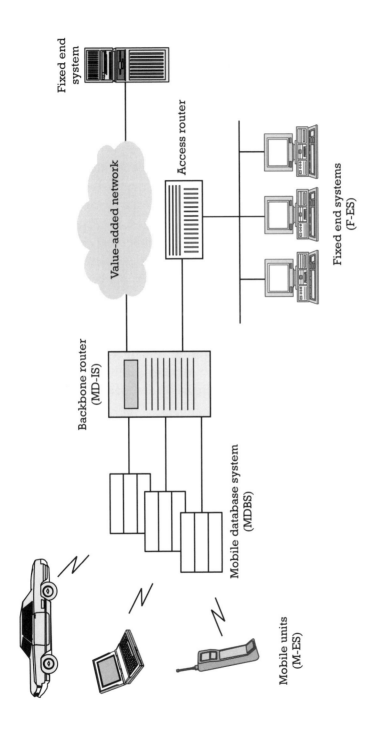

Figure 4.10 Main elements of a wireless IP network.

between the backbone router and a mobile ddatabase system (MDBS) is transmitted using a data link layer (DLL) protocol. Communications on the other side of the backbone router are handled using internationally recognized protocols.

The relay between the cellular radio system and the digital data component of the CDPD network is provided by the MDBS. The MDBS communicates with the mobile units through radio signals. Up to 16 mobile units in a sector can use the same cellular channel and communicate as if they were on a LAN. This communications technique is known as digital sense multiple access (DSMA). After the MDBS turns the cellular radio signal into digital data, it transmits the data stream to its backbone router, using frame relay, X.25, or the Point-to-Point Protocol (PPP).

Although the physical location of a mobile-end system, or mobile unit, may change as the user's location changes, continuous network access is maintained. At the network sublayer and below, mobile units and backbone routers cooperate to allow the equipment of mobile subscribers to move transparently from cell to cell, or roam from network to network. This mobility is accomplished transparently to the network layer and above. The backbone routers terminate all CDPD-specific communications with mobile units and MDBSs, producing only generic IP packets for transmission through the backbone network.

As noted, the requirement that mobile units support IP is meant to ensure that existing applications software can be used in CDPD networks with little or no modification. However, new protocols below the network layer have also been designed for CDPD. These protocols fall into two categories—those required to allow the mobile unit to connect locally to an MDBS, and those required to allow the mobile unit to connect to a serving backbone router and the network at large.

Digital sense multiple access is the protocol used by the mobile unit to connect to the local MDBS. DSMA is similar to the carrier sense multiple access (CSMA) protocol used in Ethernet. It is a technique for multiple mobile units to share a single cellular frequency, much as CSMA allows multiple computers to share a single cable. The key difference between the two, apart from the data rate, is that CSMA requires the stations on the cable to act as peers contending for access to the cable in order to transmit, whereas in DSMA the MDBS acts as a referee, telling a mobile unit when its transmissions have been garbled.

A pair of protocols permit communications between the mobile unit and the backbone router. The Mobile Data-Link Layer Protocol (MDLP) uses media access control (MAC) framing and sequence control to provide basic error detection and recovery procedures; the Subnetwork Dependent Convergence Protocol (SNDCP) provides segmentation and head compression.

In addition to segmentation and header compression for transmission efficiency, other important features of SNDCP include encryption and mobile unit authentication. While the cellular network provides a certain amount of protection against eavesdropping because of its channel-hopping techniques, the applications expected to be used on the CDPD network require explicit security—competing businesses must have the confidence that their information cannot be seen by competitors. SNDCP encryption uses the exchange of secret keys between the mobile unit and the backbone router to ensure that there can be no violation of security when transmitting over the airwaves. The authentication procedure guards against unauthorized use of a network address.

4.4.3.3 Mobility management

Traditionally, the network address of the end system has been used to determine the route used to reach that end system. CDPD is unique in allowing mobile units to roam freely, changing their subnetwork point of attachment at any time—even in mid-session. To find the best route for transmitting data to an end system, CDPD mobility management definitions describe the creation and maintenance of a location information database suitable for real-time discovery of mobile unit locations. Three network entities—the mobile units, the home backbone router, and the serving backbone router—participate in mobility management.

Mobile units are responsible for identifying their unique network equipment identifiers (NEIs) or network layer addresses to the CDPD network. As the mobile unit moves from cell to cell, it registers itself with the new serving backbone router. Each NEI is permanently associated with a home backbone router. The serving backbone router notifies the home backbone router of a mobile unit when it registers itself in the new serving area. Mobility management makes use of two protocols: the Mobile Network Registration Protocol (MNRP) and the Mobile Network Location Protocol (MNLP).

MNRP is the method mobile units use to identify themselves to the network. This information is used to notify the network of the availability of one or more NEIs at a mobile unit. The registration procedure includes the information required by the network for authenticating the user's access rights. The MNRP is used whenever a mobile unit is initially powered up and when the mobile unit roams from cell to cell. In either case, the mobile unit automatically identifies itself to the backbone router so its location can be known at all times.

MNLP is the protocol communicated between the mobile serving function and mobile home function of the backbone routers for the support of network layer mobility. MNLP uses the information exchanged in MNRP to facilitate the exchange of location and redirection information between backbone routers, as well as the forwarding and routing of messages to roaming mobile units.

Wireless IP is an appealing method of transporting data over cellular voice networks because it is flexible, fast, compatible with a vast installed base of computers, and has security features not offered with other wireless data services.

4.4.4 Satellite

Today's fiber optic, copper, and wireless networks define mere pockets of coverage, primarily in urban/suburban areas of developed countries. Truly ubiquitous networks are now under development by companies worldwide that are investing in fleets of low-earth-orbit (LEO), middle-earth-orbit (MEO), and geostationary-earth-orbit (GEO) satellites, enabling them to provide a variety of wireless phone, data, fax, paging, and multimedia services on a global basis.

The first satellites for communications use—Echo 1 and Echo 2—were launched by the United States in the early 1960s. They were little more than metallic balloons, which simply reflected microwave signals from point A to point B. Although highly reliable, these passive satellites could not amplify the signals. Reception was often poor and the range of transmission limited. Ground stations had to track them across the sky and communication between two ground stations was only possible for a few hours a day when both had visibility with the satellite at the same time. Geosynchronous (or geostationary) satellites overcame this problem. They were high enough in orbit to rotate along with the earth to service a large geographical area on a continuous basis.

Types of satellites   Satellites are categorized by type of orbit and area of coverage as follows:

▶ GEO satellites orbit the equator in a fixed position about 22,000 miles above the earth. Three GEO satellites can cover most of the planet, with each unit capable of handling 20,000 voice channels. Because of their large coverage "footprint," these satellites are ideal for radio and television broadcasting and long-distance domestic and international communications.

▶ MEO satellites circle the earth at about 6,100 miles up. It takes about 12 satellites to provide global coverage. The lower orbit reduces power requirements and transmission delays that can affect signal quality and service interaction.

▶ LEO satellites circle the earth only 600 miles up (see Figure 4.11). As many as 200 satellites may be required to provide global coverage. Since their low altitude means that they have nonstationary orbits and they pass over a stationary caller rather quickly, calls must be handed off from one satellite to the next to keep the session alive. The omnidirectional antennas of these devices do not have to be pointed at a specific satellite. There is also very little propagation delay. And the low altitude of these satellites means that earthbound transceivers can be packaged as low-powered, handheld devices that cost less money.

4.4.4.1   Iridium

Among the dozen or so new MEO/LEO satellite networks in various stages of planning and implementation is Motorola's Iridium, a 66-satellite constellation that started service in November 1998. The LEO network provides wireless personal communications services—voice, data, fax, and paging—using a combination of frequency division multiple access and time division multiple access (FDMA/TDMA) signal multiplexing to make the most efficient use of limited spectrum. The L band (1616–1626.5 MHz) serves as the link between the satellites and subscriber equipment. The Ka band (19.4 to 19.6 GHz for downlinks, 29.1 to 29.3 GHz for uplinks) serves as the link between the satellite and the gateways and earth terminals.

Intersatellite crosslinks make it possible for Iridium satellites to hand off calls between satellites in the same or adjacent orbiting planes. These

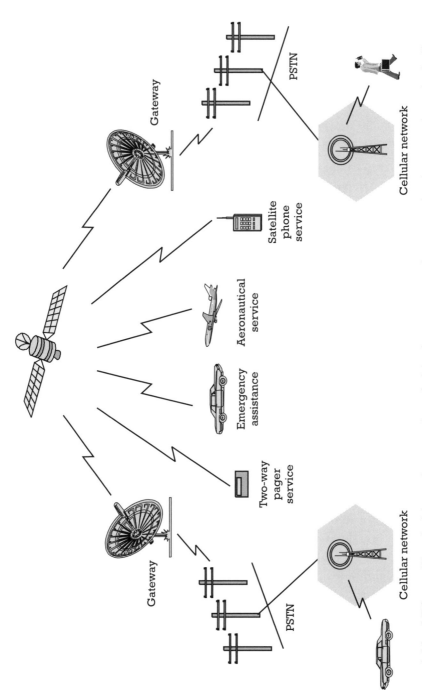

Figure 4.11 LEO satellites hold out the promise of ubiquitous personal communications services, including telephone, pager, and two-way messaging services worldwide.

intersatellite links are essential for providing truly global communication coverage. As the only proposed mobile satellite service using intersatellite links, the Iridium system will allow users to make and receive calls from anywhere on earth, including polar and ocean regions. Intersatellite crosslinks transmissions will take place in the Ka frequency band between 23.18 and 23.38 GHz.

One of the most important value-added features in the future will be roaming, the ability to use a wireless telephone while travelling abroad. The Iridium system offers a simple solution to global roaming. Even if an Iridium subscriber's location is unknown, the system will provide global transmission by tracking the location of the telephone handset. A single Iridium telephone number enables worldwide access to subscribers. The caller is offered a full complement of advanced calling features including enhanced call completion, call forwarding, voice mail, call waiting, and call conferencing.

GSM-based Iridium gateways located in strategic global locations interconnect the satellite constellation to land-based wireless systems and PSTNs, making communication possible between Iridium telephones and any other telephone in the world. The gateways also manage regional traffic and settlement procedures. The gateways are owned and operated by Iridium investors that have procured the rights to service various territories. Gateway operators are responsible for selecting service provider organizations to distribute Iridium services. The service providers are responsible for sales, billing, customer care, and other wireless provider activities.

A system control facility serves as the central management component for the Iridium network. It operates in conjunction with the master control facility, located outside of Washington, D.C., in northern Virginia (U.S.), which performs satellite control and network management. In addition, three Telemetry, Tracking, and Control Centers (TTACs) located in Hawaii and Canada are linked directly with the master control facility. The TTACs facilitate communication with the satellite network to, among other things, regulate the positioning of satellites during launch placement and subsequent orbit.

Both single-mode satellite phones and dual-mode satellite adapters are available for use with the Iridium World Satellite Service, providing users with the ability to make and receive calls anywhere in the world. The dual-mode handsets work over AMPS and GSM 900 terrestrial cellular services, enabling subscribers to use the Iridium World Roaming Service when they are in urban areas with cellular coverage.

For Iridium system access, the handsets include a Subscriber Identity Module (SIM) card, which implements the authorization procedure. For terrestrial access, moving between local networks is accomplished by inserting a different radio cassette—each one compatible with a different local cellular standard, such as CDMA, AMPS, and NAMS 800.

Pagers are available for use over the Iridium network for alpha and numeric messaging. The devices feature four-line, 80-character displays, international character sets, phone book memory, battery gauge, dual time zone clock (home and local time), and travel alarm.

4.4.4.2 Teledesic

In March 1997, Teledesic became licensed by the FCC to build and launch the Teledesic Network consisting of 288 LEO satellites, divided into 12 planes, each with 24 satellites (see Figure 4.12). Dubbed the "Internet-in-the-Sky," when completed, Teledesic will operate in a portion of the high-frequency Ka band—28.6 to 29.1 GHz on the uplink and 18.8 to 19.3 GHz on the downlink. Initial service will begin in 2002.

A multiple access scheme implemented within the terminals and the satellite serving the cell manages the sharing of channel resources among

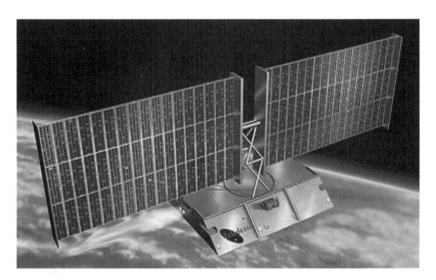

Figure 4.12   Artist's rendering of a Teledesic satellite.

terminals. Within a cell, channel sharing is accomplished with a combination of multi-frequency time division multiple access (MF-TDMA) on the uplink and asynchronous time division multiplexing access (ATDMA) on the downlink.

The satellite planes orbit north-to-south and south-to-north. To make efficient use of the radio spectrum, frequencies are allocated dynamically and reused many times within each satellite footprint. Within any circular area of 100-km radius, the Teledesic Network can support over 500 Mbps of data to and from user terminals. The Teledesic Network supports bandwidth-on-demand, allowing a user to request and release capacity as needed. This enables users to pay only for the capacity they actually use, and for the Teledesic Network to support a much higher number of users.

The Teledesic Network is designed to support millions of simultaneous users. Most users will have two-way connections that provide up to 64 Mbps on the downlink and up to 2 Mbps on the uplink. Broadband terminals will offer 64 Mbps of two-way capacity. This represents access speeds up to 2,000 times faster than today's standard analog modems. For example, transmitting a set of X-rays may take 4 hours over one of today's standard modems. The same images can be sent over the Teledesic Network in 7 seconds.

Teledesic represents the vision of its chairman, telecommunications pioneer Craig McCaw, and is backed by Microsoft chairman Bill Gates. The Boeing Company, the world's largest aerospace company, is leading the international industrial team to build and launch the $9 billion satellite constellation. End-user rates will be set by service providers, but Teledesic expects rates to be comparable to those of future urban wireline services for broadband access.

4.4.5 Digital power line

Among the objectives of the Telecommunications Act of 1996 is to accelerate the entry of public utilities into the telecommunications marketplace. At the same time, deregulation in the utility industry has prompted energy companies to move into other fields to improve overall revenues. Opportunities range from automation services, telephony and cellular, to Internet projects. To date, the impact of utility companies in the converging voice-data communications marketplace has been minimal. This could change soon with the introduction of digital power line technology to the United States, which economically delivers information access services through the existing power distribution infrastructure of electric utility companies.

For many utility companies, leveraging assets is now a critical business objective—especially in the new deregulated environment. Utilities are well positioned for many of the new developments in the telecommunications marketplace, and have numerous strengths and assets that can be brought to bear in the telecom business. Some of these assets can be leased to established telecom service providers or used directly to offer new services to customers.

4.4.5.1 Utility company assets

Among the strengths and assets that utility companies can bring to the telecom business are:

- Extensive rights-of-way and poles, ducts, conduits, and physical assets for routing cables;

- Substantial space in substations and other properties for new equipment and towers;

- Space on existing towers, masts, and poles for the addition of new antennas and equipment;

- Maintenance operations including vehicles, personnel, and associated expertise;

- Established customer relationships and a direct link with every potential telecom customer;

- Billing and customer information systems;

- Reputation for reliable service and associated brand name recognition;

- Telecom technology experience for internal purposes;

- Fiber optics installation, operations, and maintenance experience.

In offering new telecom services, either directly or through alliances, utility companies can open new sources of revenue, improve their relationship to customers, increase efficiencies, and discourage electric power business competition by offering value-added service.

Utility companies that lack expertise in data and voice services can acquire companies with that expertise or partner with established service providers. There are even outsourcing firms that cater specifically to the needs of utility companies, providing expertise in such areas as local number portability (LNP), signaling system 7 (SS7), rating systems, commission systems, customer care, service activation, wholesale and retail billing, convergent billing, and network management. The unit also provides automated mapping/facilities management and geographic information systems.

4.4.5.2 Transmission technology

Although the direct impact of utility companies in the converging voice-data communications marketplace has been minimal, the availability of DPL technology will enable utility companies to become major competitors in the telecommunications industry.

Developed in the United Kingdom by Nortel in the early 1990s, DPL offers 1 Mbps of bandwidth to support multiple applications. In addition to basic Internet access, it can be used for telephony (IP voice), multimedia, smart applications/remote control, home automation and security, home banking/shopping, data backup, information services, telecommuting, and entertainment.

The technology turns the low-voltage signals going between the customer premises and local electricity substation into a local area network. Multiple substations are then linked by fiber optic circuits to Internet switching points.

By giving customers access to the Internet through their existing electricity supply system, the technology is available to virtually anybody. It offers permanent online connection with the potential for lower charges.

Although the technology's developers—Northern Telecom and British utility Norweb Contracting—have focused their attention in European and Asian markets, they are now turning their attention to building partnerships with utility companies in the United States. To promote DPL, Nortel and Norweb Contracting have established a company called Nor.Web. The company has already signed up 10 utility companies, mainly in Europe, to deploy DPL service.

4.4.5.3 System configuration

The digital power line solution consists of a DPL 1000 main station, a base station, a coupling unit, and a communications module. The total solution provides a competitively positioned access network, which integrates seamlessly into today's WAN networks.

At the customer premises, a standalone unit connects to the power supply by standard coaxial cable. The unit is then connected to the computer by a standard Ethernet cable. It supports laptops as well as PCs and Macintoshes and supports a wide variety of platforms. It also supports the USB, enabling several machines to be connected to a single DPL box simultaneously to create a LAN in the home or office.

The DPL 1000 has the added advantage of allowing advanced network management by the utility company for the purpose of monitoring and maintenance as well as providing additional services to satisfy growing customer requirements. The system also unleashes the future potential for home automation by enabling remote operation of appliances such as the lights or the oven.

Due to the nature of this technology, it can be rolled out in discrete, targeted phases. Utilities not wishing to operate data services themselves have the option of charging a right-to-use fee to an operating company for accessing their plant. The system is managed with Nortel's Magellan ATM management platform, which allows network operators to monitor and maintain predetermined service levels and availability.

4.4.5.4 Regulation

To help utility companies get into the telecom business, per the Telecommunications Act of 1996, the FCC has granted them special status as "Exempt Telecommunication Companies" (ETCs), which relieves them from many of the rules that govern incumbent telephone companies (i.e., RBOCs).

The rules require applicants seeking such status to file a brief description of their planned activities together with a sworn statement attesting to any facts or representations to demonstrate ETC status as defined in the 1996 Act. The procedures also provide for public comment on the application, but limits comment to the adequacy or accuracy of the information presented.

CSW Communications was the first company to file an application for ETC status. CSW is a subsidiary of Central and South West Corporation, a registered public utility holding company. Within six months following the enactment of the 1996 Act on February 8, 1996, the FCC received 15 applications for a determination of ETC status and approved them all. Many of these applications included a public utility that had acquired or maintained an interest in an ETC. Since then, many more utility companies have chosen to participate as ETCs.

As noted, many utility companies are starting to offer telecommunications services, mostly in partnership with established service providers. Progress to date has not been significant enough to worry incumbent telephone companies because market entry entails a huge investment of capital. However, this situation could change dramatically if DPL technology proves as successful in the United States as it is in Europe. This would mean that utility companies could potentially reach every household and business with new voice and data services simply by leveraging their existing infrastructure, which entails only an incremental capital investment. Another technology from Media Fusion promises 2.5 Gbps of bandwidth capacity for voice and data over electrical power lines.

4.5   Conclusion

Over the years, distinct service environments developed to fulfill the specific needs of users—wireline voice, wireline data, wireless, and cable. As long as customers did not want all of the services, the separation was transparent for the customer and desirable for the carrier. Today, customers want all of these services and carriers find it increasingly difficult to justify continued investment in separate voice and data networks, as do large companies. Consequently, the telecommunications industry is making the transition from traditional voice-oriented circuit-switched networks to open, packet/cell-based architectures that support voice, data, and video services. This transition is already under way, spurred on by the growth of the Internet and its demonstrated capability to support different traffic types.

In today's networking environment, where each application has different performance requirements, a variety of services come into play. Sometimes ATM is used, other times frame relay or IP is used. Each network provides its own service benefits, as well as costs and features. Instead of trying to use one type of network to meet all application requirements, current thinking favors an integrated network that draws upon different services as needed to support data, voice, and video. A successful multiservice network is one that uses a combination of technologies that have been around for years: ATM at the core, frame relay and IP for low-speed access, and IP performing application integration functions.

In essence, each of these maturing technologies is relegated to role players in the multiservice network. Together, they combine their strengths to create a scalable, interoperable, high-performance network that suits most enterprise network needs.

Cisco Systems, for example, offers a multiservice ATM WAN solution that enables organizations to consolidate data, voice, and video traffic over a single network while significantly reducing IT management, bandwidth, and equipment costs (see Chapter 8). Advanced traffic management techniques are used to dynamically assign bandwidth as each application needs it to provide bandwidth efficiencies of up to 95% or better. The result is that organizations can reduce the total cost of ownership for their WAN backbones by as much as 30–50%, according to Cisco, when compared to TDM networks.

However, the successful evolution from the traditional circuit-switched telephony-based network to a multiservice and software intensive architecture requires a strong consensus among all the players in the telecommunications industry. In November 1998, the major international switch vendors and service providers formed the Multiservice Switching Forum (MSF) to define and implement an open systems model of ATM-capable switching systems that will expedite the delivery of new integrated broadband communications services to the marketplace.

The MSF will complement the scope and activities of existing associations by enabling the integration of leading components from multiple vendors into a single multiservice switching system. The MSF will focus on protocols and interfaces used within switching systems and will use specifications of the ATM Forum, IETF, ITU, Frame Relay Forum, and other industry associations.

The goal is to speed the arrival of an open switching platform that will enable vendors and carriers to deliver new and better services at a lower cost. An important step toward this goal is to develop next-generation networks for access aggregation, enhanced quality of service features, and to separate network intelligence from switching and routing. It is expected that multiservice switching systems will benefit from the same innovations and cost reductions that open systems in the computing world have achieved.

This is not the end of the voice-data convergence story. The ITU's IMT-2000 initiative will result not only in the wireless convergence of voice and data, but wireless access to the global telecommunications infrastructure through both satellite and terrestrial systems, serving fixed and mobile users in public and private networks. This "family of systems" framework is the topic of Chapter 7.

More information

The following Web pages contain more information about the topics discussed in this chapter:

AT&T: http://www.att.com/
Bellcore (see Telcordia Technologies)
BreezeCom Ltd.: http://www.breezecom.com/
Cisco Systems: http://www.cisco.com/
GTE: http://www.gte.com/
Home Phone Networking Alliance (Home PNA):
    http://www.homepna.org/
Home Radio Frequency Working Group (HomeRF WG):
    http://www.homerf.org/
Intel: http://www.intel.com/
International Telecommunication Union: http://www.itu.org
Internet Engineering Task Force: http://www.ietf.org/
Iridium: http://www.iridium.com/
MCI WorldCom: http://www.mciworldcom.com/
Media Fusion: http://www.mediafusioncorp.com/
Microsoft: http://www.microsoft.com/
Motorola: http://www.motorola.com/
Multiservice Switching Forum: http://www.msforum.org/
Nortel Networks: http://www.nortel.com/
Paradyne: http://www.paradyne.com/
Rockwell: http://www.rockwell.com/
Southwestern Bell: http://www.swbell.com/

Sprint: http://www.sprint.com/
Sycamore Networks: http://www.sycamorenet.com/
Telcordia Technologies: http://www.telcordia.com/
Teledesic: http://www.teledesic.com/
Tellium: http://www.tellium.com/
Universal ADSL Working Group: http://www.uawg.org/

Voice-data convergence in the home and on the road

The trend toward voice-data convergence is not confined to the corporate environment or to carrier networks—it is a trend that is apparent among America's 100 million households and 62 million mobile phone users as well. Voice-data integration within the home and on the road with portable hand-held units not only controls costs and optimizes bandwidth, but creates the foundation for integrated PSTN/Internet access; extends access to every person, home, and office; and leverages that access for multimedia applications.

As the previous chapter noted, efforts at converging voice and data over cellular networks began with CDPD, otherwise known as "wireless IP," which uses the silent periods of analog voice channels to pass data in the form of IP packets at up to 19.2 Kbps. To accomplish this a separate CDPD modem is attached to the cellular phone. Of course, the user must be in a service area that supports CDPD. This service is available in 3,000 cities in the United States, mostly from AT&T. As will be discussed later in this chapter, voice-data convergence is undergoing a renaissance with the development of new mobile phones that provide all the voice features to which users are accustomed. In addition, they provide Internet access for e-mail, Web browsing, and personal information management—all without the requirement of an external modem.

More recently, momentum has been building for voice-data convergence within the home. Among households, the number of multiple-PC homes is growing faster than the number of single-PC homes. In 1998, according to Dataquest, there were 15 million homes in the United States with at least two computers, and that number is expected to more than double to 35 million homes in the year 2000. Such explosive growth reflects the increased emphasis on computers for education as well as the desire for Internet connectivity.

In addition, there is also the continuing trend of telecommuting in which corporate employees spend one or more days each week working at home on their computer instead of at the company office. According to the Gartner Group, the number of telecommuters in the United States is around 30 million. There is also the continuing growth of home-based businesses, which now accounts for 53% of the 20 million small businesses in the United States, according to the Small Business Administration (SBA).

In view of these and other trends, the personal computer has indeed become a powerful platform in the home for work, education, communication, and entertainment, especially when connected to the Internet. Consequently, a new segment of the networking marketplace has emerged and is poised for rapid growth—home networking. With a network in the home, voice-data convergence becomes possible and offers numerous practical benefits.

Setting up a network in the home can save money by eliminating the need for duplicate peripherals, software, telephone lines, and Internet accounts that would normally be required for each computer. When three or more computers are involved, the savings can amount to thousands of dollars, plus hundreds of dollars a year in ongoing costs for separate telephone lines and Internet accounts.

With a home network, it is not only possible for household members to share resources, they can implement an in-home messaging system; centrally monitor and control security, environmental, and entertainment systems; and automate the handling of incoming calls. Of course, any of the household members can use the in-home network to place voice calls or engage in videoconferences over the Internet—even among themselves, if desired. Through connection sharing software, all of the Internet voice conversations and video sessions can occur simultaneously, just as if the users were all browsing the Web simultaneously.

5.1   Applications

Among the key applications driving the demand for home-networked devices are peripheral sharing, file and application sharing, and Internet sharing.

5.1.1   Peripheral sharing

Today, in a home with multiple PCs, each computer needs its own set of peripherals, such as printer, modem, CD-ROM, tape drive, and scanner. With a home network, the limited budget can stretch farther for one higher-end shared peripheral rather than many low-performance units. Any device that can be attached to one computer can be shared by all computers.

In the future, more peripherals will be designed to connect directly to the network, simplifying installation and eliminating dependence on computers for network connections. Already many "network ready" printers are available with a 10/100Base-TX port, for example. Cable modems also come with an Ethernet port. There is at least one file server on the market, the Snap! Server from Meridian Data, that provides an extra 8 GB or 16 GB of storage capacity simply by attaching it to the network via its RJ45 port. There are ISDN/modem combo units that also can be attached to the network in this manner.

5.1.2   File and application sharing

Through a network, multiple users in the home can easily move files, share applications, and back up data. Each user's directory tree, unless password-protected, is available to any other user, along with its files. In terms of applications, only one copy is necessary for one computer. Assuming a single-user license, when the software is not being used by the owner, other networked users can take turns using it. If the software license permits multiple usage, it need only be installed on one computer to be accessed by all users.

For file backup, only one computer need have a tape, ZIP, or CD-RW drive—internal or external. All other users can take control of the storage unit to write to or read files from their own tape cartridge, ZIP disk, or optical disk.

5.1.3 Internet sharing

A major market driver for home networking is sharing an Internet connection over a single line and ISP account. Previously, each home user required a separate line, modem, and ISP account. The alternative was to wait for a turn to get onto the Internet. Home networks provide significant savings and greater utility by enabling shared access to a single Internet connection with only one ISP account. Instead of paying for multiple low-speed lines, the money is better spent on a single "always-on" high-speed CATV, xDSL, or satellite connection that everyone can use simultaneously.

5.1.4 Other applications

Other applications that are facilitated by a network in the home include:

- ▶ Entertainment: family members can play popular multiplayer network games, either within the home or over the Internet.

- ▶ Small office: for those who work part-time or full-time out of the home, the ability to network multiple computers and share peripherals provides numerous opportunities to increase productivity, introduce efficiencies, and save money on software, hardware, phone lines, and Internet accounts.

- ▶ Home automation: a home network facilitates automation applications, such as environmental control and security systems, which can be monitored through a PC or accessed through the Internet.

▶ Messaging: a home network offers the opportunity to implement a family messaging system. Things like Post-It notes, pink slips for phone messages, voice mail, and e-mail can be sent electronically to the computers of family members. This saves trips up and down stairs or from having to yell throughout the house.

▶ Voice- and video-over-IP: cable TV systems are being upgraded all over the country for two-way capability. In addition to eliminating the need for a separate analog modem to establish the Internet connection and send URL requests, this will allow cable companies to offer voice and video services over their networks. Cable modems are already available that handle voice calls.

There are many ways to implement a network in the home. Among the choices are Category 5 cable, phoneline, coaxial CATV cable, powerline, wireless, and USB cable. As discussed below, some are better than others in supporting the convergence of voice and data.

5.2   Category 5 cable

Category 5 cable, consisting of four pairs of wires, has long been a standard in implementing corporate LANs—both Ethernet and Token Ring.[1] It is routinely used to support data rates of 10 Mbps, 100 Mbps, and more recently, 1,000 Mbps as in Gigabit Ethernet. In terms of pure capacity, Category 5 cable offers the best medium for heavy-duty applications and thus for voice-data convergence. Of the two most common networks—Ethernet and Token Ring—the former is by far the least expensive and the easiest to configure and will be the network of choice for most homes.

5.2.1   Installation

A frequent criticism of Category 5 cable is that it is too difficult for the average homeowner to install. After all, the home is not a very friendly

---

1. A special kind of Category 5 cable, called plenum cable, offers fire resistance. Although plenum cable is mostly found in the corporate office environment, it is available for the home office as well. It is no more difficult to install than ordinary Category 5 cable, but is priced slightly higher.

place—typically, there are no wiring closets, patch panels, riser conduits, cableways, and other easements that make network installation fairly easy and straightforward in commercial buildings. Nevertheless, a determined do-it-yourselfer can wire a home simply by following the existing paths of electrical, CATV, and plumbing systems. The object is to run a segment of cable from each computer location to a hub, which provides the local interconnectivity and, through an attached modem (i.e., analog, CATV, ADSL, ISDN), provides wide area connectivity through the Internet (see Figure 5.1).

Another criticism of Category 5 cable for the home is that special tools are required (i.e., RJ45 crimp and punch-down) and that, once completed, the walls are marred by gaping holes where the cables come out to the computers and hub. The truth is that vendors offer connectorized cabling in lengths up to 100 feet, modular RJ45 jacks and wall plates, and other components that are specifically aimed at the home network market, enabling homeowners to achieve professional results with only a razor blade and screwdriver.

Often it is not possible to snake Category 5 cables through existing routes occupied by electrical wire, CATV cable, or water pipes with the connectors attached. If this is the case, the connector can be cut off. Once the cable segments are routed and pulled through the wall at the desired locations, the bare end must be connected to the back of an RJ45 wall plate. With a razor blade, a two-inch length of the cable's outer jacket is stripped off to expose the 8 color-coded wires. These wires are connected to a special wall plate with 8 color-coded screw-down terminals (see Figure 5.2). This kind of wall plate is available at any Radio Shack store.

With the wires connected to the wall plate, the wall plate can then be screwed into place on the wall. One end of a connectorized segment of Category 5 cable plugs into the RJ45 port of the computer's network interface card (NIC) and the other end into the wall plate's RJ45 jack.

At the hub end, where several Category 5 cables terminate, a different procedure is used to achieve a clean, professional installation—again, with no special tools. At this end, a 2-, 4-, or 6-port wall plate is used, along with modular jacks that snap flush into the plate. Not any modular jack will do, however. There is a two-part modular jack, the bottom half

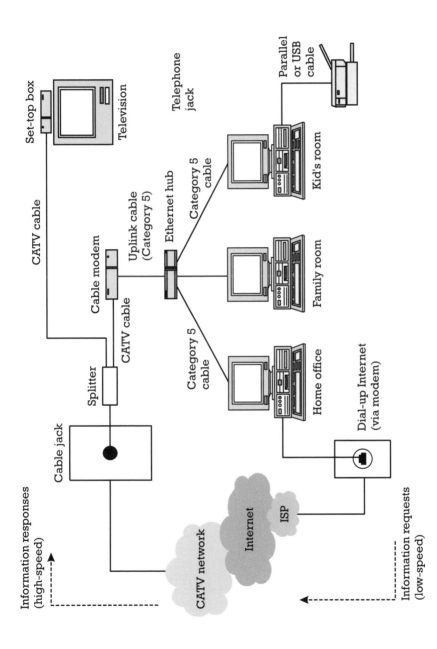

Figure 5.1 A typical in-home Ethernet configuration.

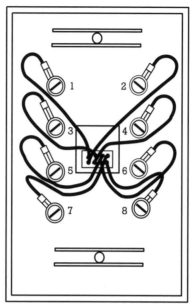

RJ45 wall jack
(rear view)

Figure 5.2   An RJ45 wall plate with 8 color-coded screw-down terminals.

of which holds the wires in a particular order so that the top half can close down on them to establish the electrical connections.[2]

As before, each cable will be stripped to expose its 8 color-coded wires. This bundle of wires is inserted through the sleeve of the bottom piece of the modular jack, and each wire is pressed down into the holding grooves in the specific order as follows (see Figure 5.3):

1. Solid brown;

2. Braided brown;

3. Solid green;

4. Braided blue;

---

2. Actually, there are several kinds of two-part modular jacks, which assemble in different ways, depending on manufacturer. What they have in common is that they terminate Category 5 cable wires without the need for crimp or punch-down tools.

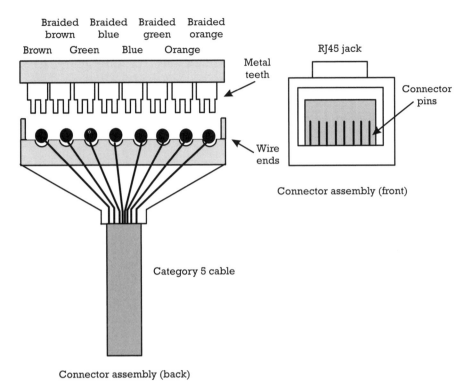

Connector assembly (back)

Figure 5.3 A two-part modular jack organizes the individual wires of Category 5 cable (bottom) and makes the electrical connections when the teeth (top) cut through them as the two pieces are snapped together.

5. Solid blue;

6. Braided green;

7. Solid orange;

8. Braided orange.

Once the wires are held firmly into place within the grooves, any excess is trimmed off, so that none of the wire ends extends beyond the unit. The top half of the modular jack contains 8 metal teeth, each with a sharp groove. When this half is snapped into the bottom half of the jack, the individual wires held in position by the plastic grooves are forced into the grooves of the metal teeth, which cut into the insulation to expose the wires, thereby making an electrical connection. This snap-together

modular jack makes special RJ45 crimp and punch-down tools unnecessary, saving about $75 on the cost of installation. With the modular jacks
fully assembled, they are simply snapped into position through the back
of the multiport wall plate. This results in each RJ45 jack being flush-
mounted to the front of the wall plate. Short segments of connectorized
Category 5 cable make the connection from the wall plate to the hub's
RJ45 ports.

5.2.2 Network interface cards

For each computer that will be networked in the home, an NIC is
required, just like at the office. This is an adapter that plugs into a vacant
ISA or PCI slot inside the computer. The card provides an RJ45 port into
which the connector on a segment of Category 5 cable will be plugged.
Many NICs are optimized to work in Windows environments—specifically PCs running Windows 95, 98, or NT. Windows provides the software
drivers for most brands of NICs, making installation and configuration
essentially a plug-and-play affair.

Once the card is installed and the computer is restarted, Windows will
recognize the new hardware and configure it automatically. If the driver
or a driver component is not found, however, it may be necessary to get
out the Windows CD-ROM or the NIC vendor's 3.5-inch installation disk
so Windows can find the components it needs.

After Windows has recognized and configured the NICs, and each
computer is connected to the hub, a green LED on the hub goes on to indicate that the connection is active. As each computer is connected, another
green LED goes on at the hub. To actually share files and peripherals over
these connections, each computer must now be set up for file and printer
sharing and for Internet access. This is accomplished by following the
directions offered in the Windows Help file.[3]

5.2.3 Hub

As alluded to thus far, hubs consolidate LAN segments at a central point.
A collapsed backbone within the hub interconnects the various segments.

---

3. A more detailed explanation of the network configuration process for Windows 95/98/NT
   is beyond the scope of this book. This information is immediately available in the
   Windows Help file and is included in the documentation that comes with nearly every
   network device marketed for home usage, including NICs, hubs, and cable modems.

This simplifies wiring and facilitates troubleshooting. Through the hub, the network is segmented so the failure of one segment—shared among several devices or dedicated to just one device—does not impact the performance of the other segments.

Work group hubs come in various port configurations—usually 4 to 12 ports are sufficient for home use. These so-called stackable hubs provide a dedicated uplink port, which allows the hub to be connected to other hubs as a means of accommodating future growth. The uplink port looks like any other RJ45 port, but may require a special crossover cable that reverses the transmit and receive pairs. Sometimes the uplink port is used for the connection to a cable, DSL, or ISDN modem.

Depending on the specific hub, there are two types of cables that are used to make the connection between the cable modem and hub: straight-through (normal) and cross-over. The right selection depends on the designation found on the hub's uplink port. A straight line indicates that straight-through or normal cable should be used, whereas an "x" indicates that special crossover cable should be used. With this type of cable, the send/receive pairs within the cable are crossed. If no designation appears on the uplink port, normal straight-through cable should be used.

For small networks, the hub usually will be a "dumb" device that is not manageable with SNMP, so no additional software is installed. Once the computers are properly configured for networking and connected to the hub, the network is operational. In addition to files and applications, devices inside of any of the computers can be shared by any of the other computers, including hard drives, tape drives, ZIP drives, and optical drives (e.g., CD-ROM, CD-RW, DVD). Likewise, any external devices connected to the computers via parallel, serial, SCSI, or USB ports can also be shared. The result is a peer-to-peer network that is easy to use and navigate through the standard tree structure visible through Windows explorer.

5.2.4 Starter kits

One of the easiest ways to set up a work group LAN is to buy a starter kit. Some vendors offer starter kits for both 10BaseT and 100BaseT networks. This is a convenient, affordable solution that comes complete with the necessary hardware and cabling needed to create a 2-node network. The typical kit includes a 4- or 5-port stackable hub, two auto-sensing

Ethernet 10/100 NICs, and two Category 5 cables of 15 or 25 feet. A 10BaseT starter kit typically sells for less than $100, while a 100BaseT starter kit costs less than $200. Additional NICs can cost between $25 and $60. An auto-sensing 10/100 PC Card for portable computers costs less than $80. There are also starter kits for switched Ethernet for under $250. A 2-port switch offers two auto-sensing 10/100BaseT ports, two NICs, and cables. The starter kits are ready to run with all major network operating systems. Among the vendors that offer such kits are D-Link, Kingston, Linksys, NetGear (a business unit of Bay Networks, a subsidiary of Nortel Networks), and 3Com.

Of course, there are other products besides Windows that implement peer-to-peer networks. One of the most popular is Artisoft's LANtastic, which includes software for Windows and DOS platforms, allowing users to mix and match the installation to fit changing needs. LANtastic allows hundreds of PCs to share resources on the same network. It includes drag-and-drop network administration and interfaces easily with Microsoft and Novell network operating systems. The network configuration process with these and other products, including phoneline networks, is similar to the one described above.

5.2.5 Cable modems

A CATV coaxial cable modem is not required for the operation of a home network, but it is a common method by which the home network is connected to the Internet for simultaneous multi-user access. ISDN and ADSL can also be used, but these services are not nearly as ubiquitous or economical—or offer as much bandwidth—as CATV. Over 70% of American homes are capable of receiving television programming through a cable company, and many cable operators now offer Internet access in partnership with an ISP as an option in their service package.

The CATV network may provide uni-directional or bi-directional access to the Internet. With the older uni-directional service, an analog modem is included within the cable modem, which establishes the initial Internet connection via a stored telephone number. The analog modem is used for sending low-speed information upstream toward the ISP, while the cable modem's demodulator is used for receiving data on the downstream from the Internet. For this arrangement, a dedicated telephone line must be connected to the cable modem. This phone line cannot be used for telephone calls while the Internet connection is in use. Next, the cable modem is connected to the uplink port of the hub with a segment of Category 5 cable (see Figure 5.4).

Bi-directional service means that the CATV network is used for sending data in both the upstream and downstream directions. In this case, the cable modem does not have an analog component. About 20% of American homes with cable service have two-way communications. Most cable operators are upgrading their networks for bi-directional communication. With both uni- and bi-directional services, information can be downloaded at speeds of 10 Mbps and potentially 28 Mbps in the future.

Whether the service is uni- or bi-directional, the cable modem connects directly to the same coaxial cable that delivers television programming via standard gauge RG-59 coaxial cable with an F-type connector. Using a splitter attached to the main cable, one segment of coaxial cable goes to the modem and the other segment goes to the television's set-top box. Another piece of cable links the set-top box to the television.

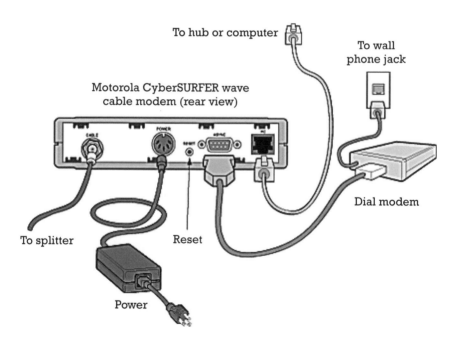

Figure 5.4 Connections for a Motorola CyberSURFR Wave Cable Modem. The modem can attach to a single computer or to a hub, in which case several computers can use it for simultaneous Internet access.

The same cable can support both data channels and TV channels simultaneously. This is because one or more of the TV channels have been assigned to carry data rather than the usual video and audio programming. All channels go through both the TV set and the cable modem, but the TV set tunes only to the video channels, while the cable modem tunes only to the data channels.

Once connected to a hub, the cable modem's firmware may have to be updated to enable access by multiple users. This usually can be done through an online configuration process. When multiple computers are used, each must have its own IP address for purposes of communicating with the cable modem. The cable modem has it own IP address, which is the one presented to the Internet via the CATV network. The assignment of IP addresses may look like this:

- Cable modem = 192.168.100.1;

- Computer #1 = 192.168.100.2;

- Computer #2 = 192.168.100.3;

- Computer #3 = 192.168.100.4;

- Computer #4 = 192.168.100.5.

Since the cable modem is always on, the network connection is always open. Getting on the Internet is as simple as opening the Web browser (see Figure 5.5). Unlike dialup Internet access, there is no waiting for a slow log on procedure.

Cable modems provide economical, high-speed access to the Internet via local CATV networks. The devices cost about $150 and are usually purchased or leased through the cable operator who decides what vendor it is going to use. About 60% of cable modems in use today are from Motorola.

5.2.6 ISDN LAN modems

Similar considerations apply to ISDN and DSL modems that may be used to provide shared access to the Internet. And having been developed in large part for the small office and remote office markets, ISDN and DSL modems are easily connected to the local area network with Category 5 cable. These devices often include a multiport hub for implementing the

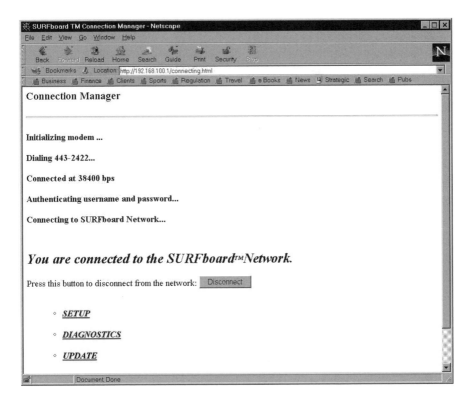

Figure 5.5 With a cable modem, accessing the Internet is simply a matter of opening the browser interface of the modem and clicking on the "Connect" button. After that, the network connection is always open until the user clicks on the "Disconnect" button.

internal Ethernet LAN. A built-in Web browser provides configuration and management assistance.

With regard to ISDN, for example, a Web wizard provides automatic network addressing to the attached PCs. The user connects the ISDN LAN modem to the ISDN BRI line, and attaches the PCs to the LAN modem's integrated 10BaseT Ethernet hub. Using a standard Web browser, or the one provided with ISDN LAN modem, the Web wizard guides the user to set up the unit. The Web wizard automatically detects the type of ISDN line and supplies the service profile identifier (SPID) or order code needed by the carrier to complete the service activation process.

An example of an SPID for a telecommuter might be H6, in which case, the following services are associated with the ISDN line:

- ▶ Circuit switched voice on one B-Channel;

- ▶ Circuit switched data on the other B-Channel;

- ▶ Voice features include:

  - ▸ Call appearance call handling (CACH);

  - ▸ Call hold;

  - ▸ Call transfer (flexible calling);

  - ▸ Caller ID (calling number identification);

  - ▸ Conference calling—3 way (flexible calling);

  - ▸ Redirecting number delivery;

  - ▸ Secondary telephone number;

- ▶ Data features include:

  - ▸ Caller ID (calling number identification);

  - ▸ Redirecting number delivery;

- ▶ Number of directory numbers per switch type:

  - ▸ DMS 100 = 3;

  - ▸ 5ESS = 1;

  - ▸ EWDS = 1.

There are about 70 different SPIDs. The Web wizard automates the task of configuring the SPID. When the SPID is sent to the carrier's central office, the ISDN connection is automatically initiated and put into service.

5.2.7 DSL modems

There are several versions of DSL being implemented around the country. To take advantage of ADSL Full Rate, for example, the user needs to place an order with the local broadband service provider. This can be the telephone company or a company that specializes in broadband services. The service itself will run over an existing phone line that comes into the home or business. Full rate service provides up to 8 Mbps downstream and 1 Mbps upstream.

At the telephone company's central office is a digital subscriber line access multiplexer (DSLAM). The function of the DSLAM is to terminate all the remote ADSL modems into a single location and send the data traffic out to the Internet or another remote destination like a VPN. With ADSL, the user has a private, secure channel of communications with the service provider, unlike a CATV cable that is shared with others in a neighborhood and poses potential security problems.

At the customer premises, the service provider typically installs an external splitter, which separates the voice and data traffic.[4] The splitter is required for simultaneous use of telephone and data access when operating in full rate mode. An ADSL modem or router connects to the network hub through its RJ45 port and to a phone through its RJ11 port. Sometimes the ADSL modem will have an integral 10BaseT Ethernet hub, eliminating the need for a separate network hub. As needs grow, additional hubs can be connected to the router's integral hub, so many users or devices can share the same ADSL circuit.

The ADSL device is configured at a PC using a Web-based management application and setup wizard. A built-in server provides for network address translation (NAT) and port address translation (PAT), and supports the DHCP, eliminating the need to manually manage IP addressing. A built-in Domain Name System (DNS) proxy server permits shared access to the Internet connection.

5.3 Phoneline networking

Although a Category 5 cable network is the most labor-intensive network to install, it offers the highest speed—up to 100 Mbps, plus the possibility of migrating to 1,000 Mbps when the pricing for Gigabit Ethernet hubs and cards drop to commodity levels. For homeowners who prefer not to be bothered with installing Category 5 cable, a simpler way to network home computers comes from the Home Phoneline Networking Alliance (PNA). Products that adhere to the HomePNA standard permit the creation of simple, cost-effective home networks using existing phone wiring.

Two phoneline technologies are available. The HomeRun technology developed by Tut Systems is the basis for the 1-Mbps specification

---

4. ADSL Lite eliminates the need for a splitter; the trade-off is that the bandwidth is reduced to a maximum of 1.5 Mbps downstream and between 384 Kbps to 512 Kbps upstream.

developed by the Alliance, while Epigram's InsideLine technology is the basis for the 10 Mbps specification. At this writing, Epigram is working on a 100-Mbps phoneline technology, which is likely to be incorporated into a future HomePNA specification.

5.3.1 Adapters

The use of phone wiring for this purpose eliminates the need for cable installation and a hub, yet provides connectivity among computers and allows shared Internet access with one ISP account—all without interfering with regular phone service. All that need be installed are phoneline-compliant NICs in each computer. The NICs have RJ11 ports instead of RJ45 ports. Connectorized telephone extension wire makes the connection between the NIC and the nearest phone jack. However, if additional extension wiring is required to link computers around the house, the same procedure described earlier for Category 5 cable can be used.

Adding data to voice over existing phone wiring does not pose interference problems because different frequencies are used for each. Standard voice occupies the range from 20 Hz to 3.4 KHz in the United States (slightly higher internationally), while phoneline networking operates in a frequency range above 2 MHz. By comparison, DSL services like ADSL occupy the frequency range from 25 KHz to 1.1 MHz. The frequencies for each type of service are far enough apart that the same wiring can support all three without any risk of crossover interference.

The first generation of phoneline products offered a top transmission speed of only 1 Mbps with up to 500 feet between nodes. Backward-compatible second-generation products boost that to 10 Mbps, making them worthy of serious consideration by the average homeowner. At this speed, the phoneline network is equivalent in performance to 10BaseT, capable of supporting any mix of voice, data, and video for multimedia collaboration, videoconferencing, and on-demand entertainment.

PCs on home phone networks must run Windows 95, 98, or NT, or some other software that supports file sharing. Once networked, the PCs can share a printer, as well as a modem connection for Internet access. Users can also work high-speed DSL, ISDN, and cable modems into the mix. In fact, HomePNA lets the consumer choose the method of WAN access, which can also include wireless services as well.

5.3.2 Modem sharing

The modem-sharing software that enables multiple users to access the Internet simultaneously usually supports any type of modem, including dial-up, cable, DSL, and ISDN. The software consists of a server component, which is installed on the computer with the modem or other Internet connection, and a client component, which is installed on the other computers on the network. The server component routes the Internet requests from the client stations by proxy through the single Internet connection. Upon installation, the client component automatically configures all client computers to ensure that Internet requests are routed to and from the server.

In addition to providing Internet connection sharing, some products such as WinGate, published by Deerfield.com, protect the in-home network with a firewall component. The firewall prevents intrusion by restricting IP addresses that can access the internal network from the Internet and by binding the ports in the operating system. The firewall accomplishes this by identifying in-bound requests for access to the network and verifying that the IP address of the request is allowed to access the internal network. It also binds the communications ports necessary for outsiders to access the internal network.

This type of software offers other useful features. WinGate logs a history of Internet activity that can be reviewed to monitor Internet usage and also allows for restricting access to specific Web sites. A "pro" version of the software includes user management utilities that can be used to restrict individual or group access to Internet services according to custom settings. In addition, a built-in cache increases the performance of the shared connection by storing frequently accessed Web content, thus minimizing the need to go out onto the Internet.

5.3.3 Phoneline kits

As with Category 5 cable products, phoneline vendors offer network kits that provide connections for two computers, plus modem-sharing software. Among the vendors offering such kits are ActionTec, Best Data Products, Boca Research, Diamond Multimedia Systems, and Linksys—most of which cost less than $100.

For example, Linksys offers its HomeLink Phoneline Network In A Box, which includes two PCI network cards, two phone line cables, and

WinGate modem sharing software for an estimated street price of $89.99. Additional phoneline network cards go for $49.99. With extra cards purchased as needed, up to 25 computers can be daisy-chained over the same telephone line. The modem sharing software allows three computers to access the Internet at the same time.

Linksys includes both RJ11 and RJ45 ports on its cards. The two RJ11 ports are used to implement a daisy chain—one provides the "in" connection and the other provides the "out" connection (see Figure 5.6). Alternatively, one port can be used for connecting a telephone and the other for connecting to the network via a wall jack. The inclusion of an RJ45 port lets users migrate to a higher-speed Category 5 cable network later without having to change cards.[5]

Boca Research's Home Area Network (HAN) kit has a suggested retail price of $109.99. Additional cards go for $59.99, and a combo card with built-in 56K modem goes for $99.99. The kit contains the usual components that connect two computers, but its network cards do not include an RJ45 port. However, the software not only allows computers to be networked together when daisy-chained over the same phone line, but allows them to be networked even if they use separate telephone lines (see Figure 5.7).

5.3.4 Market momentum

As the phoneline networking market gains momentum, we will likely see more products built to HomePNA specifications. In addition to adapters, other products—including printers, modems, access devices, cable modems, PC Cards, and more—will be home phoneline network-enabled as well. Compaq, for example, was the first computer maker to offer a home phoneline network adapter in select models of its Presario 5600 Series of high-performance desktop computers. The computer maker offers Diamond Multimedia Systems' HomeFree Phoneline card as part of its "Built for You" program under which consumers can select their own components and have them installed at retail stores nationwide before taking the computer home.

---

5. RJ11 and RJ45 capabilities cannot be used at the same time. If one is in use, the other is automatically disabled. If the RJ45 port is used, it must be connected to a hub.

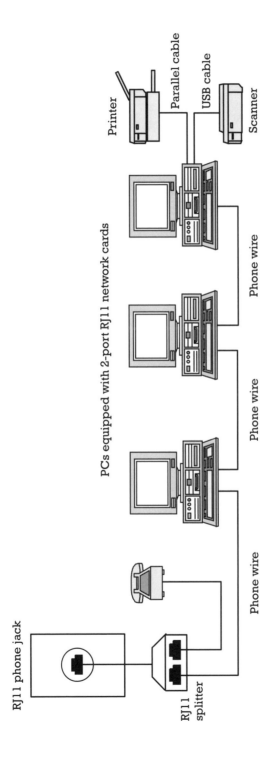

Figure 5.6 A typical home phoneline network where computers are daisy-chained together using the RJ11 ports as the in/out connections.

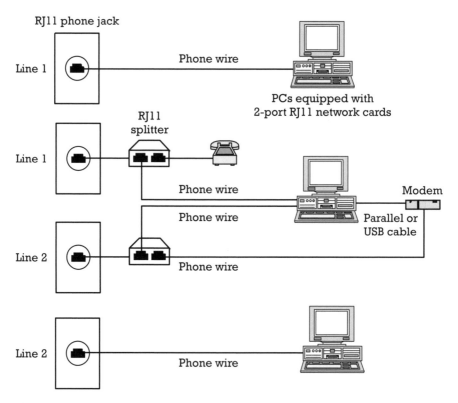

Figure 5.7   A typical in-home phoneline network spanning two separate telephone lines.

5.4   Wireless home networking

Although there are numerous wireless LAN products based on a variety of technologies for the corporate environment, there is a standard for wireless networking in the home. It is based on frequency-hopping spread spectrum and uses the unlicensed 2.4-GHz ISM (industrial, scientific, medical) band to provide data rates of 1 Mbps or 2 Mbps, depending on the type of modulation used—two- or four-phase frequency shift keying (FSK).

A consortium of vendors called the Home Radio Frequency Working Group (HomeRF WG) has developed a platform for a broad range of interoperable consumer devices. Its specification, called the Shared Wireless Access Protocol (SWAP), is an open standard that allows PCs, peripherals, cordless telephones, and other consumer electronic devices to communicate and interoperate with one another without the complexity and expense associated with wired networks.

SWAP-compliant products can carry both voice and data traffic, and interoperate with the PSTN and the Internet. The specification itself was derived from extensions of existing cordless telephone (digital enhanced cordless telephone, or DECT) and wireless LAN technologies which, when combined, enable a new class of home cordless services. SWAP-compliant products support both TDMA to deliver interactive voice and other time-critical services, and CSMA/CA (carrier sense multiple access/collision avoidance) to deliver high-speed packet data. Table 5.1 summarizes the key performance parameters of the SWAP specification.

5.4.1 Components

Using various types of nodes employing various technologies, the SWAP system has the ability to support intense high-demand packet traffic and high-quality voice traffic as well as infrequent command and control traffic. The nodes are of four basic types:

▶ Connection point: a gateway server that supports voice and data services;

Table 5.1
Performance Perameters of the SWAP Specification

Performance	Parameters
Frequency hopping network	50 hops/second
Frequency range	2.4 GHz ISM band
Transmission power	100 mW
Data rate	1 Mbps using 2FSK modulation; 2 Mbps using 4FSK modulation
Range	150 ft; covers typical home and yard
Supported stations	Up to 127 devices per network
Voice connections	Up to 6 full-duplex conversations using 32-Kbps ADPCM
Data security	Blowfish encryption algorithm (over 1 trillion codes)
Data compression	LZRW3-A algorithm
48-bit network ID	Enables concurrent operation of multiple co-located networks

Source: HomeRF Working Group.

▶ Voice terminal: communicates with a base station via TDMA;

▶ Data node: communicates with a base station and other data nodes via CSMA/CA;

▶ Integrated node: communicates with a base station using both TDMA and CSMA/CA for voice and data services.

A SWAP system can operate either as an ad hoc network—a simple network where communications are established between multiple stations in a given coverage area without the use of an access point or server—or as a managed network under the control of a connection point. In an ad hoc network, where only data communication is supported, all stations are equal and control of the network is distributed between the stations.

For time critical communications, such as interactive voice, a connection point is required to coordinate the system, assigning the nodes guaranteed bandwidth for bounded latency communication. The connection point, which provides the gateway to the PSTN, can be connected to a PC via a standard interface such as USB.

The connection point also provides the SWAP system with power management for prolonged battery life by scheduling device wakeup and polling. For example, a typical network may include a laptop whose power could be managed with the connection point buffering data until it can wake up.

The SWAP specification holds out the promise of flexible, low-cost wireless data communication within the home at around $100 per PC. For those who live in apartments or temporary residences, wireless might be the preferred network solution, if only because it can be easily dismantled and set up again when it becomes necessary to move. In the near future, SWAP-compliant products offering speeds of 30 Mbps may be possible.

5.4.2   SWAP alternatives

The HomeRF WG released Version 1.0 of the SWAP specification in January 1999, with SWAP-compliant products becoming available by mid-year.

Of course, there are non-SWAP wireless LAN products available for small office/home office (SOHO) environments. RadioLAN, for example, offers a 10-Mbps solution that operates in the 5-GHz range. Zoom Telephonics offers ZoomAir, a 2-Mbps solution that operates in the 2.4-GHz range. BreezeCOM offers the 3-Mbps BreezeNET PRO.11 wireless system, which operates in the 2.4-GHz band and complies with the IEEE 802.11 specification. None of these products is SWAP-compliant and all cost far more to equip a single desktop computer with a wireless transceiver card—street priced at between $269 and $700—than to install a cable network for three home computers.

Although components for corporate wireless LANs can be purchased for use in the home, the expense is prohibitive for most households—costing $500 or more per PC for all the LAN components. This is changing, however, as more vendors turn their attention to the potentially lucrative home market.

NDC Communications offers a wireless home network product called CableFREE, which is based on IEEE 802.11 standards-based core technology. Although several IEEE 802.11–compliant wireless LAN products have been available since early 1998 for industrial and corporate office applications, they have not been affordably priced and convenient to use for most home PC users. CableFREE uses the 2.4-GHz operating frequency band to provide a 2-Mbps data rate and coverage radius up to 250 ft. A CableFREE starter kit to connect two PCs is priced at $395. Additional ISA and PC Card adapters are priced at $199. Another company that addresses the home market is ShareWave, which has managed to reduce the cost-per-PC to $150. With its proprietary Osprey technology, PCs can be linked at up to 4 Mbps.

Another IEEE 802.11 product comes from WebGear. The company pioneered the wireless home networking market in 1997 with the introduction of the first 900-MHz wireless home solution called the Aviator. This parallel port solution was followed up with a USB version. Up to 25 nodes may be networked together, each spaced as much as 125 feet apart. WebGear also offers the AviatorPRO, which supports data rates of up to 2 Mbps using 2.4-GHz spread-spectrum technology. Up to 32 nodes may be networked together, each spaced as much as 1,000 ft apart.

5.5 CATV networking

Another way to network various devices in the home is through the same CATV cable that brings television programming into the home. As noted, the CATV cable can support shared access to the Internet by using a cable modem connected to an Ethernet hub to which multiple PCs may be connected. However, some vendors are developing networks that use the CATV cable that snakes through several rooms of the home to unite televisions, set-top boxes, audio/video components, and PCs. One vendor, Peracom Networks, calls this arrangement "entertainment networking."

The company's HomeConnex goes beyond traditional PC-to-PC networking by including home entertainment devices such as VCRs, DVD decks, satellite receivers, video cameras, and video game stations into an integrated home network system. The entire network of PCs and video devices is controlled using a wireless remote control and a graphical user interface that is viewable from any TV in the home.

This type of network offers all the benefits of traditional PC-to-PC home networking products, including file sharing, print sharing, and access to the Internet. In addition, it delivers entertainment to the user wherever they may be in the home. From any TV in the home users can:

- View and control VCRs, DVD decks, and video game stations;

- Watch a DVD movie playing from a PC;

- Move entertainment from satellite receivers or cable boxes;

- Transform existing video camcorders into baby monitors or security cameras;

- View and control the PC from any TV in the house, enabling users to run PC applications, check e-mail, surf the Web, or monitor the Web surfing activities of children;

- Use infrared transmission to move files between disconnected laptop computers and PCs on the network.

The HomeConnex system supports up to 16 simultaneous video sources, including PCs, on the network. Each video device becomes part of the network and each TV is capable of viewing every video device on the network. This is accomplished through the four basic components that comprise the HomeConnex system:

- PC Caster—a hub device that insulates the home network from the outside world, providing high quality pictures to every TV and ensuring that sensitive data on the network never leaves the home;

- MediaCaster—connects video devices to the network via standard RCA audio/video connectors;

- ▶ CableCaster—connects PCs to the network via the USB port;

- ▶ Wireless remote—combined with an easy to use graphical interface that pops up on every TV, the wireless remote controls each device on the network.

With central control at the television, for example, a different channel can be assigned to each video source, whether that source is a PC, security camera, VCR, game station, or another television in the house. Turning to that channel enables the assigned video source to be viewed and controlled. Not only can a PC's screen be viewed, for example, but any application or resource running on the PC can be viewed and controlled from the television, including Web browsing, e-mail, a multicast session, and chat. In addition, a DVD movie playing on a PC can be picked up by any television in the house, eliminating the need for separate DVD units at every television.

In the future, audio-only sources will also be interconnected throughout the home using the same CATV cable. This will permit the creation of home intercom systems, piped music, and audible alerts for wake-up calls, schedule reminders, or appliance malfunctions. As more CATV networks are upgraded for bi-directional communication, it will even become possible to place telephone calls over the cable network and receive notifications of incoming calls at the television or PC through the display of caller ID information. With the addition of a video camera, videoconferences can be set up with family, friends, or colleagues at the office.[6]

5.6 Powerline networks

In addition to Category 5, phoneline, wireless, and CATV networks, there is a technology called powerline, which uses a home's existing electrical wiring for the network connections. Creating the network is as simple as plugging special modules into electrical outlets throughout the home and then plugging a computer or printer into each module. This type of

---

6. The ability to receive caller ID information and conduct videoconferences at the television are currently available through separate add-on products. What CATV networks do is integrate these and other capabilities so they can be accessed or controlled from any system that is connected to the network.

network currently is limited to providing 350-Kbps connections between devices, but improved technology can boost the connections to 2 Mbps and, later, 10 Mbps.

Powerline networks are the easiest to install—by merely connecting the computer and printer into modules that are plugged into electrical outlets, the devices are network-ready. Through the plug-in modules, each device on the network automatically recognizes the others.

The first vendor to offer a powerline network kit is Intelogis, a company spun-off from Novell. It began shipping its PassPort Plug-In Network in mid-1998. It uses a home's existing electrical wiring to create an instant network that allows users to share a single Internet connection, printers, and files or play multiuser games between PCs. The configuration procedure for sharing resources is similar to that used for Category 5 and phoneline networks, except that no additional wiring, internal network interface cards, or hubs are required.

The PassPort Plug-In Network operates with Microsoft Windows 95/98/NT. The suggested retail price is $199.99, which includes two Pass-Port PC Plug-In adapters, one PassPort printer Plug-In adapter, and Internet-connection sharing software. Additional PC and printer Plug-In adapters can be added to the network for $79.99 and $59.99, respectively.

The first release of PassPort was disappointing in that it could only network computers at a paltry 350 Kbps. However, the company's second-generation product operates at 2 Mbps. A third-generation product will offer 10 Mbps. The company intends to push speeds of future generations of PassPort products in excess of 10 Mbps, which will allow power lines to carry streaming audio and video.

5.7 USB cable networking

One of the newest and easiest ways to link multiple PCs in the home or small office is through USB cabling. With the use of a USB network adapter and the installation of appropriate driver software, users can connect USB-equipped desktops or laptops to form a 10BaseT Ethernet network without having to open their PCs (see Figure 5.8). The connections built from USB cable can support the same kinds of applications as an Ethernet LAN built with Category 5 cable, including data, voice, and video—subject to bandwidth limitations. In addition, the USB connections provide the same opportunities for peripheral and resource sharing.

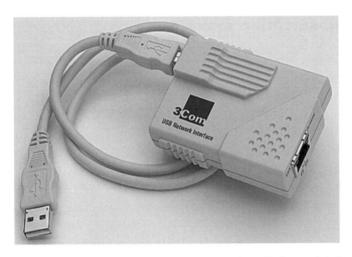

Figure 5.8   3Com's USB 10BT Network Interface links multiple PCs through their USB ports.

The external USB adapter plugs directly into the USB port on PCs and laptops. It functions like a conventional NIC, providing an immediate 10BaseT connection to other computers. The USB adapter usually comes with a 12-inch cable attached. This end plugs into the USB port of the host computer. The adapter has an RJ45 port into which a Category 5 cable connector is inserted. The other end of the Category 5 cable plugs into an Ethernet hub.[7]

The USB interface adapter is bus-powered, drawing power from the USB itself, making an external power supply connection unnecessary. As a USB bus-powered device, the adapter draws less than 500 mA from the PC. In keeping with the USB specification 1.0, the adapter provides support for energy-saving suspend and resume operations. LEDs on the adapter indicate network and traffic status for simple troubleshooting.

While USB interfaces offer an easy way to implement a PC network, due to the speed limitations of the USB ports, the connections are limited to providing a 10BaseT Ethernet network. The actual speed of the connections is considerably less—sustained throughput is 6 Mbps.

---

7. A USB hub is not unlike a network hub used to implement an Ethernet LAN. It is simply a device for expanding a computer's USB port capacity so extra devices can be connected such as scanners, digital cameras, joysticks, and mice. A USB hub can also be used to extend the length of a USB connection beyond the current 16-foot limitation.

5.8 Subscription services

Despite the impressive growth in the number of households with multiple PCs and the meteoric rise of the Internet, a significant number of Americans do not want to own a computer, much less go through the time and expense of setting up a home network. This number might be as high as 20%, according to Forrester Research. These are the households that will find subscription services like that offered by U S West more appealing.

In partnership with Network Computer, Inc. (NCI), U S West offers an innovative service called @TV, which integrates telephone, television, and Internet services without requiring a PC platform.[8] The service lets customers send and receive e-mail, place and answer telephone calls, view caller ID, and alternately surf TV channels and the Web—or surf the Web and TV channels at the same time—using high-speed DSL or dial-up connections to their televisions.

Subscribers use a television set-top box equipped with a speakerphone and NCI software to receive and make telephone calls as well as to access Internet-based features. Some of these features include programming guides, electronic commerce, news, and electronic mail. In addition to access over conventional connections, U S West provides support for high-speed DSL technologies that use existing copper-wire networks offering data transmission speeds up to 200 times faster than dial-up connections.

On the client side, NCI TV Navigator software offers support for integrated telephony and enhanced television, allowing U S West customers to display Internet content and applications while watching television and the ability to make and receive telephone calls (see Figures 5.9 and 5.10). The platform uses NCI Connect ISP Suite server software to manage the set-top boxes, provide security, and administer the network.

NCI's IQView technology formats standard Internet content for the television automatically, ensuring high-quality graphics and crisp on-screen text. Color correction is performed and real-time flicker is reduced to further enhance the image quality. Intelligent graphic scaling adapts Web sites and eliminates the need for horizontal scrolling.

---

8. Although the interactive television offering of WebTV Networks, Inc. integrates television viewing with electronic mail and Web browsing, it does not include the capability to make and receive telephone calls.

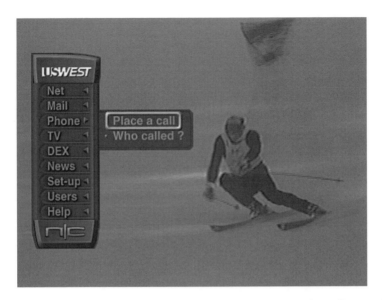

Figure 5.9 With a television program in progress, subscribers of U S West's @TV service can select from a list of Internet-based value-added features, including "Phone."

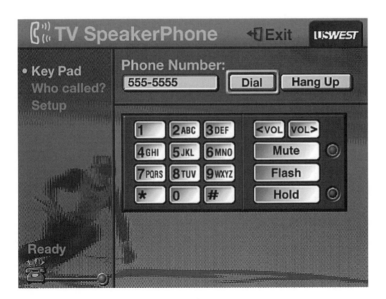

Figure 5.10 Upon selecting "Phone," the @TV service pushes a TV SpeakerPhone interface to the television screen, which displays a telephone pad. To place calls and implement features, the user points and clicks with the remote device.

5.8.1 Standards

A vendor group, called the Advanced Television Enhancement Forum (ATVEF), has defined protocols for television programming enhanced with data, such as Internet content. The goal is to allow content creators to design enhanced programming that may be delivered over any form of transport—analog or digital TV, cable, or satellite—to all types of broadcast receivers that comply with the specification.

The broad capabilities of Internet authoring standards on which the ATVEF specification was developed allow content producers and advertisers to create interactive content once without having to customize it for viewing on every type of receiver. Among the industry standards that are supported in the ATVEF's specification are:

- HTML—this is the same language used to create Web pages, including image maps, a popular way of making graphic objects on an HTML-based page "clickable."

- Extensible architecture—ATVEF specifies the familiar extensibility scheme of "plug-ins," which are used by most Web browsers to accommodate new capabilities.

- Java—currently, ATVEF supports ECMAScript, a general purpose, cross-platform programming language. Provision for hooks into Java are likely to be added to the ATVEF specification.

- Document object model (DOM)—this platform- and language-neutral interface allows programs and scripts to dynamically access and update the content, structure, and style of documents. The document can be further processed and the results of that processing can be incorporated back into the presented page.

- JavaScript—ECMAScript plus DOM is equivalent to JavaScript 1.1.

Enhanced television offers new entertainment and commerce opportunities, such as sports events with complementary information on players and teams, news with additional details and related stories, and advertisements that allow consumers to order merchandise with the click of a button. Although the ATVEF's goal is to enhance a broad spectrum of differentiated services and receivers from which consumers can make selections, consumers will have the choice whether to turn on the enhanced elements or not. Once they turn them on, there can be two levels of interactivity—content only as an enhancement, or content with links to additional information for two-way interactivity.

A common specification such as that developed by the ATVEF promises to accelerate the creation and distribution of enhanced television programs and allow consumers to access such programs cost-effectively and conveniently, regardless of which transport or broadcast receiver they use. The specification also allows content providers and distributors to choose from a variety of enhanced television business models and delivery methods.

5.9   Multimedia home information gateway

The various standards bodies discussed so far have the overriding goal of developing specifications that will meet the performance objectives of homeowners at a price they can afford. However, they share another goal, which entails positioning their individual standards so that they complement the standards of the others. This would mean that in the future, homeowners would be able to mix and match multiple technologies to achieve an integrated network. This can be accomplished with a multimedia home information gateway.

A home multimedia information gateway would integrate a cable modem with the "no new wires" approach to in-home data networking. It would allow families with multiple PCs to do all the things previously mentioned, including share computer resources, simultaneously surf the Internet, play interactive games, and subscribe to emerging low-cost IP telephony services. The home multimedia information gateway enables household members to do all of this using any transmission medium—phoneline, electrical wiring, CATV cable, and wireless—whatever happens to be the most convenient and economical for each room.

For example, it might not be practical or economical to install new wiring to a home office above a detached garage. In this case, a HomeRF WG-compatible wireless technology could be used to communicate with the home multimedia information gateway in the main part of the house. The gateway would then be able to distribute voice and data throughout the rest of the house, perhaps over Category 5 cable or phoneline wiring, or even the electrical system. For external communication, the gateway could route data to the Internet and voice to an IP telephony service provider.

Multiple IP services would be brought directly into the residence via the same gateway. As noted in the previous chapter, however, instead of using the H.323 protocol suite, which is generally regarded as too complex and slow for call establishment, complementary alternative standards will have to be relied upon, such as the Session Initiation Protocol (SIP), Simple Gateway Control Protocol (SGCP), or Multimedia Gateway Control Protocol (MGCP). Of these, SGCP is the first industry-proposed protocol that has been demonstrated for the provision of residential services.

Meanwhile, cellular equipment vendors have made great strides in integrating voice and data. Although data communications through mobile phones and other portable devices have been around for several years, they required an awkward hookup to an external modem. Today's devices provide integrated voice-data capabilities through a single handheld unit, enabling mobile users to take advantage of many of the same computing and communications capabilities they left behind at home or at the office.

5.10 Voice-data convergence through mobile phones

One of the key features available to subscribers of PCS services, for example, is messaging, which allows users to receive numeric and/or text messages of up to 80 characters in length on the phone display. For many users, this could eliminate the need for a separate pager, since messages can be sent to the phone instead.

In the case of AT&T Wireless's Digital PCS Service,[9] for example, a message is sent by calling a subscriber's PCS telephone number. If that person is not available, the caller hears a recorded voice greeting, instructing them to leave a voice message or press 5 to enter a call-back number

---

9. AT&T also offers an AMPS-based mobile phone called PocketNet Phone, which contains both a cellular circuit-switched modem and a CDPD modem, enabling users to access information on the Internet and corporate intranets as well as two-way wireless messaging services.

using a touch-tone phone. Callers have the option to leave a voice message, a call-back number, or both.

For alphanumeric service, 80-character messages are sent by using AT&T's MessageFlash software for modem-equipped PCs or Macintosh computers. The software is included in the AT&T Digital PCS service subscription. In addition, with the separately-available operator-assisted text dispatch service, callers can dial 1-888-ATT-5432 and dictate their messages to a live operator who will then send the message to the subscriber's phone.

When a message is received, the phone alerts the user with an envelope indicator that appears on the screen. If the phone is turned off, or if the user is out of the coverage area when a message comes in, the AT&T network will store the message up to 72 hours. The message will be delivered when the user returns to the coverage area and powers up the phone within that 72-hr period.

Whether the user should give up a pager service depends on several factors. First, since messages can be received on PCS phones, users may no longer need to carry both a phone and a separate pager. Users will receive one bill for both services and have one point of contact for customer service.

Second, traditional paging networks make only one attempt to deliver messages to the pager. If the pager happens to be turned off, or the user is out of the paging coverage area, they will miss the message. Since PCS messaging stores messages for up to 72 hours, there is less chance of missing an important message. As long as the user returns to the coverage area and turns on the phone within that 72-hr period the message will be delivered to the phone.

Third, the coverage area may differ between the current paging service and the PCS provider's network. Therefore, users must evaluate their travel and usage patterns to determine which service will best suit their needs. Finally, PCS messaging handles messages of only 80 characters in length, whereas some paging services are capable of supporting messages of 240 characters or more.

Mobile telephone networks based on the GSM standard support short message service (SMS), which allows the user to send and receive messages of up to 160 characters in length on cellular phones, PDAs, and other portable appliances equipped with a liquid crystal display (LCD). With the touch of a button, the same message can be sent to a group of people simultaneously. Receipt notification allows mailboxes of all types—voice, fax, and text—to send an acknowledgment to a user's mobile telephone, indicating that the message has been delivered.

The main benefits of the SMS are that messages can reach the recipient even if he or she is already busy with a voice or data call. If the recipient is in a meeting and has enabled the silent service—no ringing tone on incoming call, visual call alert only—he or she can still receive and respond to messages with as little disturbance as possible to others.

For mobile users who need to send and receive messages of more than 80 or 160 characters, there are wireless e-mail services. The primary advantage of a wireless e-mail service is that it eliminates the need for mobile professionals to find a phone and attach a coupler to it. All that is needed is a portable computer equipped with a radio modem, which can take the form of a PC Card for a notebook computer. Depending on the service provider's gateway capabilities, mobile computer users can send and receive e-mail over a wide range of private and public networks, including corporate intranets and the Internet.

Should the mobile user roam into areas that are beyond the coverage range, the wireless service provider holds all incoming messages in its system and delivers them instantly when contact is reestablished. At the same time, outgoing messages composed at any time will be transmitted as soon as contact with the system is established. And if the radio frequency signal is interrupted, such as when a user drives through a tunnel, the transmission simply picks up where it left off when contact is reestablished on the other side.

All e-mail products offer the basic functions of creating messages, locating addresses of recipients, sending and receiving mail, and sorting received mail into folders. They differ in terms of the graphical user interface, mail-processing capabilities, level of customization, and integration with other applications. They also differ in their ability to support the unique needs of mobile users who rely on wireless services to stay in contact with customers and co-workers.

Some vendors offer graphical front-ends that make e-mail over wireless services easier to use. Products such as Lotus cc:Mail are designed to facilitate wireless e-mail transmission and management (see Figure 5.11). They include filtering mechanisms that let users prioritize incoming e-mail and selectively download e-mail after viewing only the headers. Because wireless connections are still far more expensive than wireline

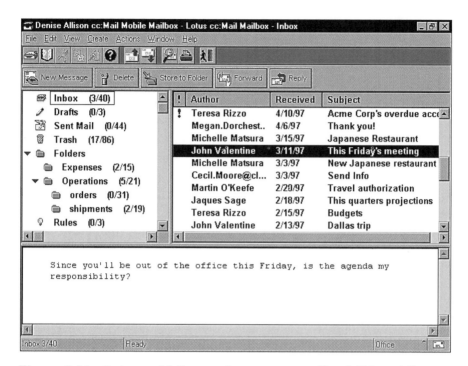

Figure 5.11 Lotus cc:Mail not only runs on an office LAN, mobile professionals can load it on their laptops and take it with them on the road. LAN and mobile mailboxes can be opened at the same time and synchronized with a mouse click. cc:Mail locates the laptop's modem port, detects modem speed, and sets up the right modem file automatically.

connections, these filters can play a key role in containing the cost of airtime.

Other vendors have adapted existing e-mail packages to function in both wireless and wireline environments. In such cases, some extra code goes on top of the existing application software, which brings up an extra menu selection for sending e-mail. After the e-mail message is prepared, the user can choose an appropriate sending option, such as "wireless" or "facsimile" or "printer." This provides users with the familiarity of their favorite e-mail software and the flexibility to communicate from nearly anywhere over a variety of services.

To make it easier for users to get started with wireless e-mail, the software, hardware, and wireless service are often bundled together and offered at a discounted monthly rate. Companies like Intel, Microsoft, and BellSouth Mobile Data (formerly known as RAM Mobile Data) combine their products and expertise to offer total e-mail solutions to their customers. Intel offers its Wireless Modem that connects to the serial port of any Intel-architecture laptop or PC. The modem includes the hardware, software, and support needed to make it easy for users of Microsoft Mail Remote for Windows to send and receive messages wirelessly over the BellSouth Wireless Data network. After installing the hardware and e-mail software, users can get online by simply filling out an on-screen form and sending it off to BellSouth Wireless Data, which will add the user to its network within a few hours.

Even billing for wireless services is getting simpler. Taking advantage of all communications options has usually meant having different companies provide different services, each sending a separate bill with different rates. Many providers now offer consolidated billing for all of their services. MCI WorldCom, for example, offers consolidated billing for telephone calls, cellular calls, paging, Internet access, and e-mail. For small businesses, invoices can be itemized by individual, work group, or project. For larger businesses, invoices can be itemized by department as well.

Internet-enhanced cell phones potentially represent an important communications milestone, assuming the availability of compelling content. One vendor that has been particularly active in this area is Nokia, the world's biggest maker of mobile phones. The company's model 7110 is indicative of the types of new mobile phones that about 70 other manufacturers are aiming at the world's 200 million cellular subscribers. It displays Internet-based information on the same screen used for voice functions. It also supports SMS and e-mail and includes a calendar and phone book as well.

The phone's memory can also save up to 500 messages—SMS or e-mail—sorted in various folders such as the inbox, outbox, or specially defined folders. The phone book has memory for up to 1,000 names, with up to five phone and fax numbers and two addresses for each entry. The user can mark each number and name with a different icon to signify home or office phone, fax number, or e-mail address, for example. The phone's built-in calendar can be viewed by day, week, or month, showing details of the user's schedule and calendar notes for the day. The week view shows icons for the jobs the user has to do each day. Up to 660 notes in the calendar can be stored in the phone's memory.

Nokia has developed several innovative features to make it faster and easier to access Internet information using a mobile phone:

- Large display—the screen has 65 rows of 96 pixels (see Figure 5.12), allowing it to show large and small fonts, bold or regular, as well as full graphics.

- Microbrowser—like a browser on the Internet, enables the user to find information on the Internet by entering a few words to launch a search. When a site of interest is found, its address can be saved in a "favorites" folder, or input using the keypad.

- Navi Roller—this built-in mouse looks like a roller (see Figure 5.13) that is manipulated up and down with a finger to scroll and select items from an application menu. In each situation, the Navi Roller knows what to do when it is clicked—select, save, or send.

Figure 5.12 Display screen of the Nokia 7110.

Figure 5.13 Close-up of the Navi Roller on the Nokia 7110.

▶ Predictive text input—as the user presses various keys to spell words, a built-in dictionary continually compares the word in progress with the words in the database. It selects the most likely word to minimize the need to continue spelling out the word. If there are several word possibilities, the user selects the right one using the Navi Roller. New names and words can be input into the phone's dictionary.

However, the phone cannot be used to access just any Web site. It can only access Web sites that are meant to give users access to important information that has value when they are mobile. For the most part, these would be Web sites that comply with the Wireless Access Protocol (WAP), a standard for Internet-enhanced cell phones that is based on Unwired Planet's Handheld Device Markup Language (HDML). CNN and Reuters are among the content providers that are developing WAP-specific news and information services for delivery to cell phones. The WAP Forum envisions cell phone access to news, weather reports, stock prices, flight schedules, and wireless banking, plus access to corporate and ISP e-mail.

The WAP standard supersedes HDML in favor of its own Wireless Markup Language (WML). The strength of WAP is that it spans multiple airlink standards and, in the true Internet tradition, allows content publishers and application developers to be unconcerned about the specific delivery mechanism. The sites that built content with HDML can easily convert their content to conform with the WML format, enabling their sites to be accessed by the new breed of Internet-enabled phones.

Another device that offers access to Internet content is 3Com's PalmPilot VII. While it provides wireless access to information, such as flight schedules and news headlines, and can be used to conduct online transactions, such as movie-ticket purchases or online stock trades, it lacks the voice component. This and other similar devices also do not have nearly the penetration that mobile phones do. Phones with data are a low-cost addition to an already-inexpensive consumer product. The Palm VII and similar devices are intended to appeal to those who are data-centric and feel the need to go to any Web site.

To support the Palm VII, content providers are not required to adhere to a special development format. Instead, a content provider needs only to build a small, downloadable "query application" with HTML to make standard Web content available to the Palm VII users. To help Palm users organize their Web activities, portals are emerging to aid navigation. A

Web portal is a site on the Internet or a corporate intranet that aggregates content, applications, or commerce services in a single place, making it convenient for users to find what they need without having to engage in time-consuming searches of producer-centric Web sites.

QualComm has very effectively bridged the gap between data-centric and voice-centric cellular usage. The company's all-in-one digital phone and pen-based organizer, pdQ Smartphone, comes in two versions: an 800-MHz dual-mode CDMA and analog unit, and a 900-MHz CDMA PCS unit. The pen-based organizer is actually the Palm computing platform adopted for the cell phone form factor, enabling the Smartphone to run personal information management (PIM) applications. As with the standalone Palm device, users can transfer and synchronize information between the Smartphone and a personal computer. A wireless modem built into the pdQ Smartphone allows users access e-mail, browse the Web, and perform other online activities.

As a full-function e-mail client that uses standard Internet mail protocols (i.e., POP/SMTP), the Smartphone's pdQmail application supports the following features:

▶ Displays HTML content embedded in messages; clicking on a URL launches the pdQbrowser;

▶ Recognizes attachments;

▶ Screens messages by downloading only the first few lines;

▶ Sorts messages by date, sender, subject, priority, and threaded subjects;

▶ Looks up e-mail address from Palm address book;

▶ Filters for selective downloading and filing.

The Smartphone's pdQbrowser is a Web browser for the Palm computing platform that supports the following features:

▶ Navigates to any Web page;

▶ Click on any hyperlink;

▶ Sophisticated bookmark management;

▶ Supports HTML forms;

▶ Click on "mailto:" launches pdQmail.

The pdQ smartphone also includes support for the SMS and adds a customizable alert manager. The alert manager automatically launches the particular application with which it is associated. For example, a voice mail alert will dial a voice mail number automatically to retrieve messages.

The Smartphone is indicative of the new types of integrated wireless data and voice communications devices. Since the device is based on the Palm computing platform, more than 1,000 productivity applications are available to users, including communications, enterprise information management, contacts, scheduling and groupware, and business and personal productivity applications.

5.11 Conclusion

From the consumer perspective, voice-data convergence manifests itself in both the mobile phone and the home network. In terms of mobile phones, there are many choices to make regarding network operator, services provided, options, coverage areas, costs, and of course, the mobile phone itself. Likewise, there are many choices to make with regard to the home network.

While a network based on Category 5 cabling takes more time and effort to install, the advantage of doing so is compelling. With its current speed advantage of up to 100x that of other types of networks—and the possibility of a 1,000x in the future, as Gigabit Ethernet products drop to commodity prices—a Category 5 cable network can easily handle sophisticated applications, including voice, as well as the additional load of resource sharing. In addition, its performance makes it the preferred type of network for any kind of interactive game playing.

While important, resource sharing among PCs in the home is only one aspect of the total convergence story. Other key aspects include:

- Using the same transmission medium for data transfer and voice calls simultaneously;

- Interconnecting previously separate systems—telephone, computer, television, game station, security, and environmental controls—so they can be accessed over a single network;

- Integrating wireline and wireless technologies.

It is not enough that these things are possible from a technical standpoint, they must offer homeowners significant practical value at an affordable price, be easy to set up and use, and offer connectivity with the outside world for both voice and data applications—over both the Internet and the PSTN.

The addition of a mobile phone that achieves these goals enables users to take advantage of the benefits of voice-data convergence even while away from the home or office. These benefits will be available wherever the user roams, even to international locations under the new global wireless communications infrastructure, called IMT-2000. The subject of the next chapter, IMT-2000 is an initiative of the ITU that will result in wireless access to the global telecommunication infrastructure, including the Internet, through both satellite and terrestrial systems, serving fixed and mobile users in public and private networks.

More information

The following Web pages contain more information about the topics discussed in this chapter:

3Com: http://www.3com.com/
ActionTec: http://www.actiontec.com/
Advanced Television Enhancement Forum: http://www.atvef.com/
AT&T: http://www.att.com/
BellSouth Mobile Data: http://www.data-mobile.com/
Best Data Products: http://www.bestdata.com/
Boca Research: http://www.bocaresearch.com/
BreezeCOM: http://www.breezecom.net/
Cable Television Laboratories http://www.cablelabs.com/
Cisco Systems: http://www.cisco.com/
D-Link Systems: http://www.dlink.com/
Deerfield.com: http://www.deerfield.com/
Diamond Multimedia Systems: http://www.diamondmm.com/
Epigram: http://www.epigram.com/
Home Phone Networking Alliance (Home PNA):
    http://www.homepna.org/
Home Radio Frequency Working Group (HomeRF WG):
    http://www.homerf.org/

Infonetics Research: http://www.infonetics.com/
Intel: http://www.intel.com/
Intelogis: http://www.intelogis.com/
International Data Corp.: http://www.idc.com/
International Telecommunication Union: http://www.itu.org/
Kingston: http://www.kingston.com/
Linksys: http://www.linksys.com/
Lotus Development: http://www.lotus.com/
MCI WorldCom: http://www.mciworldcom.com/
Meridian Data: http://www.meridian-data.com/
Microsoft: http://www.microsoft.com/
NDC Communications: http://www.ndclan.com/
NetGear: (see Nortel Networks)
Network Computer: http://www.nc.com/
Nokia: http://www.nokia.com/
Nortel Networks: http://www.nortel.com/
Novell: http://www.novell.com/
Peracom Networks: http://www.peracom.com/
QualComm: http://www.qualcomm.com/
RadioLAN: http://www.radiolan.com/
RAM Mobile Data (see BellSouth Mobile Data)
ShareWave: http://www.sharewave.com/
Small Business Administation (SBA): http://www.sba.gov/
Texas Instruments: http://www.ti.com/
Tut Systems: http://www.tutsys.com/
Unwired Planet: http://www.unwiredplanet.com/
U S West: http://www.uswest.com/
WebGear: http://www.webgear.com/
Zoom Telephonics: http://www.zoomtel.com/

CHAPTER

6

Contents

Next-generation networks

To combine voice and data successfully over the same network and provide new multimedia services that are economical and of the same quality that customers have become accustomed to in their experience with plain old telephone services and data-only networks, traditional carriers are reinventing themselves and changing the way they do business. At the same time, new carriers have entered the long-distance market to challenge the likes of AT&T, MCI WorldCom, and Sprint with multigigabyte-capacity fiber optic backbone networks capable of handling far more voice and data traffic than their legacy counterparts.

Not only have the traditional carriers responded with integrated networks of their own, but through mergers and acquisitions, they now dominate the major Internet routes in terms of ownership and control. In addition to laying the foundation for voice-data convergence in the new millennium, these developments portend a period of innovation in multimedia service creation and rollout that will unfold at an unprecedented pace, fueled by the robust economy of the United States.

6.1 Nextgen service providers

A next-generation network is one which provides integrated voice-data services over an IP-based fiber optic infrastructure. The services are aimed at businesses and consumers, and are more attractively priced than the comparable services of incumbent interexchange carriers, such as AT&T, MCI WorldCom, and Sprint. Among the growing-number of so-called "nextgen" service providers are Delta Three, Frontier, ICG Communications, and Qwest LCI. Even Tier I national Internet service providers, such as PSINet, are transforming themselves into full service communications firms capable of supporting any business service—voice and data—over their extensive IP backbones. Although some of these companies lease bandwidth from other service providers, their goal is to eventually have full ownership of their networks. Ownership reduces their backbone cost to about one-tenth of the cost of equivalent leased bandwidth.

Nextgen carriers offer a variety of services, but they usually focus on those that have the most market potential. Examples include IP telephony, facsimile, and VPNs. The use of IP for these and other services makes global distribution more efficient and economical than using traditional PSTNs, while permitting their integration with legacy data and applications.

The traditional long-distance telephone companies are not standing still. After all, failure to compete aggressively with the nextgen service providers can result in huge losses. Various industry estimates indicate that telephone companies could lose between $8 and $10 billion in revenue worldwide over the next few years to IP telephony. AT&T, MCI WorldCom, and Sprint are pursuing different strategies to offer IP services—as much to stave off competition and minimize revenue loss as to enter new markets with innovative services that can protect their customer base and shore up the bottom line.

Compared to incumbent interexchange carriers, nextgen service providers currently lack a broad customer base and offer a narrower range of services. Customer support is also not up to par with the incumbents. With strategic relationships, however, the nextgen carriers could quickly overcome these limitations. Accordingly, traditional carriers cannot count on retaining this competitive edge for very long.

What follows are brief portraits of the notable nextgen service providers. These portraits provide a clear indication of how networks and services are rapidly evolving to meet the voice, data, and applications requirements of consumers and businesses in the twenty-second century. Although not all of these companies currently offer IP telephony, they are certainly building and expanding their infrastructures in ways that position them to offer the service when they believe the time is right.

6.1.1 Frontier Corporation

Frontier Corporation offers integrated communications services—including Internet, IP and data applications, long distance, local telephone, and wireless—to business customers nationwide. The company offers services characteristic of the nextgen carriers, but also resells local service in 32 states and uses its own facilities to offer services in 23 major metropolitan markets nationwide. Its nationwide long-distance footprint ranks the company as the fifth largest interexchange carrier, offering integrated local, long-distance, and data services within reach of approximately 70% of the United States business population. Frontier also resells wireless services and manages a cellular joint venture with Bell Atlantic.

Frontier acquired a significant presence in the Internet market with its purchase of GlobalCenter in 1997. GlobeCenter, an Internet provider for medium and small businesses, had previously merged with Primenet, an ISP and Web hosting provider with a nationwide ATM backbone for the small office/home office market.

Frontier's IP Packet Over SONET/ATM network interconnects with all major public exchanges and private networks—including AT&T, MCI WorldCom, and Sprint—and has more than 8,000 business customers. Through its six domestic and international media distribution centers and national backbone, the company provides direct Internet connections, Web hosting, and collocation services.

The fully redundant IP backbone network employs both DS3 (45 Mbps) and SONET OC-3 to OC-12 (155 Mbps to 622 Mbps) links, which can move data over IP, frame relay, or ATM. The network is comprised of more than 60 high-speed private peering network connections to major Internet carriers and high-speed links to eight Internet exchanges—network access points (NAPs) and metropolitan area exchanges (MAEs)—with over 100 public peering arrangements. The network also has 23 IP hubs and eight media distribution centers (domestic and international) to help improve content performance and connection efficiency on the Internet. The network is accessed from more than 90 local POPs.

For reliability, Frontier has deployed 13 bi-directional line switched rings (BLSRs) to protect customer data from network outages. A BLSR is a method of SONET transport in which half of the working network is sent counter-clockwise and the other half is sent clockwise over another fiber. If a data path is blocked by a fiber cut, this arrangement allows customer data to be rerouted to another SONET ring within 50 ms.

In 1999, Frontier doubled the capacity of its IP backbone to 2.5 Gbps (OC-48). The company has two OC-48 circuits linking Los Angeles to New York City and will expand capacity to 10 Gbps (OC-192) in the year 2000. Frontier's use of advanced optical fiber with dense wavelength division multiplexer (DWDM) gives its network virtually unlimited capacity and positions the company to meet customers' applications needs well into the future.

While it was upgrading its network, Frontier Corporation merged with fiber optic network provider Global Crossing Ltd. in a transaction valued at $11.2 billion. The combined network would span 71,000 route miles in 159 cities in 20 countries and provides voice, Web hosting, private line, ATM, and Internet services. Global Crossing then bought the Global Marine business of Cable & Wireless Group for about $850 million. Global Marine specializes in laying undersea fiber optic cable systems with a fleet of 13 ships and 21 subsea vehicles.

Frontier also offers high-speed local access to its IP backbone through DSL facilities provided by NorthPoint Communications, a data-oriented CLEC. NorthPoint's DSL technology provides the two-way symmetric bandwidth needed for services such as IP telephony, intranets, virtual private networks, e-mail, electronic commerce, and Web hosting. The combination of Frontier's fiber optic network and NorthPoint's DSL connectivity to the desktop will facilitate the development and delivery of new multimedia applications such as unified messaging services.

6.1.2 PSINet

As the first and the world's largest independent commercial Internet service provider, PSINet offers high-speed LAN/WAN connectivity services to corporations as well as secure Internet, intranet, electronic commerce, and Web hosting services. The company also provides bandwidth to other carriers and ISPs. PSINet has over 55,000 commercial customers.

The company's international frame relay–based, IP-optimized network connects to ISDN, ATM, SMDS, and wireless/satellite systems. It has over 225 POPs in the United States out of more than 450 worldwide, which can also be accessed via dialup and leased lines. The managed, redundant backbone is connected to all the major NAPs and MAEs.

At this writing, PSINet is installing more than 10,000 route miles of optical fiber, which will operate at up to 9.95 Gbps (OC-192). PSINet boasts that its complete ownership and operation of all aspects of the switching and routing infrastructure—from the physical fibers and the low-level transmission gear to the IP routing devices and the high-level information servers—guarantees the best operational and technical solutions for corporate internetworking requirements.

PSINet offers PSIVoice, a suite of voice-over-IP services that includes iPEnterprise, iPEnterprise Plus, and iPGlobal, which address the differing communications needs of businesses.

6.1.2.1 IPEnterprise

This service enables businesses with private branch exchanges (PBXs) to use PSINet's IP-optimized network for carrier-grade voice services between their distributed locations at a lower price than is available from existing carriers. For a flat-rate price, PSINet offers a fully managed, turn-key service that provides customers with voice services that are equal in performance to traditional tie-line or dedicated voice services between PBXs. By purchasing the iPEnterprise solution with other value-added PSINet services, such as IntraNet and iPFax, businesses can reduce their internal communications costs by as much as 50%.

6.1.2.2 IPEnterprise Plus

This service adds voice communication to the capabilities of multi-company extranets, enabling businesses—as well as their select partners, customers, and suppliers—to share simplified dialing codes and enhanced features such as desktop faxing, conference calling, and unified messaging services.

6.1.2.3 IPGlobal

This voice-over-IP service is implemented through H.323-compliant IP/PSTN gateways deployed throughout PSINet's worldwide network. H.323 compliance enables voice-over-IP services to be accessed from a broad range of technologies, including dial, DSL, and cable, and over a range of protocols including IP, ATM, and frame relay. However, in keeping with its strategy to be a carrier's carrier, PSINet will provide iPGlobal primarily on a wholesale basis to its carrier and ISP customers. These customers are expected to use the cost savings generated by iPGlobal to improve their branding and marketing in order to compete in the voice market against more expensive carriers that use less efficient networks.

PSINet offers DSL service in partnership with Covad Communications. In addition to standard DSL service for businesses—with speeds ranging from 144 Kbps to over 1.5 Mbps—PSINet offers an enhanced DSL service that provides a seamless integration with its frame relay and VPN offerings. PSINet also has a wireless component. Its InterSky offering is a business-grade wireless Internet access service that uses the unlicensed 2.4 GHz spread spectrum band. Speeds of up to 512 Kbps are available, with the potential for up to 2 Mbps in the future.

6.1.3 Qwest LCI International

Qwest started out in 1992 as a construction company that built fiber optic lines for others, most notably GTE, WorldCom, and Frontier. The company uses a patented rail-plow fiber deployment process that enables it to lay down fiber deeper (42 to 56 inches below ground) and faster than anyone else. About 85% of the company's fiber is buried along the national rail system rights-of-way company chairman Phil Anschutz had access to from previous ventures related to Southern Pacific Telecom (SP Telecom), which built fiber networks for Sprint and MCI, among others. But for every fiber Qwest put down for a customer, the company installed another for itself. In effect, the company's customers—who are also its long-distance competitors—have covered much of Qwest's initial network construction costs.

Today, Qwest LCI International, the merged company of Qwest and LCI International, is the fourth-largest long-distance provider in the United States, right behind AT&T, MCI WorldCom, and Sprint. Qwest LCI has a 21,000-mile high-capacity fiber network and over two million business and residential customers.

LCI International brought to the 1998 merger its full array of voice and data transmission services that it offers to businesses, residential customers, and other carriers through its 8,500-mile fiber optic network. The company also provides long-distance service to over 230 international locations and local telephone service to commercial customers in 41 U.S. markets.

Qwest brought to the merger its high-capacity fiber optic network that delivers high-quality data, video and voice services to businesses, consumers, and other communication's service providers. Upon its completion in mid-1999, Qwest's domestic 18,449-mile network began serving over 130 cities, which collectively represent approximately 80 percent of the data and voice traffic originating in the United States. The bi-directional self-healing network is deployed in an OC-192 SONET ring architecture.

The company also has a 1,400-mile backbone segment that extends into Mexico and—with its acquisition of EUnet International, an Amsterdam-based Internet service provider—a transatlantic fiber link that serves 13 countries in Europe. Qwest is also participating in a consortium of communications companies that is building a submarine cable system connecting the United States to Japan. The 13,125-mile cable will provide connections to other digital lightwave submarine cable systems reaching numerous points in Asia.

The submarine cable system planned for the Pacific Rim is designed as a self-healing ring. The synchronous digital hierarchy (SDH) network will consist of four-fiber-pair cable, which will ultimately possess the capability to transmit information at the rate of 640 Gbps, making it the most reliable and capacious transpacific cable available. Once this cable is online, Qwest LCI will have the capability to transmit data, multimedia, and voice traffic from as far as Eastern Europe, across the United States, into Mexico, and on to Asia.

Qwest LCI believes the demand from interexchange carriers and other communications entities for advanced, high-bandwidth voice, data, and video transmission capacity will increase over the next several years due to regulatory and technical changes and other industry developments. These anticipated changes and developments include:

- Continued growth in capacity requirements or high-speed data transmission, ATM and frame relay services, Internet and multimedia services, and other new technologies and applications;

- Continued growth in demand for existing long-distance services;

- Entry into the market of new communications providers;

- Reform in the regulation of domestic access charges and international settlement rates which Qwest expects will lower long distance rates.

Among the company's many voice and data services is Q.talk, an IP telephony service available from 15 cities that has no connection fees, per-call charges, or for minimum call lengths. State-to-state calls are billed at a flat $0.075 per minute any time of the day. In-state calls are less, depending on the state.

In one of its most innovative marketing arrangements conceived to date in this new era of IP telephony, Qwest tried to link its high-speed backbone with the local loops of Ameritech and U S West in partnerships that would offer residential and small business customers a single-source package of local and long-distance services.

In the five-state Ameritech region, customers were able to sign up for CompleteAccess, which includes Ameritech local service and custom calling features, as well as Qwest's Q.talk long-distance service at the rate of seven cents per minute for evenings and weekends and $0.15 per minute during weekdays for all state-to-state and in-state calls. For small business customers, CompleteAccess featured Ameritech's ValueLink Extra-Select volume discount local services and Qwest LCI's long-distance service at a flat rate of $0.095 per minute, 24 hours a day, seven days a week for all state-to-state and in-state calls.

In the 14-state U S West region, customers were offered a package called Buyer's Advantage that included local services from U S West and an option of long-distance service from Qwest at a flat $0.10 per minute for residential customers. Rates for business customers varied depending on usage.

However, this type of marketing arrangement was strongly opposed by AT&T and MCI Worldcom, which argued to the Federal Communications Commission (FCC) and in lawsuits filed in two Federal District courts that it illegally sidestepped the FCC's procedures for determining the eligibility of the regional carriers to enter the long-distance market based on the amount of local competition they offer. Qwest had declared that efforts to halt the marketing agreements would have the effect of limiting competition and preventing companies from implementing new programs that provide customers with low cost, value-added services for which they have long asked and which the Telecommunications Act of 1996 ostensibly encourages.

In the end, the FCC's decision to negate the agreements prevailed. The commission found that Qwest's marketing arrangements with dominant local telephone companies violated the Telecommunications Act and would not be allowed until their local markets are open to competition, as measured against the FCC's own 14-point checklist. Consumers will continue to have choices for long-distance service, including those of Qwest. The FCC ruling simply meant that Qwest cannot offer its long-distance service bundled with the local services of the dominant telephone companies.

Although this attempt at bundling long-distance IP telephone services with local telephone service failed, it indicates the direction in which the industry plans to move in the future, once the FCC is satisfied that local competition exists and the former Bell companies are free to offer long-distance services. Instead of the Bells building their own long-distance infrastructures at great cost, they need only partner with nextgen carriers like Qwest to provide long-distance services to their customers. The Bell companies might even merge with nextgen carriers, thereby gaining a national footprint instantaneously from which they can launch virtually any conceivable information, entertainment, multimedia, and electronic commerce service.[1] All of these offerings could be bundled together with local and long-distance telephone service at attractive discounts that are designed to retain and enlarge their customer base and, just as importantly, discourage potential competitors. The FCC would like to put off this situation as long as it can by making sure the local loops are fully opened to competitors before giving the telephone companies permission to offer long-distance services.

---

1. Qwest merged with U S West in 1999 to form a $65 billion worldwide company.

BellSouth has invested about $3.5 billion for a 10% equity stake in Qwest. BellSouth and Qwest will jointly market Qwest's Internet and voice services along with BellSouth's local networking services when BellSouth gets approval to enter the long-distance market. Until Bell-South and the other former "Baby Bells" convince regulators that they have opened their markets to competition for local calling service, they are limited to 10% ownership in any long-distance company and may not offer services on a bundled basis. But once regulatory approvals permit the Baby Bells to enter the long-distance market, they will be free to link up with the nextgen carriers to package a comprehensive set of high-speed data, image, and voice communications for business and residential customers.

Qwest offers DSL services in partnership with Rhythms NetConnections and Covad Communications. These agreements provide Qwest with market overlap and multiple connection points in key U.S. cities and the necessary bandwidth at the local level to improve customer connections between their homes and offices to the Qwest network. Aside from DSL connectivity, Qwest is also in the process of assembling 19 metropolitan area networks in major U.S. cities that will allow large businesses to directly connect to the Qwest network.

6.1.4   Other players

There are a number of other service providers that are rapidly building high-speed fiber optic networks capable of handling packetized voice and data. Some are smaller Tier I or major Tier II backbone providers.

6.1.4.1   AGIS

This company provides connectivity services to millions of users through its extensive customer base of carriers, LECs, CLECs, ISPs, content providers, and large multinational corporations. AGIS has private interconnections with the industry's largest Tier I Internet backbone service providers and maintains a broad range of interconnections through public peering arrangements at the major NAPs and MAEs.

AGIS offers ATM and frame relay VPNs, and dial and dedicated Internet access connectivity services. It owns and operates a nationwide self-healing SONET network that extends over 10,000 route miles and offers dedicated connectivity at speeds ranging from 768 Kbps to the OC-3 rate of 155 Mbps. The AGIS network is accessible from over 200 POPs.

In partnership with Novell, AGIS provides directory-enabled Internet services to its enterprise customers. Novell Directory Services (NDS) provides a managed VPN solution for deployment at the customer premise. This directory-enabled approach frees business customers of AGIS from much of the effort that had previously been required to connect with vendors, customers, employees, and the public Internet.

AGIS is expanding its service portfolio with a full suite of next-generation communication solutions such as DSL, videoconferencing, IP telephony, and a variety of value-added and integrated applications. At the same time, it is upgrading its network to OC-48 (2.488 Gbps), giving it a migration path to OC-192 (9.95 Gbps) in the future. Shortly after the millennium mark, the company expects to have its network span the entire globe.

6.1.4.2 CAIS Internet

Operating a nationwide ATM backbone that uses DS-3 (45 Mbps) links, this company is the underlying Internet access and services provider for more than 50 ISPs in the United States and around the world. In addition to connections to the major NAPs, CAIS has 44 POPs nationwide. The company guarantees 99.999% service availability on its fully redundant, continuously monitored network.

CAIS offers a variety of services to ISPs and consumers, including:

▶ DSL access in selected areas at speeds ranging from 640 Kbps to 7.1 Mbps;

▶ Dedicated access for businesses that require high capacity, high performance Internet connections;

▶ High-speed dial-up access;

▶ Web hosting for businesses;

▶ Dedicated servers for direct national backbone access.

CAIS also offers a service called OverVoice, a high-speed Internet technology that relies on existing in-building telephone wiring to provide consumers with simultaneous phone service and Internet access. Unlike ADSL, cable modems, and other high-speed technologies, the OverVoice solution is based on the 10BaseT Ethernet. The technology enables users to connect to the Internet at speeds of up to 10 Mbps and talk on the phone at the same time over the same line. This is accomplished without the need for expensive external modems or in-room powered electronic components. All that is required is a standard Ethernet card for PCs and laptops for connecting them to the LAN.

For this service, CAIS seeks distribution and licensing agreements with commercial real estate developers, and owners and operators of apartments, condominiums, and hotels, as well as schools, other Internet service providers, and value-added resellers.

6.1.4.3   Concentric Network

Concentric designed and operates its own low/fixed latency, scalable WAN with performance that is adaptable on a per-call basis. Based on ATM, the continuously monitored backbone in the United States and Canada consists of 16 SuperPOPs in major metropolitan calling areas, supplemented by approximately 136 traditional POPs located in smaller markets. Designed to optimize local call access, the high-density Super-POPs broaden call coverage for users within a 150–200 mile radius. SuperPOPs house thousands of modems that process traffic aggregated from a surrounding area by competitive local exchange carriers and other telecommunications service providers.

The company offers an array of services including VPNs, wireless Internet access at up to 384 Kbps, dedicated access, dedicated and shared hosting services, and electronic commerce solutions. Concentric partners with several companies to provide customers with DSL access to its nationwide ATM network. These partners include Covad Communications and NorthPoint Communications.

Concentric also offers long-distance and international IP voice service for carriers and resellers, permitting their customers to call phone-to-phone to more than 200 countries. To take advantage of this service, a carrier or reseller installs Concentric's dedicated IP access facilities and all the necessary voice gateway hardware at their switch location. The carrier or reseller maintains full management and control of the international long-distance services provided to their end-consumers. As part of the service, they receive a monthly call detail report itemizing all the IP voice calls routed by Concentric.

Like many of the larger backbone providers, Concentric has grown through acquisitions. In mid-1997, Concentric acquired Delta Internet Services, also known as DeltaNet, which enabled it to get into the business market. In early 1998, Concentric acquired InterNex Information Services, a provider of hosting and dedicated access services. Concentric Network combined InterNex's public peering arrangements with its dedicated private peering and transit connections to high-volume network providers.

The same year, Concentric and Telemedia International (TMI)—a subsidiary of Telecom Italia, the world's sixth largest telco—launched MondoNet, the first global IP network designed and built expressly to support virtual private network services. Built upon TMI's infrastructure, MondoNet offer VPNs in more than 30 cities in 24 countries located throughout Europe, South America, and the Pacific Rim. TMI also provides Concentric customers with connectivity in 104 cities in 45 nations worldwide.

6.1.4.4 Covad Communications

Established in 1996, Covad is a packet competitive local exchange carrier (PCLEC) that focuses on corporations and Internet service providers. The company wants to be the dominant national provider of packet network services over telephone lines enhanced with DSL technology for the delivery of data, voice, and video.

The company's TeleSpeed access services are available in 8 bandwidth increments from 144 Kbps to 1.5 Mbps. The services are sold through leading carriers and ISPs. Another service, TeleSpeed Remote, provides nationwide high-speed access service for secure interconnection between corporations, remote branch offices, and teleworkers at a fraction of the cost of dial-up, ISDN, frame relay, or leased lines. The reach of the TeleSpeed Remote service is made possible by Covad's agreements with AT&T and Qwest, which provide redundant, continuously monitored ATM backbone links. The backbone networks provide a complete, local, regional, and national data connectivity solution for a wide range of customers.

The network architecture provides user access to the network through Covad connections. The data is then sent across the United States on either an AT&T or Qwest backbone and delivered to the Covad network at the remote location. The end-to-end service eliminates the need to juggle multiple carriers for local and long-distance data connectivity. Covad provides all necessary connectivity from site-to-site. Capacity planning is handled by Covad and is completely transparent to the customer.

Unlike routed or VPN services, each TeleSpeed remote client uses an individual Layer 2 virtual circuit. No encryption, authentication, or tunneling hardware or software is required, although customers can implement these independently if added security is desired.

Covad is also looking at using ATM to provide voice service over DSL. This technology has the capability to provide up to 16 channels of toll-quality voice over a single DSL line. The company's goal is to deliver a product with the same functionality of today's business-class telephone service. The ATM-based solution not only combines digital voice and data in the local loop, it also provides the same quality and full functionality of today's telephone service, including features such as caller ID and call forwarding.

6.1.4.5 Delta Three

This subsidiary of RSL Communications Ltd. offers three types of IP telephone service that offer global reach: PC-to-PC, PC-to-phone, and phone-to-phone. PC-to-PC calls are made from one multimedia computer to another. PC-to-phone calls are made from a PC to a regular telephone. Phone-to-phone calls are made from a regular telephone to another regular telephone but are routed over IP networks. According to the market research firm Frost and Sullivan, Delta Three routes 17% of all IP telephony traffic worldwide.

The company has 57 POPs in 29 countries. Its parent company fills the gaps where the Delta Three IP network does not reach. Through RSL's traditional telephone network and termination services Delta Three can complete calls to over 200 countries. The company's IP/PSTN gateway sites are either wholly owned by Delta Three, or are joint ventures with international companies, ISPs, and PTTs. Delta Three controls over 50 gateways. The Delta Three sites are redundantly interconnected via a managed international intranet composed of frame relay and dedicated leased lines. The infrastructure, the Internet, and all POPs in the Delta Three global network are monitored on a 24-hr basis.

The company's Communications Portal offers interactive voice, fax, e-mail, paging, voice-mail, roaming services, messaging, and other services which can be accessed via regular telephones or Delta Three's site on the Web. Since subscribers pay in advance for their services, the company offers account status information online and provides the means for subscribers to recharge their accounts via credit card (see Figure 6.1).

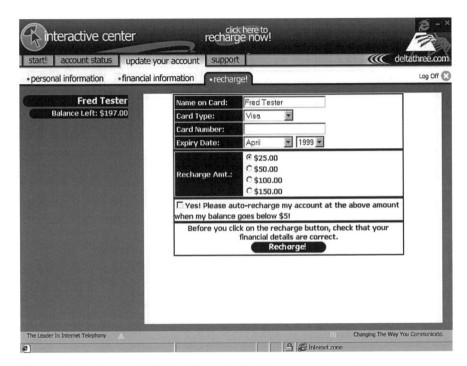

Figure 6.1 Through Delta Three's Web page, subscribers can access the company's Interactive Center to check account status or recharge their accounts.

Delta Three's technology development partnership with Ericsson resulted in the IP Telephony Solution for Carriers (IPTC). The IPTC is installed and operational in the Delta Three global network on four continents. With the Ericsson platform, carriers and ISPs can join the Delta Three network through the Network Partners Program.

6.1.4.6   Electric Lightwave

A Tier I Internet provider, Electric Lightwave serves businesses, ISPs, and long-distance carriers across the United States by consolidating voice-data traffic over its ATM-based OC-3 fiber optic broadband network. The company operates in 100 municipalities. In the western United States, Electric Lightwave is a full service communications provider, offering business customers a suite of integrated products and services, including local phone service, switched and dedicated long distance, private networks, advanced data and Internet access services, nationwide videoconferencing, and prepaid long-distance services. Nationwide, the company offers data, Internet access, and broadband transmission services. It owns and operates 21 data POPs, maintains connections to the primary NAPs, and continuously monitors its infrastructure through a network operations center.

In 1998, Electric Lightwave began an ATM network upgrade. The company is building additional SONET-based long haul routes and expanding its metropolitan area networks (MANs). When completed, Electric Lightwave's western SONET network will consist of more than 4,500 miles of fiber. The company's move to OC-3 has enabled it to become one of the top 15 data networks in the country. In addition, the company is nearing completion of an upgrade to its coast-to-coast data-only network to OC-12.

Electric Lightwave offers service level guarantees on packet delivery, latency, and service provisioning. A "looking glass" feature available on the Web reveals the real-time performance of its fiber optic data network (see Figure 6.2). Users can even conduct custom data tests of speed and accuracy between cities that matter most to them.

6.1.4.7 ICG Communication

ICG offers extensive switched fiber optic networks and provides local, long-distance, and enhanced telephony and data services to small and medium-sized businesses in California, Colorado, Ohio, and the southeastern United States. With its acquisition of ChoiceCom, the company has a presence in Texas. ICG also provides high-speed data network capabilities and services to Internet service providers throughout the United States over its nationwide network. Through other subsidiary operations, ICG designs and installs copper, fiber, and wireless infrastructure for buildings and campuses, and offers international satellite voice and data services.

At year-end 1998, ICG had 354,482 local lines in service and was connected to 5,397 buildings. It had 4,255 operational fiber route miles, 28 voice switches, and 16 data switches. The company serves ISPs using its nationwide data network of 236 POPs, offering ISPs such services as modemless remote access services (RAS), expanded originating service (EOS), and DSL.

ICG Communications offers an IP-based long-distance service called IPLD, which lets users make prepaid long-distance phone calls for as low

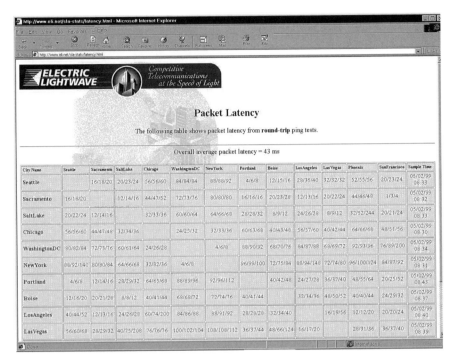

Figure 6.2 Packet delivery and latency statistics shown on the Web are the result of pings directly between Electric Lightwave's routers using the extended ping feature in Cisco's Internetwork Operating System (IOS).

as $0.059 per minute to an expanding list of U.S. cities and $0.089 per minute everywhere else. With an ordinary phone, users dial a local access number, which gets them onto the IP network of ICG, and then the long-distance number. However, the IPLD service will not work if the user's phone has Caller ID Blocking. Caller ID Blocking can be disabled for individual calls by dialing *82 before dialing the IPLD local access number. Users can access their accounts via the Web by entering their account number and PIN, or administrative account and password.

Through an agreement with NorthPoint Communications, ICG has expanded its national DSL footprint with 75,000 DSL lines to gain access to more than two-thirds of the businesses and more than half of the residential customers in the top 25 U.S. markets. ICG offers NorthPoint's symmetrical digital subscriber line (SDSL) service as a part of its suite of bundled services aimed at businesses and ISPs.

6.1.4.8 Level 3 Communications

Level 3 supports a full range of information and communications services including local and long-distance voice and data transmission as well as other enhanced services, such as Web hosting, Internet access, and virtual private networks. Through its collocation program, Level 3 offers companies and ISPs space in its own facilities in which they can put telephone and Web-hosting equipment for connection over its IP network.

The company leases capacity over 8,300 miles of network from Frontier Corporation's national network and over 7,000 miles of IXC's national networks. In addition, the company has local network access lease arrangements in 18 markets. Leasing capacity allows Level 3 to begin to offer services and build a customer base. Level 3 will move the traffic over to its own networks as they are built.

At year-end 1998, network development was under way in 25 of 50 U.S. cities in which it plans to offer local service. Of those, gateway sites were completed in 17 cities. The first Level 3-owned local loops were operational in three cities and loop construction has started in five additional cities. A typical city network consists of a gateway site connected to a fiber network. This network interconnects with the other carriers and directly to potential customers within a city, ultimately allowing Level 3 to offer end-to-end services. Ultimately, the company plans to offer local service capabilities over its own fiber facilities in all 50 U.S. cities. The company is also building its own 16,000-mile long-distance national network to interconnect these cities. The network will also connect approximately 150 cities where the company can terminate and provide long-distance services.

In addition, the development of local city networks and a Pan-European network were under way in five European countries, while transoceanic capacity agreements were signed for three systems—one to participate in the build of the Japan-U.S. Cable Network, an undersea cable system that will connect Japan and United States; and two for transatlantic capacity. The first transatlantic cable system, expected to be in service in late 2000, will be upgradeable to a total capacity of 1.28 terabits per second (Tbps), enough bandwidth to transport the contents of the U.S. Library of Congress across the Atlantic in less than 16 seconds.

Level 3 has engaged in acquisitions to supplement its own network build program. For example, it acquired BusinessNet in the United

Kingdom, a leading provider of IP based services in London. BusinessNet provides a variety of services to over 300 clients, including Internet access, secure intranets, interactive voice and video services, and media streaming.

6.1.4.9 NorthPoint Communications

This PCLEC provides services in 28 metropolitan areas, offering a range of value-added services to ISPs and small to medium-sized businesses. The data connectivity needs of business are served with dedicated SDSL at speeds from 144 Kbps to 1.5 Mbps for such applications as Web hosting, intranets, VPNs, electronic commerce, and multimedia. Those who opt for entry-level services can move up without any hardware changes to higher speed premium services as their business needs evolve.

The company establishes its network footprint by co-locating equipment in telephone company central offices to provide Internet access to end users in the local service area, and funneling that traffic over high-speed connections to a centrally located hub. Service providers gain access to their subscriber base through DS3 links to the central facility.

Although NorthPoint does not currently offer any voice services of its own, it has positioned itself for this market in the future. When the company decides to offer voice service, it will do so over a data network rather than the circuit-switched network traditionally used by telephone companies. The equipment co-located in phone company central offices will take calls from the traditional phone network and turn them into signals for transport on an IP or ATM data network. Meanwhile, ISPs and companies that use NorthPoint's DSL services can offer end-to-end voice via their own backbone networks.

These thumbnail portraits provide only a small part of the voice-data convergence story. There are dozens of larger and smaller firms using the latest technologies to build and operate fiber optic backbones, and new companies continually emerging—all intending to provide Internet capacity in the near term and position themselves for more advanced services later. This trend shows no indication of slowing down and the traditional carriers have been responding by pouring more capital into alternative IP/ATM infrastructures than into conventional circuit-switched infrastructure.

6.2   Responses from traditional carriers

One of the responses of traditional carriers to the emergence of nextgen carriers has been to gobble up key Internet backbone routes and elevate themselves to the status of Tier I national backbone providers.

Tier I backbone providers occupy the top of the Internet hierarchy in the United States, providing bandwidth, connections, and services to regional and local ISPs and other types of carriers. Just as there are multiple long-distance telephone companies in the United States, there are multiple Tier I backbone providers that compete with each other for traffic and customers. Despite this competition, the Tier I service providers cooperatively interconnect with each other at various high capacity NAPs and MAEs—so the voluminous traffic of their customers can be routed in the most expeditious manner possible anywhere on the global Internet. Through mergers and acquisitions, the giants in the telephone industry in recent years have managed also to become the major Tier I Internet backbone providers: AT&T, Cable & Wireless, GTE, MCI WorldCom, and Sprint.

The backbones of Tier I providers typically are comprised of SONET-based optical fiber that carry traffic as fixed-length ATM cells or IP packets. On major networking routes, including those to NAPs/MAEs, the links provide 622 Mbps (OC-12) of capacity each, while the links between other hub locations provide at least 44.736 Mbps (DS3) of capacity each. Customers connect to the backbones of the Tier I providers at these hub locations. All the connections and equipment are continuously monitored by the backbone provider's network operations center.

Tier I national backbone providers charge a monthly fee to the local and regional ISPs that connect to them. The amount of the fee typically is based on the bandwidth of the connection. Depending on amount of traffic the ISP intends to deliver to the Tier I backbone provider, access bandwidth can be purchased in increments as low as 56/64 Kbps all the way up to 155 Mbps (OC-3). Transoceanic connectivity—as well as high-speed regional connectivity within Europe, Asia, and South America—is typically available at up to 44.736 Mbps. Once the fixed-bandwidth leased-line connection is in place, the local or regional ISP can pump and receive as much data as it can, up to the capacity of the connection, at no additional cost.

By definition, a Tier I provider operates its own national Internet network with a capacity of at least 45 Mbps (DS3) carrying traffic that is exchanged at all the NAPs. A Tier I provider must also maintain a network operations center 24 hours per day, seven days a week to manage their network. Tier I providers are the core of the fabric that constitutes the Internet. All other Internet providers and users must connect through these Tier I operators to access the Internet.

The backbones of all of these providers are interconnected at the following major NAP/MAE switch points (see Figure 6.3):

- MAE East (Washington, DC);

- MAE Central (Dallas, TX);

- MAE West (San Jose, CA);

- Ameritech Chicago NAP (Chicago, IL);

- Pac Bell San Francisco NAP (San Francisco, CA).

In addition to serving local and regional ISPs, these backbone providers also serve the networking needs of small to large businesses with such services as VPNs, extranets, IP multicast, Web hosting, and remote access. All of these backbone providers even compete with their local and regional ISP customers by offering dialup Internet access services directly to consumers.

The trend of mergers and acquisitions in the telecommunications industry has resulted in an elite club of companies controlling most of the Internet's major backbone routes.

6.2.1 AT&T

With its $11 billion merger with TCG CERFnet in mid-1998, AT&T became one of the top 10 Internet service providers for businesses.

Prior to the merger, AT&T's WorldNet consisted of 16 backbone nodes on a DS3 fiber network. More recently AT&T upgraded the backbone to OC-3 and from there to OC-12 and OC-48. The entire upgrade, completed in 1999, included a conversion to an ATM switch fabric that interconnects 80 OC-48 SONET rings. All routers and switches are deployed in a redundant configuration. The hub nodes, which utilize network routers and ATM switches, are connected to the IP backbone via OC-3 links. The network has redundant management systems, which are used to monitor performance.

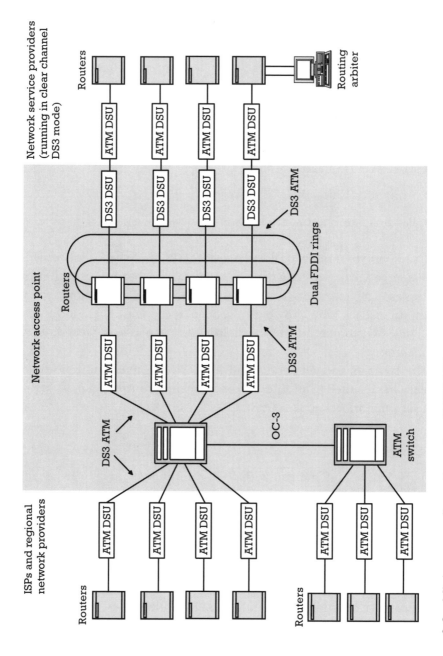

Figure 6.3 Minimum configuration of an NAP/MAE.

Dedicated access is provided from 45 POPs. Customer circuits had been back-hauled to one of the 16 backbone nodes for connection to the network. However, with the recent addition of 100 concentrator nodes that connect with the backbone nodes, customer circuits are now back-hauled to either the nearest concentrator node or backbone node for connection to the network.

Dial-up access is provided from 27 POPs. The dial platform contains 375 local dial access numbers as well as Megacom 800 service for toll-free access.

Several AT&T organizations are involved in providing IP services over the WorldNet backbone:

▶ AT&T Networked Commerce Services is the umbrella organization for all AT&T business IP-based service products.

▶ AT&T Consumer Markets is responsible for AT&T WorldNet Service consumer Internet dial access.

▶ AT&T WorldNet Managed Internet Service (MIS) provides dedicated connectivity to the Internet. Connectivity is available via frame relay or private line, at speeds ranging from 56 Kbps to 45 Mbps. In addition, flat-rate tiered T3 connections, ranging from 10 Mbps to 45 Mbps, are available.

AT&T has catapulted itself into the leadership position with its $5 billion acquisition of IBM Global Services, which has 503 local POPS in the United States and 850 local POPs internationally, for a total of 1,353 POPs. The IBM Global Network serves the networking needs of several hundred large global companies, tens of thousands of mid-sized businesses, and more than 1 million individual Internet users in 59 countries. About 5,000 IBM employees have joined AT&T as part of the transaction.

The acquisition of IBM's global data network accelerates AT&T's ability to deliver IP-based services to global customers, while giving it a sophisticated new platform for revenue growth. By providing customers with more attractive global services, the acquisition enables AT&T to compete more effectively with strong rivals for the provisioning of global managed data network services, including IP.

The addition of IBM's high capacity global network complements the 100-city, IP-based network that is being built as part of the $10 billion global joint venture announced by AT&T and BT in mid-1998. The purpose of the joint venture is to develop an intelligent, centrally managed IP-based global network that will support services such as global electronic commerce, global call centers, and new intranet applications. At the same time, the two companies agreed to invest a total of $1 billion, split equally between them, in U.S. businesses involved in high technology and emerging communications markets.

Like MCI WorldCom and Sprint, AT&T offers a managed service that combines voice, frame relay, ATM, private line, and IP traffic on the same access lines. Its Integrated Network Connect (INC) offering requires that users deploy an AT&T-owned ATM multiplexer at their sites to convert the different traffic types to ATM cells. The cells are then sent over a dedicated T1 line to an ATM multiplexer in an AT&T switching office. There, the combined traffic is split out and sent over the appropriate AT&T facilities (see Figure 6.4).

While AT&T customers could previously combine different kinds of traffic on their access lines using a service called static integrated network access (SINA), INC provides the benefits of AT&T owning the equipment used to blend the traffic, and consolidated billing for the different types of services. In addition, INC offers the following other advantages:

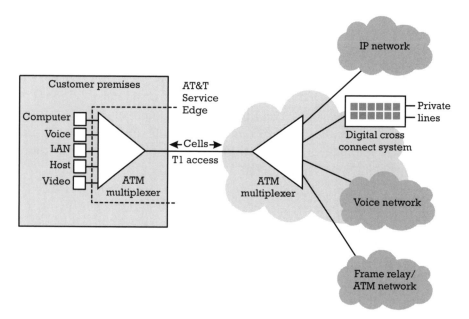

Figure 6.4 Basic configuration of AT&T's INC service.

▶ Dynamic bandwidth utilization—INC uses a combination of ATM statistical multiplexing and voice compression technologies to more efficiently transport voice traffic and to enable customers to add more data traffic. It replaces multiple access arrangements with one integrated voice, data, and IP access technology. It also allocates bandwidth-on-demand and minimizes last mile access bottlenecks for bursty data.

▶ End-to-end network management—AT&T serves as the single provider for all services and manages and maintains on an end-to-end basis the ATM multiplexer and the T1 access link.

▶ Interoperability—INC allows customers to add locations to their network as their needs evolve, and still maintain full connectivity to all other sites in their network. INC-equipped locations are fully compatible with a customer's existing premises equipment, eliminating the need to make any additional purchases.

▶ Reduced communications costs—Because INC lets businesses put more traffic on a single T1 (e.g., up to 40 voice calls simultaneously with 512 Kbps of data and IP traffic), they can expect savings of between 10–20%.

Since its merger with cable operator TCI in 1998, AT&T has been laying the groundwork for advanced communications services aimed at consumers. These services include Internet access, local, and long-distance telephone service, and multimedia applications that are delivered into homes by TCI and its affiliates, along with television programming. The hybrid fiber/coax network allows AT&T to offer its customers the ability to watch TV, send and receive faxes, use the Internet, and talk on the phone simultaneously. The telephone service features include multiple phone lines per household, along with options such as conference calling, call waiting, call forwarding, and individual message centers for family members. TCI's cable plant passes more than 17 million homes in the United States, while TCI's affiliates add another 5 million homes. AT&T has also acquired MediaOne, giving it a total of 26 million customers.

Despite its extensive cable reach, AT&T is also pursuing DSL and wireless as part of its end-to-end IP connectivity strategy. These technologies will be used to provide wholesale and retail services, giving customers fast on-ramps to its expanding global IP network in support of converged voice-data services.

AT&T operates a global IP telephony interoperability lab, an industry test bed that promotes the consistent implementation of standards for global IP telephony and other advanced IP services among vendors and service providers. The facility demonstrates AT&T's recognition that if next-generation IP services are to be truly successful, they must be as reliable and interoperable as the telephone network it helped build over the past 100 years.

6.2.2 Cable & Wireless

Cable & Wireless USA operates the Global Digital Highway, a private international data network, and the Cable & Wireless Internet Exchange (CWIX), the company's global IP backbone. The fully meshed DS3/OC-3 ATM backbone in the United States consists of 37 backbone nodes in major cities with intelligence distributed at its global network access points (GNAPs), which minimizes network delays or connection with other traffic types. The GNAPs are connected with NxT1/E1 circuits.

In mid-1998, the company acquired MCI's internetMCI division for $1.75 billion, which made it a Tier I Internet backbone provider. The MCI assets included 22 domestic nodes, 15,000 interconnection ports, routers, switches, modems, e-mail servers, and more than 40 ongoing peering agreements. Cable & Wireless also acquired MCI's approximately 3,300 dedicated access customers, 1,300 ISP customers, 250,000 consumer dial-up access customers, 60,000 business dial-up access customers, and 100 corporate and Web-hosting and firewall customers. At the time of the acquisition, Cable & Wireless created two new business divisions: Commercial Internet Services and Consumer Internet.

The company is building OC-12 connections to the major NAPs and installing new ATM switches at the MAEs. Additionally, Cable & Wireless will build OC-3s and OC-12s to its largest direct connect peers and is installing at least 25 new DS3s to handle traffic for other new and existing peers. This effort includes upgrading the connections to major ISPs at its public peering points, to direct connections.

The new OC-12 capability provides the equivalent of almost 48 DS3s, doubling the company's public peering capacity. With these enhancements, Cable & Wireless will be able to allocate bandwidth to each ISP, aiding in effective capacity planning. The benefits to Cable & Wireless customers include faster access, elimination of packet loss, and a better ability to plan for future growth for ISPs that peer with the company's upgraded backbone.

Starting in 1999, Cable & Wireless will spend about $670 million over the next few years to upgrade and expand its network in the United States to offer data and voice services to businesses and consumers nationwide. When completed in 2001, the carrier's Cable & Wireless USA Internet backbone will span more than 60 U.S. cities, providing throughput at rates of up to OC-192 (9.6 Gbps). It will tie into existing Cable & Wireless networks in operation in Europe and Asia.

6.2.3 GTE Internetworking

GTE Internetworking, a unit of GTE Corp., includes the former BBN Corp., which 30 years ago developed the ARPANET, the forerunner to today's Internet, and GTE Intelligent Network Services, which provides Internet services to consumers and small businesses.

GTE Internetworking has a redundant OC-12 and OC-3 layered architecture with DS3 links connecting additional cities. The network utilizes both its own network POPs, which number 375, and interconnected frame relay and switched multi-megabit facilities provided by local exchange and interexchange carriers to make local Internet access available to its customers throughout the United States.

GTE Internetworking offers consumers to Fortune 500 companies, a full spectrum of integrated Internet services, including dial-up and dedicated Internet access, high-performance Web hosting, managed Internet security, network management, systems integration, and Web-based application development for integrating the Internet into business operations.

As of year-end 1998, GTE Internetworking's global network infrastructure (GNI) consisted of 12,000 miles of its planned 17,000-mile dark fiber network, which offers OC-192 (9.953 Gbps) of bandwidth capacity. The GNI supports the operations of GTE's business units and their millions of customers, and provides enough capacity to support a broad array of advanced voice, video, and data services including fax and voice over the Internet, unified messaging, broadband videoconferencing, and distance learning.

For instance, GTE Internetworking uses the GNI to support its nationally available IP Fax and Internet call waiting services, as well as other enhanced IP-based services. Separately, GTE Communications Corporation, the national, sales, service, and marketing arm of GTE Corp., currently utilizes the GNI to deliver long-distance voice and data traffic, and expects the high-capacity network to carry more than one billion minutes of use by the end of 1999.

In mid-1999, the GNI spanned more than 100 major markets from coast-to-coast, providing underlying capacity for GTE business units so that they can offer a variety of services including ATM, frame relay, long-distance, and private line. These services, in turn, support a cross-section of applications such as Internet access, Web hosting, virtual private networks, dial-up access, and electronic commerce.

The GNI supports both domestic and global needs by connecting to GTE Internetworking's European POPs through undersea fiber. In the future, undersea fiber extensions will provide high-speed connections to Latin America and Pacific Rim countries, including many markets that are already served by GTE Internetworking.

6.2.4 MCI WorldCom

WorldCom and MCI merged in 1998. The new company, MCI World-Com, retained MCI's position as the second largest long-distance carrier in the United States. MCI WorldCom offers a full range of voice and data services and is the second largest carrier of international voice traffic. Before selling its extensive Internet assets to Cable & Wireless to facilitate regulatory approval of its merger with WorldCom, MCI was also a leading global Internet provider. However, because WorldCom already owned UUNET, ANS Communications, and CompuServe Network Services, the merger enabled MCI WorldCom to increase its leadership position in this market, despite the sale of its internetMCI unit to Cable & Wireless.

In 1999, MCI WorldCom combined the operations of its UUNET and Advanced Networks (i.e., ANS and CNS) subsidiaries into one organization, offering a full range of Internet services to business customers worldwide. Currently, the combined organization offers Internet services to more than 70,000 businesses and ISPs in 114 countries and operates a global network with more than 1,000 voice and data POPs throughout North and South America, and in Europe and Asia.

MCI Worldcom operates eight MAEs in the United States. MAEs East, West, and Central are considered Tier 1 or national NAPs, while the other five are Tier II or regional NAPs. The company also operates 23 smaller hub locations in the United States.

Prior to the merger with MCI, WorldCom had inherited MAE sites in Vienna, VA, and San Jose, CA, when it acquired MFS Communications in late 1996. These MAEs serve as major gateways for international traffic. In mid-1998, WorldCom put into operation an MAE in Dallas, TX. It became the first MAE to be based solely on ATM. It consists of Cisco Systems' StrataCom BPX ATM switches. WorldCom upgraded its MAEs in Vienna and San Jose as well with the ATM switches at a cost of $5 million. Combined, all three MAEs used 16 of the BPX switches. The two older sites had previously used FDDI GIGA switches from Digital Equipment Corp. (now owned by Compaq Computer Systems). However, these switches had proven to be inadequate for handling large traffic volumes and dropped as much as 40% of their packets under heavy loads. Throughout 1998, WorldCom spent another $5 million upgrading the three MAEs with 16 additional BPX switches.

MCI WorldCom offers a unique service called PeerMaker. This Web-based tool improves overall performance of the Internet by enabling ISPs to dynamically manage their peering relationships at the Internet's major network access points. PeerMaker enables ATM customers to provision permanent virtual circuits (PVCs) in near real time at MCI WorldCom's MAEs.

The company offers the tool to ISP customers at no charge, allowing them to set up, reconfigure, or delete PVCs through a secure Web interface. An ISP customer logs onto a secure Web site and fills out a request detailing the bandwidth for a specific destination. Port speeds are available in 45 Mbps, 155 Mbps, and 622 Mbps. To guard against the possibility of over-subscribing a port, PeerMaker automatically surveys a customer's port capacity and the capacity of the destination port, such as the customer's peering partners, before building the PVC. When the customer's peering partner agrees to the changed bandwidth, the PVC is implemented immediately.

MCI WorldCom's On-Net integrated access services allow customers to combine a variety of voice, data, and Internet services over a dedicated private line, or over an ATM or frame relay service. Customers can take advantage of a number of services, all integrated onto a single contract, invoice, and discount plan. For smaller companies and consumers,

MCI WorldCom offers DSL access as a cost-effective, high performance alternative to leased lines for voice as well as data. For large companies, DSL enables them to extend local bandwidth in remote locations for broadband access to MCI WorldCom On-Net services. Customers can take advantage of access speeds ranging from 128 Kbps to 7 Mbps, depending on their locations and particular applications.

6.2.5 Sprint

Sprint operates an Internet backbone network called SprintLink (not to be confused with its Integrated On-demand Network, or ION, which also offers IP-based services). The company also operates one of the four official network access points designated by the National Science Foundation. Referred to as the New York NAP, it is actually located in Pennsauken, New Jersey. Additionally, Sprint interconnects with other network backbones through the major NAPs/MAEs.

SprintLink, launched in mid-1992, originated out of Sprint's participation in building and implementing the federal government's FTS-2000 communication system. As such, Sprint became the first long-distance voice telephone carrier to offer a commercial Internet service.

Today, SprintLink service is available nationwide from all of Sprint's 320 POPs and at various circuit speeds from 56/64 Kbps to 1.544 Mbps, NxT1 service, fractional T3 service, 45 Mbps T3, and SONET OC-3 service. In mid-1998, Sprint upgraded its Internet backbone from OC-12 (622 Mbps) to OC-48 (2.5 Gbps) through the use of DWDM. One OC-48 connection could support more than 175,000 dial-up users sending and receiving files at the same time. The OC-48 route is deployed between Fort Worth, TX, and Kansas City, MO. Further installations are planned on an ongoing basis throughout the Sprint network.

At the core of Sprint's upgrade are the deployments of Cisco Systems' 12000 series gigabit switch routers (GSRs) and Ciena Corp.'s DWDM technology on SprintLink's 15 backbone nodes. Cisco and Ciena have combined forces in IP-over-DWDM technology to develop a high-capacity optical IP backbone. This high-speed deployment by Sprint ties the Cisco 12000 GSR directly to Ciena's long-haul DWDM systems to create an expanded optical layer to Sprint's all-digital, all-fiber network. The Cisco 12000 connects directly into Ciena's DWDM system interfaces at 2.5 Gbps, without the need for additional intermediate network elements such as SONET terminal multiplexers. Not only does the IP-over-DWDM capability allow Sprint to substantially increase the data load on its existing fiber in a very efficient manner, it allows Sprint to provide seamless integration between SprintLink and its new ION network.

Sprint has not purchased any circuit switches for its long-distance network for the past few years. Instead, it has been emphasizing its new packet-switched broadband network, ION, which uses ATM to consolidate all kinds of traffic, regardless of protocol, over a nationwide fiber optic backbone.

ION is similar in concept to the integrated access offerings of AT&T and MCI WorldCom. Access points called broadband metropolitan area networks (BMANs), deployed in 60 top markets, allow Sprint ION to pass within proximity of 70% of large businesses. For smaller businesses, telecommuters, small/home office users, and consumers who may not have access to BMANs, ION supports other broadband access technologies, including DSL, two-way cable, and fixed wireless.

ION supports multimedia applications over a single seamless, end-to-end, high-speed, packet-switched broadband network that is extended to the customer at Sprint's expense. To take advantage of this service, customers allow Sprint to install an integrated service hub (ISH) in their building for which the carrier is entirely responsible. This system, which is considered part of the Sprint ION network, converts all the communications protocols used in customers' local data and voice networks into ATM cells for transport over network via the BMANs. Sprint takes care of all necessary hardware or software upgrades—including replacing the entire hub itself, if necessary.

Although Sprint has chosen ATM for its new network, it fully supports IP, so it can offer public-switched telephone voice quality without losing the advantages offered by IP-based applications. Sprint ION supports all networks—frame relay, IP, voice, and data—and reduces them to a single platform.

Sprint offers a family of services to large businesses, including data services, Internet services, and voice services—all of which are integrated through a common access facility. Initially, Sprint ION services have focused on WAN and managed service solutions. As the infrastructure is expanded and enhanced, Sprint will introduce value-added services such as collaboration, e-commerce, supply chain management, and customer relationship management services. While many of these services will come from Sprint, others will come from its strategic partners who will develop solutions specifically designed to run on Sprint ION.

Sprint ION is priced differently than traditional circuit-switched services. With legacy networks, a circuit is tied up for the duration of a voice or data call, during which no other user can access it. In the ION network—and other packet-based networks—bandwidth is not used until there are packets actually moving through it. The facility is dedicated to a particular user only for an instant and then freed up for other applications to use. With dramatically more sharing of the network facilities, there must be a new model for billing. Accordingly, Sprint has decided to price access at a fixed monthly charge, which ranges in price, depending on the local loop charges in a given location. Transport charges are priced according to the number of packets sent and the quality of service they want for the applications.

6.3   Performance issues

The Tier I providers differ in some very important ways, such as the number of POPs they have on their backbone networks. These local access points allow users, particularly traveling workers and telecommuters, to access intranets, extranets, or the Internet for the cost of a local phone call. The more comprehensive the coverage, the less users must pay long-distance rates or additional "roaming" fees to secondary service providers when coverage is required to locations outside the backbone provider's home market.

Since the performance of all types of Internet service providers differ in a variety of ways, companies and individuals should make use of the valuable resource offered by Inverse Network Technology (and the comparable services of other providers), which posts monthly performance comparisons on the Internet. The "ISP BenchMark Report" is issued by the company as part of its Internet Measurement Service, which rates the major backbone providers and national ISPs according to such key metrics as:

- ▶ Call failure rate;

- ▶ Call failure rate for evening hour;

- ▶ Call failure rate for business hour;

▶ Call failure rate by time of day;

▶ Call failure rate by top 42 demographic POPs;

▶ Call success rate summary with statistics:

▶ Time to log in;

▶ Dial-up connection speeds (modem connect speed);

▶ Domain name server lookup time;

▶ DNS failure and time-out rate;

▶ Web latency;

▶ Web throughput;

▶ Web download failure and time-out rate;

▶ Historical trends (when engaged in a contract and where appropriate).

Results are collected monthly and put into graphical format (see Figure 6.5). The providers are issued a letter grade of A+, A, B, C, or D based on each of these key metrics.

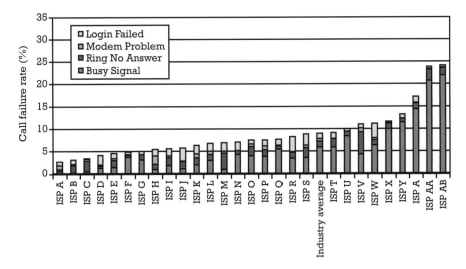

Figure 6.5 A sample call failure report issued by Inverse Network Technology that compares the performance of ISPs in terms of login failures, modem problems, ring-no-answers, and busy signals.

6.4 Analysis

Throughout 1998, the Internet backbone industry experienced robust growth coupled with fierce competition and consolidation—two trends that are likely to continue in the forseeable future. The deregulated market has resulted in the predictable pattern of fewer and bigger backbones vying for business.

Tier I national Internet backbone providers offer a wide range of connectivity options for local and regional ISPs, small and large businesses, and for consumers as well. With their state-of-the-art optical fiber networks and ATM-based switching hubs, plus direct connections to the major NAPs/MAEs, they have abundant traffic-handling capacity.

As mergers and acquisitions continue, enabling the leading Tier I backbone providers to further solidify their control of major Internet routes, there is still enough competition between them to keep prices for access reasonable. Although the Tier I providers have discontinued supporting lower tier Internet service providers through free peering arrangements, this was necessary to fund the cost of expanding and upgrading the backbones in order to meet the growing demand for Internet services among consumers and businesses, as well as ISPs.

It pays to shop around when picking a Tier I Internet backbone provider. Price is only one of the variables that require comparison. In order to get the best deal—and avoid serious problems—net managers need to make their own assessment of value for the money by digging into the details of each provider's service. The topology, capacity, and redundancy of backbones and connections with other operators all give clues to the likely quality of service.

6.5 Conclusion

As the next-generation telecommunications companies start to deliver more features at lower costs, businesses large and small will take notice and realize the benefits of combining voice and data networks. It may take a while longer to eliminate all the concerns about the performance issues raised about running voice over IP, but as the voice-quality concerns fade and more mainstream carriers start offering voice-over-IP services, the market will accelerate.

As this happens, the old telecommunications model of transporting voice over a dedicated, separate switched-circuit network will give way to a new network in which services are not priced based on distance and that offers progressively lower rates to customers.

This will not be a quick transition. Although the growth rates for next-generation services such as IP telephony are impressive, the market is dwarfed by today's $90 billion U.S. market for long-distance services. And most large companies can still negotiate long-distance services for less than $0.05 per minute over the traditional PSTN via traditional virtual private networks. This means voice-over-IP will still hold less than 5% of the total long-distance voice market in 2002.

In the new millennium, there will likely be a trend toward converging operators to perhaps 10 to 15 big global operators carrying all of the international traffic. The convergence trend has already been pervasive among Internet-related ISPs and carriers throughout the 1990s. The traditional telecom operators—the ones that choose to embrace IP—will be the real winners, if only because they own the majority of the customer base today.

More information

The following Web pages contain more information about the topics discussed in this chapter:

AGIS: http://www.agis.net/
Ameritech: http://www.ameritech.com/
AT&T: http://www.att.com/
Cable & Wireless: http://www.cw-usa.net/
CAIS Internet: http://www.cais.net/
Ciena: http://www.ciena.com/
Cisco Systems: http://www.cisco.com/
Compaq Computer: http://www.compaq.com/
Compuserve: http://www.compuserve.com/
Concentric Network: http://www.concentric.net/
Covad Communications: http://www.covad.com/
Delta Three: http://www.deltathree.com/
Digital Equipment Corp. (see Compaq Computer)
Electric Lightwave: http://www.eli.net/

Federal Communications Commission: http://www.fcc.gov/
Frontier Corp.: http://www.frontiercorp.com/
GTE: http://www.gte.com/
IBM Global Services: http://www.ibm.com/globalnetwork/
ICG Communications: http://www.icgcomm.com/
Inverse Network Technology: http://www.inversenet.com/
  products/ims/benchmark.html
International Telecommunication Union: http://www.itu.org/
Level 3 Communications: http://www.L3.com/
MCI WorldCom: http://www.mciworldcom.com/
NorthPoint Communications: http://www.northpointcom.com/
Pacific Bell: http://www.pacbell.com/
PSINet: http://www.psinet.com/
Quest LCI International: http://www.quest.com/
Rhythms NetConnections: http://www.rhythms.net/
Sprint: http://www.sprint.com/
Stratacom (see Cisco Systems)
U S West: http://www.uswest.com/
UUNET: http://www.uunet.com/

Contents

Third-generation wireless networks

In recent years, mobile telecommunications systems have been implemented with great success all over the world. Many are still first-generation systems—analog cellular systems such as AMPS, Nordic Mobile Telephone (NMT), and the Total Access Communication System (TACS). Most systems are second-generation (2G), which is digital in nature. Examples of digital cellular systems include GSM communications, Digital AMPS (DAMPS), and Japanese Digital Cellular (JDC). Although both first- and second-generation systems were designed primarily for speech, they support low bit-rate data services as well.

The spectrum limitations and various technical deficiencies of second-generation systems have led to the development of third-generation (3G) systems. The proliferation of standards and the potential fragmentation problems they could cause in the future led to research on the development and standardization of a global 3G platform.

279

The ITU and regional standards bodies came up with a "family of systems" concept that would be capable of unifying the various technologies at a higher level to provide users with global roaming and voice-data convergence, leading to enhanced services and innovative multimedia applications.

The result of this activity is International Mobile Telecommunications 2000 (IMT-2000), a modular concept that takes full account of the trends toward convergence of fixed and mobile networks and voice and data services. The third-generation platform represents an evolution and extension of current wireless systems and services available today, optimized for high-speed packet data-rate applications, including high-speed wireless Internet services, videoconferencing, and a host of other data related applications.

Vendor compliance with IMT-2000 enables a number of sophisticated applications to be developed. For example, a mobile phone with color display screen and integrated 3G communications module becomes a general-purpose communications and computing device for broadband Internet access, voice, video telephony, and conferencing (see Figure 7.1). These applications can be used by mobile professionals on the road, in the office, or at home. The number of IP networks and applications are growing fast. Most obvious is the Internet, but private IP networks (i.e., intranets and extranets) show similar or even higher rates of growth and usage. With an estimated 300 million Internet users worldwide, there exists tremendous pent-up demand for 3G capabilities.

3G networks will become the most flexible means of broadband access because they allow for mobile, office, and residential use in a wide range of public and non-public networks. Such networks can support both IP and non-IP traffic in a variety of transmission modes including packet (i.e., IP), circuit switched (i.e., PSTN), and virtual circuit (i.e., ATM).

3G networks will be able to benefit from the standards work done by the IETF, which has extended the basic set of IP standards to include QoS capabilities that are essential elements for mobile operation.[1] Developments in new domain name structures also are taking place. These new structures will increase the usability and flexibility of the system,

1. The IETF's work on QoS-enabled IP networks has led to two distinct approaches: the Integrated Services (int-serv) architecture and its signaling protocol RSVP, and the Differentiated Services (diff-serv) architecture.

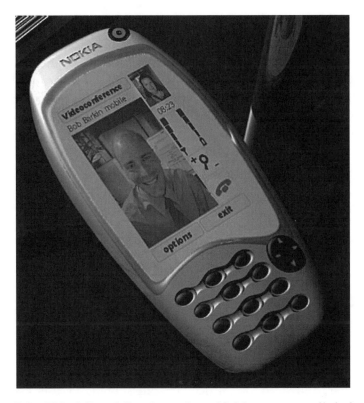

Figure 7.1 This 3G mobile phone from Nokia supports digital mobile multimedia communications, including video telephony. Using the camera eye in the top right corner of the phone, along with the thumbnail screen below it, the local user can line up his or her image so it can appear properly centered on the remote user's phone.

providing unique addressing for each user, independent of terminal, application, or location.

The promises of 3G networks could easily have gone unrealized. The objective of a 3G platform is to harmonize these and other technologies to provide users with seamless global connectivity and enhanced services and features at a reasonable cost. This is complicated by the fact that these and other technologies are not fully compatible with each other, and harmonization necessitates compromise. Until recently, progress toward the 3G platform had been hampered by a number of obstacles. Regulators wanted to protect their own country's market, operators wanted to protect their capital investments in one technology or another, while some vendors sought to protect their patents.

The last of these hurdles was cleared in March 1999 with the agreement of Ericsson and Qualcomm to end their two-year patent and licensing disputes. Now all the industry players and standards organizations can safely focus on the development and implementation of 3G networks. Although 3G networks are in various stages of planning and development all over the world—especially, the United States, Europe, and Japan—IMT-2000 will be phased in over the next decade. It is expected that IMT-2000–compliant 3G networks will start becoming available commercially by the year 2002 and that enhancements will continue to embrace new technologies and concepts to address future applications and market demands.

7.1   Current 2G networks

First-generation analog cellular technology operates in the 800-MHz frequency band.

Cellular systems conforming to the AMPS standard divide the 12.5 MHz of available bandwidth into 30-KHz channels, with one user assigned to each channel. Newer technologies such as TDMA and CDMA can support more users in the same region of the spectrum as well as provide a number of advanced features. In fact, TDMA and CDMA, along with GSM, are the primary technologies contending for acceptance among cellular and personal communications service (PCS) operators worldwide.

7.1.1   Time division multiple access

TDMA increases the call handling capacity of AMPS by dividing the available 30-KHz channels into three time slots, with each user having access to one time slot at regular intervals in round-robin fashion. This time-slot scheme enables several calls to share a single channel without interfering with one another. In using the time slots for short bursts, as many as three calls can be handled on the same 30-KHz channel that supported only one call on an AMPS system.

The current version of TDMA, IS-136, is a revision of the original version of TDMA, IS-54, which was based on the technology available in the 1970s, which had limited system performance. The revised IS-136 standard was published in 1994, and took into account such later developments as digital control channels.

TDMA IS-136 is offered in North America at both the 800-MHz and 1,900-MHz bands. IS-136 TDMA normally co-exists with analog channels on the same network. One advantage of this dual-mode technology is that users can benefit from the broad coverage of established analog networks while IS-136 TDMA coverage grows incrementally, and at the same time take advantage of the more advanced technology of IS-136 TDMA where it is already available.

IS-136 TDMA, also known as DAMPS, specifies the addition of a DQPSK digital control channel (DCCH) to the existing FSK-based control channel used in AMPS and dual-mode (IS-54B) cellular. The FSK control channel is now referred to as the analog control channel (ACC) and the FM voice channel is referred to as the analog voice channel (AVC). Signaling has been added to the AMPS channels to allow a dual-mode mobile terminal to switch between digital and analog channels to find the one that offers the best service in a given area.

At the physical layer, the digital traffic channel (DTC) is very similar to the DCCH in that it is a slotted 48-Kbps DQPSK channel. The DTC was introduced with IS-54B, but has been greatly improved in IS-136 with a new voice coder and enhanced signaling capabilities. As before, the subdivided 30-KHz TDMA traffic channel allows three simultaneous conversations. IS-136 has also added new control messages and expanded previous messages on the DTC, providing new services and supporting transparent extension of cellular services into the PCS band. In addition, IS-136 allows the mobile terminal to go into low power mode (i.e., "sleep" mode) which extends standby battery life.

From the operator's perspective, the TDMA digital channels provide a 3× capacity improvement over AMPS. Among the factors that contribute to this increase in capacity are:

▶ Three TDMA DTCs use the same spectrum as one AMPS voice channel.

▶ TDMA supports a wider range of power levels.

▶ Service selection incorporates information about service capability in addition to signal strength.

▶ Digital communication allows more dense reuse of cellular spectrum.

Service is also improved as the result of the mobile terminal's capability to constantly monitor signal strengths and provide these measurements to the base station when requested.

7.1.2 Code division multiple access

Instead of dividing the available radio spectrum into separate user channels by time slots, CDMA uses spread-spectrum technology to separate users by assigning them digital codes within a much broader spectrum. This results in higher channel capacity and provides each channel protection against interference from other signals. Like TDMA IS-136, CDMA operates in the 1,900-MHz band as well as the 800-MHz band.

With spread-spectrum technology, the information contained in a particular signal is spread over a much greater bandwidth than the original signal. CDMA assigns a unique code to each user and spreads the transmissions of all users in parallel across a wide band of frequencies. Individual conversations are recovered by the respective mobile terminals based on their assigned code. All the other conversations look like random noise and are ignored. In the IS-95 implementation of CDMA, the bandwidth of a single channel is 1.25 MHz, which is claimed to support 10 to 20 times as many users as the equivalent spectrum dedicated to analog cellular.

From its introduction into commercial communications systems in the mid-1980s, the great attraction of CDMA technology has been the promise of extraordinary capacity increases over narrowband cellular technologies. Early performance models suggested that the capacity improvement could be as much as 40 times that of the existing narrowband cellular standards, such as AMPS in North America, NMT in Scandinavia, and TACS in the United Kingdom. However, such performance claims were based on the theory of operation under ideal conditions, rather than on measurements from real-world installations. In reality, cell coverage areas are highly irregular, offered load varies greatly by time-of-day, and system engineering is often subject to uncontrollable influences, such as terrain and zoning laws, which may result in the final installation exhibiting suboptimal performance.

Even though the idealized performance claims of CDMA have not stood up in the real world—operators rarely achieve more than a 12×

performance gain over first-generation analog systems—the technology offers other advantages. For example, the CDMA standard's handoff procedure improves reliability by minimizing the chance of dropped calls.

In all cellular systems, communication between base stations and mobile stations is established by a negotiation upon call origination. In first-generation systems, once communication is established between the base and mobile stations, movement of the mobile station is detected and the service is handed over from one base station to another for the duration of the call. In the CDMA standards, the handoff concept is extended to a multiway simultaneous "soft" handoff, during which the base station exhibiting the strongest signal will take the call.

7.1.3 CDMA versus TDMA

Throughout their histories, there have been conflicting performance claims for TDMA and CDMA. Advocates of CDMA believe it is a dramatically superior digital technology that will carry the cellular industry forward into the next century. The advantages of CDMA are summarized as follows:

- *Call clarity:* CDMA's digital encoding improves call quality by eliminating static and reducing background noise.

- *Network capacity:* CDMA offers an initial call capacity advantage of up to 10 times that of analog technology, reducing the need for additional cell sites, thereby minimizing equipment and maintenance costs.

- *Service provisioning:* CDMA provides greater opportunities to offer economical, personal communications services along with advanced digital features such as voice mail alert and digital messaging, as well as an array of wireless data applications.

- *Privacy:* CDMA's unique coding for each phone conversation makes unauthorized interception of a call extremely difficult, providing users with greater privacy than first-generation systems.

- *Reliability:* With its soft handoff capability, CDMA promises fewer dropped calls than first-generation systems. CDMA handsets are "smart" phones.

▶ *Environmental:* Because existing cell sites can be upgraded to serve 10 times the customers they do today, the need for new cell site towers will significantly diminish.

Not surprisingly, advocates of TDMA claim that CDMA confers no advantage over TDMA-based technology; specifically, the call quality offered by TDMA is indistinguishable from that of CDMA and that the greater channel capacity claimed for CDMA is misleading. The principal arguments for TDMA are summarized as follows:

▶ *Proven technology:* TDMA is a field-proven technology that meets the stated objectives of the industry for future growth.

▶ *Economy:* While analog networks grow by adding cells and equipment, TDMA enables the same equipment to be shared by multiple users. This results in the kind of efficiencies needed for mass-market penetration.

▶ *Evolutionary approach:* TDMA offers an evolutionary growth path, mitigating the capacity claims for CDMA.

Initially, TDMA and CDMA systems provided only basic voice services. The future direction for both types of systems includes enhanced data services and inclusion into the global IMT-2000 family of systems.

7.1.4 GSM

Global System for Mobile Communications (GSM)—formerly known as Groupe Spéciale Mobile, for the group that developed it—was designed from the beginning as an international digital cellular service. GSM's air interface is based on TDMA technology. It was intended from the start that GSM subscribers should be able to roam across national borders and find that their mobile services and features crossed with them.

The European version of GSM operates at the 900-MHz frequency as well as the newer 1,800-MHz frequency. In North America, GSM is used for PCS 1900 service, which is currently available mostly in the Northeast and in California and Nevada. Since PCS 1900 uses the 1,900-MHz frequency, the phones are not interoperable with GSM phones that operate on 900-MHz or 1,800-MHz networks. However, this problem can be overcome with multiband GSM phones that operate in all the frequencies, eliminating the need to rent phones while traveling in different countries.

7.1.4.1 GSM development

In the early 1980s, the market for analog cellular telephone systems was experiencing rapid growth in Europe. Each country developed its own cellular system independently of the others. The uncoordinated development of national mobile communication systems meant that it was not possible for subscribers to use the same portable terminal when they traveled throughout Europe. Not only was the mobile equipment limited to operation within national boundaries, but there was a very limited market for each type of equipment, so cost savings from economies of scale could not be realized. Without a sufficiently large home market with common standards, it would not be possible for any vendor to be competitive in world markets. Furthermore, government officials realized that incompatible communication systems could hinder progress toward achieving their vision of an economically unified Europe.

With these considerations in mind, the then 26-nation Conference of European Posts and Telegraphs (CEPT) formed a study group called the Groupe Spéciale Mobile in 1982 to study and develop a future pan-European mobile communication system. By 1986 it was clear that some of the current analog cellular networks would run out of capacity by the early 1990s. CEPT recommended that two blocks of frequencies in the 900-MHz band be reserved for the new system. The GSM standard specifies the frequency bands of 890 to 915 MHz for the uplink band, and 935 to 960 MHz for the downlink band, with each band divided up into 200-KHz channels.

The mobile communications system envisioned by CEPT had to meet the following performance criteria:

- Offer high quality speech;
- Support international roaming;
- Support hand-held terminals;
- Support a range of new services and facilities;
- Provide spectral efficiency;

▶ Offer compatibility with ISDN;

▶ Offer low terminal and service cost.

In 1989, oversight responsibility for developing GSM specifications was transferred from CEPT to the European Telecommunication Standards Institute (ETSI). ETSI was set up in 1988 to set telecommunications standards for Europe and, in cooperation with other standards organizations, the related fields of broadcasting and office information technology.

ETSI published GSM Phase I specifications in 1990. Commercial service was started in mid–1991. By 1993 there were 36 GSM networks in 22 countries, with 25 additional countries having already selected or begun consideration of GSM. Since then, GSM has been adopted in South Africa, Australia, and many Middle and Far East countries. In North America, GSM is being used to implement PCS. By year-end 1998, 323 GSM networks in 118 countries were serving 138 million subscribers.

7.1.4.2 Network architecture

A GSM network consists of the following architectural elements (see Figure 7.2):

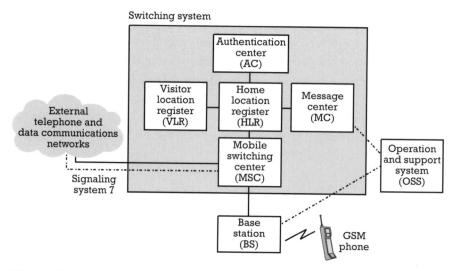

Figure 7.2 Architecture of the GSM switching system.

▶ *Mobile station:* This is the hand-held device carried by the subscriber, which is used for voice and/or data calls. It includes the removable smart card, or subscriber identity module (SIM), which contains subscriber and authentication information.

▶ *Base station subsystem:* This subsystem controls the radio link with the mobile station and monitors call status for handoff purposes.

▶ *Network subsystem:* The main part of this subsystem is the mobile services switching center (MSC), which sets up calls between the mobile and other fixed or mobile network users, and provides management services such as authentication.

▶ *Air interfaces:* The Um interface is a radio link over which the mobile station and the base station subsystem communicate. The A interface is a radio link over which the base station subsystem communicates with the MSC.

Each GSM network also has an operations and maintenance center which oversees the proper operation and setup of the network.

7.1.4.3 Channel derivation and types

Since radio spectrum is a limited resource shared by all users, a method must be devised to divide up the bandwidth among as many users as possible. The method used by GSM is a combination of TDMA/FDMA.

The FDMA part involves the division by frequency of the total 25-MHz bandwidth into 124 carrier frequencies of 200-KHz bandwidth. One or more carrier frequencies are then assigned to each base station. Each of these carrier frequencies is then divided in time, using a TDMA scheme, into eight time slots. One time slot is used for transmission by the mobile device and one for reception. They are separated in time so that the mobile unit does not receive and transmit at the same time.

Within the framework of TDMA two types of channels are provided: traffic channels and control channels. Traffic channels carry voice and data between users, while the control channels carry information that is used by the network for supervision and management. Among the control channels are the following:

▶ *Fast associated control channel (FACCH):* This channel is created by robbing slots from a traffic channel to transmit power control and handoff–signaling messages.

▶ *Broadcast control channel (BCCH):* Continually broadcasts, on the downlink, information including base station identity, frequency allocations, and frequency–hopping sequences.

▶ *Standalone dedicated control channel (SDCCH):* This is used for registration, authentication, call setup, and location updating.

▶ *Common control channel (CCCH):* This is comprised of three control channels used during call origination and call paging.

▶ *Random access channel (RACH):* This is used to request access to the network.

▶ *Paging channel (PCH):* This is used to alert the mobile station of an incoming call.

7.1.4.4 Follow-up phases

Phase II of GSM has been implemented by most operators. This phase offers additional services such as a call charge tally, calling line identity, call waiting, call hold, conference calling, and closed user groups. Phase II+ provides support for multiple service profiles, allowing a user with a single handset to adopt different roles, such as private person and a business executive. Through the introduction of private numbering plans, users can internetwork with other staff within their organization as if they were all on the same PBX. Business users can also access Centrex services, where available, through the GSM network, giving them PBX-type facilities while on the move. Phase II+ also adds internetworking specifications so that users can communicate via DECT, DCS1800, and PCS networks.

7.1.4.5 High-speed circuit switched data (HSCSD)

HSCSD, introduced in 1998, boosts data transmission capacity to 57.6 Kbps, with the potential for higher speeds. To accomplish this, the GSM standard was modified such that 14.4-Kbps channel coding replaced the original 9.6-Kbps coding used to support mobile telephony. The four channels of 14.4 Kbps each are combined into a single channel of 57.6 Kbps—almost the speed of a fixed ISDN channel of 64 Kbps. The extra bandwidth allows GSM phones and mobile terminals to handle multimedia applications.

With this much bandwidth, users can access the Web and download pages with high graphics content in seconds. They can also take advantage of the higher speeds for accessing in-house LANs via corporate intranets. Because HSCSD has an integral bandwidth on demand capability, it will not matter what speed users need for any particular activity—the service will provide whatever speed they require, up to the maximum rate of 57.6 Kbps. With end-to-end compression between a client and a server, even higher speeds can be achieved.

7.1.4.6 General packet radio services (GPRS)

GPRS for GSM became available in 1999. Like HSCSD, GPRS provides higher-speed data services for mobile users. But as a packet-switching technology, GPRS is better suited to the highly bursty nature of most data applications than HSCSD, making it ideal for e-mail and database access services, for example, where users do not want to pay high call charges for short transmissions. GPRS also permits the user to receive voice calls simultaneously when sending or receiving data calls. Messages are delivered direct to the user's phone, even without the need for a full end-to-end connection. When they switch on their phones, users receive a notification that they have a message waiting. The user can then choose to have the messages downloaded immediately or saved for later.

GPRS also offers faster call setup than HSCSD and more efficient connectivity with networks that use the IP protocol, including corporate intranets and LANs, as well as the Internet. Through various combinations of TDMA time slots, GPRS can handle all types of transmission from slow-speed short messages to the higher speeds needed for browsing Web pages. GPRS provides peak packet data rates of over 100 Kbps. The maximum rate is 171.2 Kbps over eight 21.4-Kbps channels.

With circuit switched transmission, a channel is allocated to a single user for the duration of a call. With a packet switched network like GPRS, the available radio spectrum is shared by all users in a cell. The spectrum is used only when the user has something to send. When there is no data to be transmitted, the spectrum is free to be used for another call. Thus if the data is bursty in nature—as is LAN data—the network resources can be balanced more efficiently, because the operator can use the gaps in transmission to handle other calls.

As with HSCSD, GPRS works within the existing GSM infrastructure, allowing GPRS to be introduced quickly. GPRS provides two key benefits for providers and users of data services. First, it reduces the costs of providing connectivity, as GPRS uses radio and network resources more efficiently. With GPRS the applications only occupy the network when data is actually being transferred. Compared with today's circuit switched methodology the costs of providing connectivity are reduced. Second, it provides transparent IP support. By tunneling the IP protocol transparently from the mobile terminal to the Internet or intranet and giving the terminal the same status as IP hosts on a LAN, GPRS enables mobile access to corporate intranets—without the traditional remote modem. As demand develops, higher data rates of up to 2 Mbps in packet mode will be supported in the future through the use of ATM technology.

7.1.4.7   Enhanced data for global evolution (EDGE)

The last incremental development before the arrival of third generation will be the introduction of EDGE, which uses an alternative modulation scheme to provide data rates of 300 Kbps and beyond using the standard GSM 200 KHz carrier. EDGE will support both circuit and packet switched data and offers a solution for GSM operators who do not win third-generation licenses but who wish to offer competitive wideband services.

The enhanced modulation scheme will automatically adapt to the current radio environment to offer the highest data rates for users in good propagation conditions close to the base station sites, while ensuring wider area coverage at lower data speeds. EDGE can also be implemented over TDMA IS-136 networks.

7.1.4.8   GSM and IP

GSM and IP are easily combined for value-added applications. Nokia's GSM Intranet Office, for example, combines GSM and IP telephony in the same network. This solution can substantially cut call costs for larger corporations by using the existing LAN (see Figure 7.3).

Voice calls can be made via the company's intranet to various mobile phones, fixed phones, or PCs throughout an office by using any standard GSM mobile phone. In addition, the same mobile phones can be used outside the office premises as well, where calls are routed as normal using the operator's GSM network.

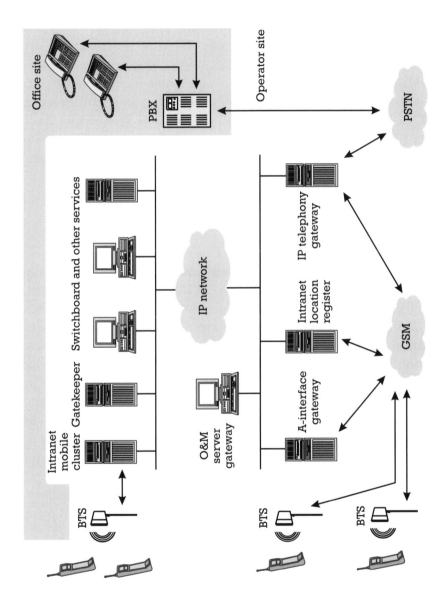

Figure 7.3 Nokia's GSM Intranet Office architecture.

By connecting a dedicated BTS—specifically, the Nokia InSite Base Station—to the corporate LAN, all intra-office phone calls can be routed locally through the corporate intranet. The InSite Base Station is a single-transceiver base station optimized for indoor pico-cellular applications. By implementing the mobile phone as the preferred phone for employees, whether in the office or traveling, a corporation can bring down its telecom costs and increase its productivity and efficiency.

GSM Intranet Office also includes a management system to monitor the performance of the IP and GSM networks. It stores information about all network element configurations and software versions. Data from all the network elements can be forwarded to other management systems, such as the Nokia NMS 2000, for further processing and analysis.

7.2   Global 3G initiative

As noted, IMT-2000 will result in wireless access to the global telecommunication infrastructure through both satellite and terrestrial systems, serving fixed and mobile users in public and private networks. This initiative is being developed as a "family of systems" framework, defined as a federation of systems providing advanced service capabilities in a global roaming offering. The initiative aims to facilitate the evolution from today's national and regional second-generation (2G) systems that are incompatible with one another toward third-generation (3G) systems that will provide users with genuine global service capabilities and interoperability soon after the year 2000. The role of the ITU is to provide direction to and coordinate the many related technological developments in this area to assist the convergence of competing national and regional wireless access technologies. Toward these ends, the ITU considered over a dozen national and regional proposals in an effort to select the key characteristics of the IMT-2000 set of radio interfaces.

7.2.1   Standards development

In 1992, the World Radio Conference (WRC) identified the frequency bands 1,885–2,025 MHz and 2,110–2,200 MHz for future IMT-2000 systems. Of these, the bands 1,980 –2,010 MHz and 2,170–2,200 MHz were intended for the satellite part of these future systems.

With the agreement on frequency bands, along with other important standards in place, work at the ITU focused next on selecting the all-important air interface technology for the system, known as the radio transmission technology (RTT). The need for additional spectrum to cope with the increasing amount of broadband and increasingly interactive traffic of third-generation systems will be addressed by the World Radio Conference in the future.

For the radio interface technology, the ITU considered 15 submissions from organizations and regional bodies around the world. These proposals were examined by special independent evaluation groups, which submitted their final evaluation reports to the ITU in September 1998. The final selection of key characteristics for the IMT-2000 radio interfaces occurred in March 1999. This will lead to the development of more detailed ITU specifications for IMT-2000.

The decision of the ITU was to provide essentially a single flexible standard with a choice of multiple access methods, which include CDMA, TDMA, and combined TDMA/CDMA—all potentially in combination with space division multiple access (SDMA)—to meet the many different mobile operational environments around the world. Although second-generation mobile systems involve both TDMA and CDMA technologies, very little use is currently being made of SDMA. However, the ITU expects the advent of adaptive antenna technology linked to systems designed to optimize performance in the space dimension to significantly enhance the performance of future systems.

The IMT-2000 key characteristics are organized, for both the terrestrial and satellite components, into the RF part (front end), where impacts are primarily on the hardware part of the mobile terminal, and the base-band part which is largely defined in software. In addition to RF and baseband, the satellite key characteristics also cover the architecture and the system aspects. According to the ITU, the use of common components for the RF part of the terminals, together with flexible capabilities which are primarily software defined in baseband processing, should provide the mobile terminal functionality to cover the various radio interfaces needed in the twenty-first century as well as provide economies of scale in their production.

The ITU noted that there are already many multi-mode/multi-band mobile units appearing on the market to meet the evolution needs of today's systems, and by early next century, there should be negligible impact in areas such as power consumption, size, or cost due to the flexibility defined within the IMT-2000 standard.

The key characteristics of the radio transmission technology by themselves do not constitute an implementable specification but establish the major features and design parameters that will make it possible to develop the detailed specifications.

However, due to the constraints on satellite system design and deployment and because it was considered that there was little to be gained at this time from harmonizing any of the satellite proposals since they were already global, several satellite radio interfaces were included in the March 1999 agreement. Commonalities among elements of the satellite and terrestrial radio interfaces, however, were sought and terrestrial/satellite commonality can be expected to further increase when the second phase of the IMT-2000 satellite component is introduced in the early part of the twenty-first century.

The flexible approach to IMT-2000 implementation provides choice among multiple access methods within a single standard that will address the needs of the worldwide wireless community. Specifically, this approach allows operators to select those radio interfaces that will best address their specific regulatory, financial, and customer needs, while minimizing the impact of this flexibility on end users.

7.2.2 Goals of IMT-2000

Under the IMT-2000 model, mobile telephony will no longer be based on a range of market-specific products, but will be founded on common standardized flexible platforms, which will meet the basic needs of major public, private, fixed, and mobile markets around the world. This approach should result in a longer product life cycle for core network and transmission components, and offer increased flexibility and cost effectiveness for network operators, service providers, and manufacturers.

In developing the family of systems that would be capable of meeting the communications demands of mobile users in the year 2000 and beyond, the architects of IMT-2000 identified several key issues that would have to be addressed to ensure the success of the third-generation of mobile systems.

7.2.2.1 High-speed

Any new system must be able to support high-speed broadband services, such as fast Internet access or multimedia-type applications. Demand for such services is already growing fast and—by some industry projections—the market for broadband services could be worth up to $10 billion by 2010. Users will expect to be able to access their favorite services just as easily from their mobile equipment as they can from their wireline equipment.

7.2.2.2 Flexibility

The next generation of integrated systems must be as flexible as possible, supporting new kinds of services such as universal personal numbering and satellite telephony, while providing for seamless roaming to and from IMT-2000 compatible terrestrial wireless networks. These and other features will greatly extend the reach of mobile systems, benefiting consumers and operators alike.

7.2.2.3 Affordability

The system must be as affordable as today's mobile communications services, if not more so. The ITU recognizes that the economies of scale achievable with a single global standard would have the benefit of driving down the price to users. While important for all consumers, affordability is vital to extending the penetration of telephony in developing countries. For third-generation equipment to be taken up quickly by consumers, it must deliver at least the same or better service than current systems, and it must be cheap. Even though economies of scale will inevitably bring prices down once sufficient volumes are achieved, if the systems are more expensive and do not initially offer much greater functionality, consumers will not buy.

7.2.2.4 Compatibility

Any new generation system has to offer an effective evolutionary path for existing networks. While the advent of digital systems in the early 1990s often prompted the shutting down of first-generation analog networks, the enormous investments which have been made in developing the world's second-generation cellular networks over the last decade makes a similar scenario for adoption of third-generation systems completely untenable.

7.2.2.5 Differentiation

In coordinating the design of the IMT-2000 framework, the ITU is also aware of the need to preserve a competitive domain for manufacturers in order to foster incentive and stimulate innovation—mindful that industrial organizations need to have the freedom to compete when it comes to technology. Accordingly, the aim of IMT-2000 standards is not to stifle the evolution of better technologies or innovative approaches, but to accommodate them.

7.2.3 Universal mobile telecommunications system

One of the major new 3G mobile systems being developed within the ITU's IMT-2000 framework is the Universal Mobile Telecommunications System (UMTS), which has been standardized by ETSI. UMTS makes use of UTRA (UMTS terrestrial radio access) as the basis for a global terrestrial radio access network. Europe and Japan are implementing UTRA in the paired bands 1,920–1,980 MHz and 2,110–2,170 MHz. Europe has also decided to implement UTRA in the unpaired bands 1,900–1,920 MHz and 2,010–2,025 MHz.

UMTS combines key elements of TDMA—about 80% of today's digital mobile market is TDMA-based—and CDMA technologies with an integrated satellite component to deliver wideband multimedia capabilities over mobile communications networks. The transmission rate capability of UTRA will provide at least 144 Kbps for full mobility applications in all environments, 384 Kbps for limited mobility applications in the macro and micro cellular environments, and 2.048 Mbps for low mobility applications particularly in the micro and pico cellular environments. The 2.048-Mbps rate may also be available for short range or packet applications in the macro cellular environment.

Because the UMTS incorporates the best elements of TDMA and CDMA, this 3G system provides a glimpse of how future wireless networks will be deployed and what possible services might be offered within the IMT-2000 "family of systems."

7.2.3.1 UMTS objectives

UMTS makes possible a wide variety of mobile services ranging from messaging to speech, data, and video communications; Internet and intranet access; and high bit-rate communication up to 2 Mbps. As such, UMTS is expected to take mobile communications well beyond the current range of wireline and wireless telephony, providing a platform that will be ready for implementation and operation in the year 2002.

UMTS is intended to provide globally available, personalized, and high-quality mobile communication services. Its objectives include:

▶ Integration of residential, office and cellular services into a single system, requiring one user terminal;

▶ Speech and service quality at least comparable to current fixed networks;

▶ Service capability up to multimedia;

▶ Separation of service provisioning and network operation;

▶ Number portability independent of network or service provider;

▶ The capacity and capability to serve over 50 percent of the population;

▶ Seamless and global radio coverage and radio bearer capabilities up to 144 Kbps and further to 2 Mbps;

▶ Radio resource flexibility to allow for competition within a frequency band.

7.2.3.2 Description

UMTS separates the roles of service provider, network operator, subscriber, and user. This separation of roles makes possible innovative new services, without requiring additional network investment from service providers. Each UMTS user has a unique network-independent identification number and several users and terminals can be associated with the same subscription. This enables one subscription and bill per household to include all members of the family as users with their own terminals.

This would give children access to various communication services under their parents' account. This application would also be attractive for businesses, which require cost-efficient system operation—from subscriber/user management down to radio system—as well as adequate subscriber control over the user services.

UMTS supports the creation of a flexible service rather than standardizing the implementations of services in detail. The provision of services is left to service providers and network operators to decide according to the market demand. The subscriber—or the user when authorized by the subscriber—selects services stored in individual user service profiles that are set up either with the subscription or interactively with the terminal.

UMTS supports its services with networking, broadcasting, directory, localization, and other system facilities, giving UMTS a clear competitive edge over mobile speech and restricted data services of earlier generation networks. Being adept at providing new services, UMTS is also competitive in the cost of speech services and as a platform for new applications.

UMTS offers a high-quality radio connection that is capable of supporting several alternative speech codecs at 2 Kbps to 64 Kbps, as well as image, video, and data codecs. Also supported are advanced data protocols covering a large portion of ISDN services. The concept includes variable and high bit-rates up to 2 Mbps.

7.2.3.3 Functional model

The UMTS functional model relies on distributed databases and processing, leaving room for service innovations without the need to alter implemented UMTS networks or existing UMTS terminals. This service-oriented model provides three main functions: management and operation of services, mobility and connection control, and network management.

▶ *Management and operation of services:* A service data function (SDF) handles storage and access to service related data. A service control function (SCF) contains overall service and mobility control logic and service related data processing. A service switching function (SSF) invokes service logic—to request routing information, for example. A call control function (CCF) analyzes and processes service requests in addition to establishing, maintaining, and releasing calls.

‣ *Mobility and connection control:* Drawing on the contents of distributed databases, UMTS provides for the real-time matching of user service profiles to the available network services, radio capabilities, and terminal functions. This function handles mobile subscriber registration, authentication, location updating, hand-offs, and call routing to a roaming subscriber.

‣ *Network management:* Under UMTS, the administration and processing of subscriber data, maintenance of the network, and charging, billing, and traffic statistics will remain within the traditional TMN.

TMN consists of a series of interrelated national and international standards and agreements, which provide for the surveillance and control of telecommunications service provider networks on a worldwide scale. The result is the ability to achieve higher service quality, reduced costs, and faster product integration. TMN is also applicable in wireless communications, cable television networks, private overlay networks, and other large-scale, high-bandwidth communications networks. With regard to UTMS (and other 3G wireless networks) TMN will be enhanced to accommodate new requirements. In areas such as service profile management, routing, radio resource management between UMTS services, networks, and terminal capabilities, new TMN elements will be developed.

7.2.3.4 Bearer services

Under UMTS, four kinds of bearer services are provided to support virtually any current and future application:

‣ *Class A:* This bearer service offers constant bit-rate (CBR) connections for isochronous (real-time) speech transmission. This service provides a steady supply of bandwidth to ensure the highest quality speech.

‣ *Class B:* This bearer service offers variable bit-rate connections, which are suited for bursty traffic, such as transaction processing applications.

‣ *Class C:* This bearer service is a connection-oriented packet protocol, which can be used to support time-sensitive legacy data applications such as those based on IBM's SNA.

▶ *Class D:* This is a connectionless packet bearer service. This is suitable for accessing data on the public Internet or private intranets.

7.2.3.5 Technology approaches

In developing the UMTS standard, there had been ongoing disagreement within the UMTS Forum about whether to use TD-CDMA (time division) or W-CDMA (wideband) for the radio interface portion of the network.

TD-CDMA uses CDMA signal-spreading techniques to enhance the capacity offered by conventional TDMA technology. Digitized voice and data would be transmitted on a 1.6 MHz-wide channel using time-segmented TDMA technology. Each time slot of the TDMA channel would be individually coded using CDMA technology, thus supporting multiple users per time slot. The proposal establishes an economical and smooth network migration for existing GSM customers to the next-generation cellular standard. At the same time, the TD-CDMA solution allows CDMA technology to be integrated into the TDMA-based GSM structure worldwide, enabling GSM operators to compete for wideband multimedia services, while protecting their current and future investments.

W-CDMA, on the other hand, not only has the advantage of providing high capacity, it is a widely deployed cellular technology. Proponents of W-CDMA had insisted that this be the air interface standard for UMTS.

In January 1998, members of the UMTS Forum, which coordinates standards development, agreed to combine key elements of both TD-CDMA and W-CDMA cellular technologies into a new solution, called UTRA. Technically, the solution is as follows: In the paired band (FDD) of UMTS, the system adopts the radio access technique advocated by the W-CDMA group. In the unpaired band (TDD) the UMTS system adopts the radio access technique advocated by the TD-CDMA group.

UTRA offers a competitive continuation for GSM evolution to UMTS and will position UMTS as a leading member of the IMT-2000 family of systems recommendations being finalized in the ITU.

Among the leading telecommunications manufacturers participating in the development of UMTS are Alcatel, Bosch, Ericsson, Italtel, Motorola, Nokia, Nortel, Siemens, and Sony.

7.2.3.6 Applications

UMTS comprises a new air interface and new radio components. The aim is to combine these in a modular way with new network components and components from pre-UMTS fixed and mobile networks. This approach will allow new entrants to establish UMTS networks and enable existing operators a smooth migration by reusing parts of their existing infrastructure to the extent possible.

For the user, UMTS will provide adaptive multimode/multiband terminals or terminals with a flexible air interface to enable global roaming across locations and with second generation systems. Software download to terminals may offer additional flexibility.

By harnessing the best in cellular, terrestrial, and satellite wideband technology, UMTS will guarantee access, from simple voice telephony to high speed, high-quality multimedia services. It will deliver information directly to users and provide them with access to new and innovative services and applications. It will offer mobile personalized communications to the mass market regardless of location, network, or terminal used.

7.2.3.7 Implementation issues

Since UMTS service is not yet in operation, users have no recourse but to stay with their current mode of communication until UMTS services start coming online in 2002. Current users of digital cellular technologies—whether based on TDMA or CDMA—will be able to readily take advantage of the capabilities UMTS provides because it is a platform into which other technologies can be plugged.

If changes to existing user terminals are necessary, the onus is on the network operators to make these changes as painless as possible for existing users. Sometimes this will entail only a software upgrade that can be implemented over the airwaves by dialing a special number. For terminals requiring changes to hardware, the network provider will have every incentive to offer an attractive trade-up policy.

For users contemplating a mobile service now, GSM (or a derivative such as PCS 1900 in the United States) would be the best choice, since GSM terminals can be upgraded via software download from the network service provider. If a digital service is not available and mobile service is a requirement right now, there is no harm in going with a locally available alternative service capable of meeting immediate communications needs. UMTS still is several years away from full implementation in Europe and it will have to prove itself there before it can successfully migrate around the world.

7.2.3.8 Market dynamics

The markets for mobility and for fixed multimedia are already large and growing rapidly. Customers will want to combine mobility with multimedia, resulting in higher demand for bandwidth and creating a significant shift towards new data services. For Europe alone, this new market is estimated to be as large in 2005 as the entire mobile market is today.

7.2.3.9 Regulatory environment

For the foreseeable future, there are still key issues to be worked out that will determine the future success of UMTS. To begin with, the regulatory environment for UMTS, including the licensing regime in different countries, must be made more harmonious and consistent so that competition will be encouraged.

Regulators in the EU member countries must commit to allowing the interconnection of UMTS networks with other telecommunication networks. Further, regulators must address restrictions in related markets such as entertainment and broadcasting, so that services in these environments can be allowed to converge. Developers must agree on ways to ensure interoperability of applications on a minimum level, while allowing a degree of competitive differentiation. Network operators must agree to share sites, infrastructure, and facilities; combine services; and operate throughout the value chain.

Standardization is, and will remain, a key factor in providing quality services at an affordable cost and enable roaming between systems. Continued close cooperation between operators, manufacturers and regulators in the UMTS standards process is essential for the success of UMTS. Creating a dynamic home market in the early years of UMTS will also go a long way toward promoting the competitiveness of European industry in world markets.

7.3 U.S. participation

U.S. proposals submitted to the ITU for consideration as the RTT in the IMT-2000 framework included wideband versions of CDMA of which there are three competing standards in North America: wideband cdma-One, WIMS W-CDMA, and WCDMA/NA. All three have been developed from second-generation digital wireless technologies, and are evolving to third-generation technologies that will best fit their networks. However, early on WIMS W-CDMA and WCDMA/NA were merged into a single proposed standard and, along with wideband cdmaOne, have been submitted to the ITU for inclusion into its IMT-2000 "family of systems" concept for globally interconnected and interoperable 3G networks. A U.S. proposal for a TDMA solution for the RTT was also submitted by the Universal Wireless Communications Consortium (UWCC).

7.3.1 CDMA proposals

As noted, initially there were three competing WCDMA standards in North America: wideband cdmaOne, WIMS (wireless multimedia and messaging services), W-CDMA, and W-CDMA/NA (W-CDMA North America). Most wireless operators have chosen one of these in building out and enhancing their networks. Competition between these three viable standards has brought innovation in technologies, features, and services, as well as low prices for consumers.

Wideband cdmaOne technology was submitted to the ITU by the CDMA Development Group (CDG) as cdmaOne-2000. The WIMS W-CDMA technology was submitted to the ITU by AT&T Wireless, Hughes Network Systems, and InterDigital Communications Corporation, among others. The North American GSM Alliance, a group of 12 U.S. and one Canadian digital wireless PCS carriers, submitted the WCDMA/NA technology to the ITU.

There had been talk of combining all of these technologies into a single, unified ITU submission. In September 1998, the W-CDMA/NA and WIMS W-CDMA proposals were merged into what was referred to as the enhanced W-CDMA/NA proposal. This aligned proposal offers value-added data capabilities, such as enabling packet data to be delivered to up to ten times as many users.

Supporters of the enhanced W-CDMA/NA declined to unify their proposal with wideband cdmaOne, claiming that the necessary changes would have caused a significant degradation in system capacity and performance, affected additional capabilities, and probably raised the price of services to customers.

7.3.1.1   Wideband cdmaOne

Under the proposed standard of the CDMA Development Group (CDG), wideband cdmaOne will use a CDMA air-interface based on the existing TIA/EIA-95-B standard to provide wireline-quality voice service and high-speed data services, ranging from 144 Kbps for mobile users to 2 Mbps for stationary users. It will fully support both packet- and circuit-switched communications such as Internet browsing and landline telephone services, respectively.

Support for wideband cdmaOne is not limited to North America. Major wireless carriers in Japan also support wideband cdmaOne. Korean carriers and manufacturers have also contributed to the development of wideband cdmaOne.

CDMA (TIA/EIA-95-B) is a spread-spectrum approach to digital transmission. Each conversation is digitized and then tagged with a code. The mobile phone is then instructed to decipher only the particular code it is tuned to, enabling it to pluck the right conversation off the air. This is similar in process to an English-speaking person being able to pick out the conversation of the only English-speaking person in a room of foreigners.

The spread spectrum entails tagging groups of bits from digitized speech with a unique code that is associated with a particular call on the network. Groups of bits from one cellular call are pseudo-randomly combined in a multiplexing process with those from other calls and transmitted across a broader band of spectrum—1.25MHz—and then reassembled in the right order to complete the conversation.

Spreading is achieved by applying a pseudo-noise code (also called a chip code) to the data bits, which increases the overall data rate and expands the amount of bandwidth utilized. Since each call has been tagged with a unique code, a specific conversation can be identified when the spread spectrum signal is recombined (i.e., despread) at the receiving end. However, for a given channel within a cell, the aggregate signal for all other conversations will be perceived as interference.

For the spread spectrum system to operate properly, the receiver must acquire the correct phase position of the incoming signal. Acquisition is accomplished by a search of as many phase positions as necessary until one is found which results in a large correlation between the phase of the incoming signal and the phase of the locally generated spreading sequence at the receiver. The receiver also must continually track that phase position so that lock loss will not occur. The processes of acquisition and tracking are performed by the synchronization subsystem of the receiver. These functions work together to ensure that incoming spread signals can be properly "despread."

Advanced services require more capacity, robustness, and flexibility than narrowband technologies can provide. CDG and its members have already completed specification of the 64-Kbps data rate service, and commercial deployment of this service is expected to begin soon. The 64-Kbps data capability will provide high-speed Internet access in a mobile environment, a capability that cannot be matched by other narrowband digital technologies, including CDMA.

The CDG believes that mobile data rates of up to 144 Kbps and fixed peak rates beyond 1.5 Mbps are possible without degrading the system's voice transmission capabilities or requiring additional spectrum. In other words, cdmaOne is expected to double capacity and provide a 1.5-Mbps data rate capability—all within the existing 1.2- MHz channel structure. At the same time, cdmaOne supports existing TIA/EIA-95-B services, including speech coders, packet data services, circuit data services, fax services, SMSs, and over-the-air service activation and provisioning.

cdmaOne uses a physical layer channel structure that shares much of the fundamental/supplemental channel structure from TIA/EIA-95-B. This design provides for simultaneous voice/data structure and procedures that are upwardly compatible with TIA/EIA-95-B.

The cdmaOne extends support for multiple simultaneous services far beyond the services in TIA/EIA-95-B by providing much higher data rates and a sophisticated multimedia QoS control capability to support multiple voice/packet data/circuit data connections with differing performance requirements.

The cdmaOne system MAC layer provides extensive enhancements to negotiate multimedia connections, operates multiple concurrent services, and manages QoS tradeoffs between multiple active services in an efficient, structured, and extensible manner. Delivery of these multiple concurrent data streams over the radio interface is accomplished by the cdmaOne Layer 1 (physical layer).

Layer 1 supports multiple supplemental channels that can be operated with varying QoS characteristics tailored to the individual service's requirements. For example, one channel can carry circuit data with low BER and low latency transmission requirements, while another channel carries packet data that can tolerate a much higher BER and relatively unconstrained latency.

The cdmaOne Physical Layer also supports a DCCH that can be utilized in a number of flexible configurations to provide for the highest level of independence between competing services (e.g., voice and data), while maintaining the highest level of performance.

High-speed data service negotiation procedures are extended far beyond TIA/EIA-95-B to include ATM/B-ISDN QoS parameters, including:

> Data rate requirements (e.g., CBR, ABR, VBR);

> Data rate symmetry/asymmetry requirements;

> Tolerable delay/latency characteristics.

The QoS negotiation procedures provide a service that is functionally equivalent to B-ISDN Q.2931 procedures. This provides for ease of implementing transparent multimedia call service via a gateway to ATM/B-ISDN networks (e.g., landline ATM networks).

Additionally, cdmaOne packet data services (i.e., IP) support QoS negotiation upper layer protocols such as the RSVPs that perform end to end service negotiation procedures to provide multimedia call support.

The cdmaOne system also provides extensive capabilities to support highly efficient and cost effective WLL implementations. Optimal radio interface capacity along with high single-user throughput is provided. Delay and cell/sector capacity can be traded off to optimize for the desired environment.

Optimized packet data service modes of operation provide Internet service that is highly competitive with wireline and other wireless environments. Voice quality is equal to or better than toll quality and high system capacity provides for a highly competitive replacement for landline voice systems. Improved capacity and single-user throughput using the same cell footprint as an existing TIA/EIA-95-B system permit the integration of WLL services with general cellular high-mobility traffic using the same infrastructure.

7.3.1.2 Enhanced W-CDMA/NA

As noted, Enhanced W-CDMA/NA represents a merger of WIMS (wireless multimedia and messaging services), W-CDMA, and W-CDMA/NA

(W-CDMA North America). The merger was relatively painless because WIMS W-CDMA had already incorporated four key WCDMA/NA elements into its technology:

▶ Chip rate of 4.096 Mcps (millicycles per second), which has since been reduced to 3.840 Mcps;

▶ Frame length of 10 ms;

▶ Support of the Adaptive Multi-rate Vocoder;

▶ Support of asynchronous base stations.

This merger offers a technology that allows the acquisition of packet data within 10 ms, which is much faster than other 3G technologies submitted to the ITU for consideration.

The enhanced W-CDMA/NA proposal incorporated two key WIMS elements into its technology: use of multiple parallel orthogonal codes for higher data rates, and a pilot/header structure that enables very rapid packet acquisition and release. This improves data performance and throughput to address the growing marketplace and the demanding requirements of multimedia and Internet based services.

Citing the significant commonality of key technical parameters of chip rate, frame length, asynchronous base station operation, and vocoders, among others, the GSM Alliance endorsed the merging of the two technologies. The GSM Alliance supports multiple 3G standards in the United States and abroad.

As noted, GSM was designed from the beginning as an international digital cellular service. Since GSM is based in digital technology, it allows synchronous and asynchronous data to be transported as a bearer service to or from an ISDN terminal. Data can use either the transparent service, which has a fixed delay but no guarantee of data integrity, or a nontransparent service, which guarantees data integrity through an automatic repeat request (ARQ) mechanism, but with variable delay.

Supplementary services are provided on top of teleservices or bearer services, and include such features as caller identification, call forwarding, call waiting, and multiparty conversations. There is also a lockout feature, which prevents the dialing of certain types of calls, such as international calls.

As noted, GSM uses a combination of TDMA/FDMA to divide up the bandwidth among as many users as possible. The FDMA part involves the division by frequency of the total 25-MHz bandwidth into 124 carrier frequencies of 200-KHz bandwidth. One or more carrier frequencies are then assigned to each base station. Each of these carrier frequencies is then divided in time, using a TDMA scheme, into eight time slots. One time slot is used for transmission by the mobile device and one for reception. They are separated in time so that the mobile unit does not receive and transmit at the same time.

Within the framework of TDMA two types of channels are provided: traffic channels and control channels. Traffic channels carry voice and data between users, while the control channels carry information that is used by the network for supervision and management.

The third generation of GSM will be an evolution and extension of current GSM systems and services available today, optimized for high-speed packet data-rate applications, including high-speed wireless Internet services, video on demand, and other data-related applications. Specifically, third-generation GSM adds the use of CDMA multiplexing at the radio interface level. This is one of the reasons why the GSM Alliance supported the merger of WIMS W-CDMA and W-CDMA/NA.

7.3.2 TDMA proposal

The UWCC submitted a TDMA IS-136 solution for the RTT portion of IMT-2000. Its proponents say that UWC-136 meets the high-speed data application requirements of IMT-2000, enabling telecommunications carriers and vendors to provide these capabilities with minimal costs and maximum benefits. The UWC-136 proposal was developed by the Global TDMA Forum, a technical forum of the UWCC.

UWC-136 is a 100% pure TDMA digital solution that provides an evolutionary path to the next generation from IS-136 to IS-136+ to IS-136HS (the high-speed component of UWC-136). IS-136+ will provide high-fidelity voice services and higher rate packet-data services, up to 64 Kbps using the existing 30-KHz bandwidth. IS-136HS provides user-data rates up to 384 Kbps for wide area coverage in all environments and more than 2 Mbps for in-building coverage.

UWC-136 is a market-driven solution for TDMA IS-136 that enables carriers to implement high-speed data, multimedia, and other applications incrementally to meet specific market demands. Carriers can retain infrastructure investments and implement data applications while providing quality service delivery to customers—with little impact on the networks and existing spectrum.

In addition to the flexibility of just-in-time build-out, the advantages of UWC-136 include in-building coverage, increased capacity, wireline voice quality, and tightly integrated voice and data services. IS-136+, combined with adaptive channel allocation, advanced modulation, and vocoder enhancements, increases capacity to 10× AMPs while maintaining high voice quality.

The innate hierarchical cell structures of UWC-136 will enable carriers to provide in-building coverage and seamless voice and data services delivery for end-users. UWC-136 supports a common physical level of compatibility across the major TDMA-based technologies for delivery of wideband services. This could lead to global convergence.

UWC-136 also creates a common technology base with GSM for the provision of wideband services and forms a foundation for the development of a world phone that provides wireless multimedia applications on both D-AMPS IS-136 and GSM networks in all bands.

7.4 Analysis

Broadband wireless technologies are being developed not only with an eye toward building third-generation networks capable of providing enhanced multimedia services to North American customers, but with the goal of meeting the requirements established by the ITU for its global IMT-2000 initiative. Both the wideband cdmaOne and enhanced W-CDMA/NA technology proposals have been developed to meet these twin objectives. When formally accepted by the ITU, wideband cdmaOne and enhanced W-CDMA/NA will be members of IMT-2000's family of systems. This means they will globally interconnect and interoperate with other systems in the family.

However, there will be more than one global standard for 3G wireless networks. In a world driven by consumer demand, there will continue to be more than one digital standard, just as there is today. This means that there will be an RTT solution that includes TDMA as well as CDMA.

One early adopter of cdmaOne-2000 is Bell Atlantic Mobile. The company is engaged in a phased introduction of new high-speed wireless data capabilities over its existing cdmaOne network. Bell Atlantic Mobile's

5.9 million current customers will not have to surrender their current cdmaOne handsets when cdma-2000 services become available. Instead, the technology enhancement will be performed with minimal upgrades to the base stations, which are supplied by Lucent Technologies.

Even though third-generation broadband wireless networks have only just begun to be implemented in the United States, there are many technical issues to be resolved before the full range of IMT-2000 capabilities can become available in North America.

As discussed earlier, most users of second-generation network services will not have to worry about migration issues because the onus is on the network operators to make these changes as painless as possible for existing users. Sometimes this will entail only a software upgrade that can be implemented over the airwaves by dialing a special number. For legacy terminals requiring changes to hardware, the network provider will have every incentive to offer an attractive trade-up policy.

7.4.1 Business usage

With national economies becoming increasingly global in nature, companies are establishing a local presence in the countries in which they do business. As a consequence, there is an urgent need for interoperable telecommunications services that offer a consistent set of features, regardless of national boundary, as well as a true roaming capability. The IMT-2000 initiative is an attempt to address this situation.

With regard to roaming, for example, IMT-2000 systems will permit a subscriber to roam from a private network, into a picocellular/microcellular public network, then into a wide area macrocellular network (which may actually be a second generation network), and then to a satellite mobile network with minimal break in communication.

Each system in the "family" also is expected to provide inherent IP support to deliver interactive multimedia services as well as other new wideband applications such as video telephony and videoconferencing.

As the demand for user data rates increases in the long term, systems will be developed to support even higher data rates—perhaps one or two orders of magnitude greater, if appropriate spectrum is allocated in the future by the WRC.

With UMTS, for example, later development phases will include a convergence with wireless LAN technologies (microwave or infrared) to provide data rates of as much as 155 Mbps in indoor environments. UMTS is also being designed to offer data rate on demand, where the network reacts flexibly to the user's demands, his or her profile, and the current status of the network. The use of packet-oriented transport protocols such as IP are being studied for UMTS to enhance these capabilities. Together, the combination of packet data and data rate on demand will remove technical barriers for the user and make operation of the system much cheaper. There will be no worries about how and when to connect to the network because the user's device will connect automatically upon power-up.

The ultimate goal is that all networks, signaling, connection, registration, and any other technology should be invisible to the user, so that mobile multimedia services are simple, user-friendly, and effective.

7.4.2 Benefits

Current wireless or mobile systems, despite their evolution, are still constrained in terms of the data rate they can offer and their flexibility to manipulate complex, yet user friendly multimedia services. In comparison with today's digital cellular services, users would have the following benefits with IMT-2000 systems:

▶ Global roaming capabilities that go beyond current GSM network coverage;

▶ Seamless global radio coverage;

▶ Use of the same terminal for all applications and environments—for example, home, office, outdoor, satellite, WLL;

▶ Broad choice and flexible user control of services;

▶ User-controlled adjustment of services to include adding, deleting, or changing features;

▶ Broader access to wireless information services such as news, audio and video entertainment, computer games, CD-quality music, mobile computing, road traffic information, and teleshopping;

▶ Low cost, enabling virtually everyone to afford personal communications services, including children.

Of course, these benefits will not accrue all at the same time. The initial step is to combine today's proven and emerging technologies with innovative new technologies. Actual realization is likely to be phased in accordance with the market needs of different countries.

7.4.3 Intellectual property rights

One potential problem that may have prevented the full benefits of IMT-2000 from being realized came to a resolution in March 1999. The problem was a contentious dispute surrounding the intellectual property rights of CDMA, to which both Ericsson and Qualcomm laid claim. The ITU has indicated that it may only be able to consider RTT technologies for IMT-2000 that are based on TDMA technology if the dispute surrounding intellectual property rights of CDMA could not be resolved.

To conform with the ITU patent policy, the holder of any known patent or any pending patent application related to any proposal made to the ITU in the process of international standards-setting must submit a written statement, either waiving their rights or committing to negotiate licenses on a nondiscriminatory basis and on reasonable terms and conditions. Failure to provide this statement ultimately excludes the proposal from the international standards-setting process.

Ericsson contended that it had patents and/or pending application(s) for patent that are essential to the two different proposed 3G standards based on W-CDMA and cdma-2000. Ericsson had indicated that it was fully prepared to grant license to these patents on fair, reasonable, and nondiscriminatory terms, subject to conditions of reciprocity, which would be required to create fairness in a multistandard environment. Ericsson was not prepared to offer licenses, however, if Qualcomm did not also apply such reciprocity in its technology licensing.

Qualcomm held intellectual property rights on 5 proposals based on CDMA technology or variants of it. On cdma2000, for example, Qualcomm was not willing to waive its rights, but was willing to negotiate licenses with other parties on a nondiscriminatory basis on reasonable terms and conditions. On most other CDMA proposals—Europe's UMTS Terrestrial Radio Access, Japan's Arib's W-CDMA, the United States' W-CDMA/NA, and Korea's TTA's Global CDMA II—Qualcomm was not willing to waive its rights or agree to negotiate licenses with other parties on an on-discriminatory basis with reasonable terms and conditions.

Resolution of these matters was required in order to ensure that the IMT-2000 standardization process could continue at a rate necessary to meet agreed industry implementation schedules and to ensure many of the benefits of 3G networks could be realized.

Ericsson and Qualcomm eventually settled their patent dispute, paving the way for a global wireless standard. Part of the settlement entailed Ericsson buying Qualcomm's terrestial CDMA wireless infrastructure business, including two of its research and development facilities, boosting Ericsson's position in the U.S. mobile phone market.

At the same time, Qualcomm obtained access to some of Ericsson's patents to GSM technology.

Meanwhile, the wireless industry's truce between Ericsson and Qualcomm has not won over all companies to CDMA. In fact, some of the biggest companies in the United States, including AT&T Wireless, BellSouth, and SBC Communications, will continue to use TDMA. AT&T Wireless, which serves about 10 million customers, will evolve its current 850-MHz and emerging 1,900-MHz cellular networks to a 3G TDMA wireless network with $1 billion in new equipment from Lucent Technologies.

Lucent's platform includes the development of a wideband or software defined radio to support EDGE, a technology for delivering high-speed wireless data services and IS-136 for today's voice and data service. In addition, Lucent's radio platform will support the Bell Labs–developed multi-beam smart antenna technology, which enables TDMA operators to increase voice capacity and substantially increase the number of users on their network.

Lucent's 3G TDMA platform also includes new software for TDMA voice service features and new network enhancements to support GPRS, a technology that will enable an operator to provide high-speed packet data services using their existing network infrastructure. This supports the vision that GPRS will evolve into an IP-based, packet core network of the future. This future packet core network will be able to support multimedia services including voice, data, and video and support both EDGE and W-CDMA air interfaces.

With this next-generation technology, AT&T intends to offer consumers reliable new services such as wireless Internet access whether on AT&T's Digital PCS TDMA network or while traveling internationally. The networks will satisfy the criteria for IMT-2000. AT&T's TDMA wireless network will incorporate high-speed data systems similar to those planned by carriers whose networks are based on GSM, offering seamless global roaming and converged voice, video, and data applications using Internet protocols.

7.5 Conclusion

IMT-2000 addresses the key needs of the increasingly global economy—specifically, cross-national interoperability, global roaming, high-speed transmission for multimedia applications and Internet access, and customizable personal services. The markets for all of these exist now and will grow by leaps and bounds through the next millennium. IMT-2000 puts into place standards that permit orderly migration from current 2G networks to 3G networks, while providing a growth path to accommodate more advanced mobile services.

A phased introduction of IMT-2000 systems is expected, with services, coverage, capability, and number of operators growing over time. This phased introduction ensures the early availability of services to users, while reducing risks for 2G operators who can choose to reuse their existing infrastructure assets and expertise. Global availability of IMT-2000 services will be ensured by providing for roaming between members of the IMT-2000 family of systems, and smooth hand-over between GSM, cdmaOne (IS-95 CDMA), and IS-136 TDMA systems.

A number of technologies are required in addition to the radio interface. Technologies from other fields, such as IP, will be used to reduce costs, increase the performance, and enhance the services of IMT-2000 systems. Although much attention is currently focused on the early years of deployment and keeping to schedules to make it all happen, IMT-2000 is being defined with a view to the long term. In time, its capabilities will extend far above those anticipated today.

The market-driven approach to standards for mobile communications advocated by the United States and subsequently adopted by the ITU offers substantial benefits. In the United States, without mandating standards or national coverage, there is nationwide coverage by networks in place or under construction that use four different 2G technologies. The result is that some carriers offer local and long-distance wireless service at flat rates as low as $0.10 per minute, which is competitive with wireline service prices. In many cases, there are no roaming fees. Rates in Denmark, by comparison, average $0.30 per minute; in France, the average is $0.60 per minute. The use of multiple technologies and competitors worldwide under IMT-2000 is expected to result in similar benefits to wireless users around the globe.

In addition, much of the innovation in the wireless communications market can be traced to the willingness of the United States to allow the market to drive standards. Of note is that CDMA technology, a leading contender as a standard for third generation technology, likely would not have been developed commercially if a single standard had been mandated in the United States because it was submitted for consideration after the adoption of TDMA technology.

More information

The following Web pages contain more information about the topics discussed in this chapter:

AT&T: http://www.att.com/
Bell Atlantic Mobile: http://www.bam.com/
BellSouth: http://www.bellsouth.com/
CDMA Development Group: http://www.cdg.org/
Ericsson.: http://www.ericsson.se/
European Telecommunications Standards Institute:
  http://www.etsi.org/
GSM Alliance: http://www.gsm-pcs.org/
GSM Association: http://www.gsmworld.com/
Hughes Network Systems: http://www.hns.com/
InterDigital Communications: http://www.interdigital.com/
Internet Engineering Task Force: http://www.ietf.org/
International Telecommunication Union: http://www.itu.org/
Lucent Technologies: http://www.lucent.com/
MCI WorldCom: http://www.mciworldcom.com/
Nokia: http://www.nokia.com/
Nortel Networks: http://www.nortel.com/

Qualcomm: http://www.qualcomm.com/
SBC Communications: http://www.sbc.com/
UMTS Forum: http://www.umts-forum.org/
Universal Wireless Communications Consortium:
  http://www.uwcc.org/

Convergence via multiservice internetworking

The current Internet has lacked the speed, reliability, and security for mission-critical voice and data applications. While private IP networks offer these characteristics, they have lacked the feature-richness, robustness, and global reach of the PSTN. However, corporations are not about to give up their very extensive investments in conventional network technologies and replace them with IP technologies, no matter how much marketing hype from vendors is brought to bear on the situation. For most organizations, network replacement is not an option—migration to IP is the more realistic strategy.

With the application of intelligence, a single multiservice network infrastructure that converges voice and data using IP as the common internetworking protocol can combine the best features of the IP and PSTN worlds, and offer substantial savings in both capital

319

expenditure and operational costs. Such networks provide increased bandwidth for high-demand applications, enable secure remote access for intranets and extranets, and give organizations the ability to manage, grow, scale, and change their voice-data requirements quickly and easily, while also offering simple and cost-effective management.

A multitude of IP-oriented products have emerged in recent years—all aimed at voice-data convergence. Many interconnect vendors are participating in alliances with other firms to develop platforms that are capable of supporting converged voice and data networks. Often, larger firms have acquired smaller firms to round out their IP product lines. In the week of April 12, 1999 alone, vendors committed nearly $3 billion to voice-over-IP (VoIP) acquisitions.[1] The biggest deal involved Cisco Systems' acquisition of GeoTel Communications, a developer of telecommunications routing software, in a transaction worth $2 billion. This frenzied activity not only adds credibility to the concept of convergence—as if more were needed—it demonstrates widespread awareness among vendors that implementing converged networks requires an infrastructure capable of running mission-critical multimedia applications. These alliances and acquisitions bring together complementary expertise, products, and distribution with that objective in mind.

So pervasive is the convergence theme among vendors that it is not necessary to document all of them to get a sense of where the industry is headed—all are pursuing similar strategies with only a few variations in implementation. Instead it is only necessary to focus on the pre-eminent supplier of the Internet infrastructure, Cisco Systems, which leads the router market with 70% share.[2]

8.1 Cisco and convergence

While its major competitors also are addressing the convergence market, Cisco has the broadest range of networking products available from any single supplier. At this writing, the company has voice-enabled 80% of its products, which include routers, LAN and WAN switches, dial and other

---

1. Reported in InternetWeek, a CMP publication, April 19, 1999, page 7.

2. Among the many other companies that are also very active in supporting voice-data convergence over IP—both alone and with partners—is Nortel Networks. In March 1999, Nortel announced that it was working with Hewlett-Packard, Intel, and Microsoft to integrate voice technology into computing solutions for open, unified communications.

access solutions, SNA-to-LAN integration solutions, Web site management tools, Internet appliances, and network management software. The company's Internetwork Operating System (IOS) ties these products together.

Cisco's voice applications partner (VAP) program provides developers and value-added resellers (VARs) with the expertise and tools necessary to develop, sell, and support voice applications on a range of Cisco platforms. With its voice application partners, Cisco provides voice solutions that span over 30 application categories, including unified voice/fax/e-mail communications, prepaid and postpaid calling card, single number services, tandem switching, operator services, voice dialing, information services, conference calling, and directory assistance.

Like its major competitors—3Com, Ascend Communications (owned by Lucent Technologies), Bay Networks (owned by Nortel Networks), Cabletron Systems, and IBM—Cisco provides network design, implementation, maintenance, and support. Among common product lines, all of these companies serve customers in three key markets:

▶ *Enterprises:* This market consists of large organizations with complex networking needs, usually spanning multiple locations and types of computer systems. These customers include corporations, government agencies, utilities, and educational institutions.

▶ *Service providers:* This market consists of companies that provide information services, including telecommunication carriers, Internet service providers, cable companies, and wireless communication providers.

▶ *Small and medium businesses:* This market consists of companies with a need for data networks of their own, as well as connection to the Internet and/or to business partners.

Cisco sells its products in approximately 115 countries through a direct sales force, distributors, value-added resellers, and system integrators. It has more than 225 sales and support offices in 75 countries. Cisco now has one of the world's largest electronic commerce sites. In 1998, more than 64% of its orders and nearly 70% of its customer inquiries were handled via the Web. That year, total sales revenue approached $8.5 billion.

Cisco's strategy for growth is to buy whatever technologies, research, and development it cannot provide in-house. For example, in late 1998,

Cisco acquired Selsius Systems, a leading supplier of network PBX systems for high-quality telephony-over-IP networks. Selsius's IP phones and call manager software initially enabled small and medium businesses and branch offices to migrate voice traffic onto packet data networks. Cisco has extended the technology to the full campus environment and plans to enhance it further to enable value-added applications such as virtual call centers and unified messaging. To help realize this vision, Cisco also purchased Amteva Technologies, which makes middleware that meshes voice mail, fax, and e-mail over IP networks. Cisco is blending Amteva's technology with its Selsius IP PBX and other data and voice products to target enterprises that want to combine voice and data over a single network. At mid-1999, Cisco had acquired over 30 companies since 1993—making it one of the most active companies in pursuing acquisitions.

Like its competitors, Cisco is implementing a strategy for addressing the converged networks market, which it fine-tunes along the way. The company's data, voice, and video integration strategy—announced in November 1998 and updated in March 1999—addresses the WAN, campus area network (CAN), and LAN, as well as remote users and branch offices, with multiservice solutions and policy-based end-to-end call management. This open multiservice architecture includes QoS, call management, infrastructure integration, and infrastructure management capabilities.

8.2   Concept of multiservice networking

Multiservice networking has emerged as a strategically important issue for both companies and carriers. It entails the convergence of all types of communications—data, voice, and video—over a packet or cell-based infrastructure. The benefits of multiservice networking are reduced operational costs, higher performance, greater flexibility, integration and control, and faster deployment of new applications and services than can otherwise be achieved over traditional voice-oriented PSTNs or even TDM-based leased-line private networks.

The interest among organizations for data, voice, and video integration is fueled by short-term cost savings and increased budget flexibility. This integration also meets medium-term requirements for the support of emerging applications, and leads to the long-term objectives of reducing complexity and network redundancy through technology convergence.

In Cisco's scheme of things (see Figure 8.1), multiservice networking can result in immediate cost savings and increased budget flexibility by allowing the telecom budget to be reduced and the savings shifted to the information systems (IS) budget. This can be achieved by consolidation in the WAN, where costs are highest and where transmission options are greatest. This cost savings and budget flexibility can also be achieved by integration on the campus and in the branch office, where PBX capital and operations costs are high, but can be reduced with the addition of IP voice technology that takes advantage of the local- and campus-area infrastructure.

At the same time, according to Cisco, multiservice networking addresses key emerging business applications by inherently supporting any type of traffic, and therefore any type of applications networking requirement. Among these applications are unified voice and e-mail messaging, computer telephony integration (CTI), and desktop video streaming and videoconferencing. The real-time, near-real-time, and nontime-sensitive nature of the various applications requires that they be mixed in a reliable way. This can be achieved in an open multiservice architecture where end-to-end QoS capabilities guarantee voice and video throughput and latency requirements throughout the network.

Other benefits can accrue as well. When the multiservice infrastructure is used to provide virtual connectivity, IP telephones connected to the LAN provide users with familiar PBX functions, at a lower cost, on a more scalable, distributed, open architecture. The mobility that IP affords reduces PBX move, add, and change costs, while the intelligence embedded in the IP-based multiservice infrastructure makes CTI seamless.

To fulfill its vision of a multiservice infrastructure, Cisco offers a range of products that support legacy data and emerging Internet applications—as well as an array of voice over packet technologies—the combination of which span the network from edge to core.

8.3   IP telephony implementation

Cisco provides full telephony capabilities over IP networks via the IP PBX developed by Selsius Systems, a company it acquired in 1998. The complete IP telephony system consists of desktop telephone, call manager

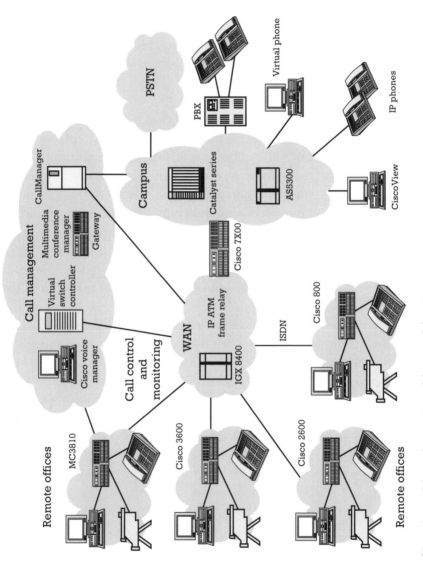

Figure 8.1   Cisco's multiservice networking architecture.

software, and a WAN/IP gateway—all of which are attached to an existing LAN/WAN infrastructure. The infrastructure can be a shared or switched Ethernet that provides the bandwidth connectivity between devices (see Figure 8.2).

The IP phones digitize and compress voice, place the resulting data into packets, and transmit them across the local- or wide-area packet network. The IP telephones communicate with each other directly and only communicate with the IP PBX when a special call processing function is required, such as transferring a call, creating a conference call, or sending a call to voicemail.

Local connectivity is provided over a standard Ethernet 802.3 LAN. Directory services are provided via the Lightweight Directory Access Protocol (LDAP). IP address assignment is provided through the DHCP. Signaling, call control, and audio compression for the delivery of packetized voice and video over IP comply with the H.323 standard, which ensures interoperability between the compliant systems of different vendors.

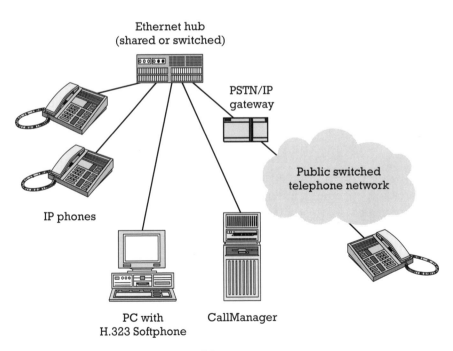

Figure 8.2 Cisco's IP PBX architecture.

8.3.1 IP telephones

By way of its acquisition of Selsius Systems, Cisco offers two fully programmable IP telephones. They provide the most frequently used business features and plug directly into the network with a standard 10BaseT Ethernet connection without the need for a companion PC. The IP phones provide toll-quality audio via their support of G.711 and G.723.1 audio compression for low-bandwidth requirements. Each model also contains an Ethernet repeater, so a single Ethernet switch port can be used for a computer and IP telephone. And because they rely on IP, the phones can be installed anywhere on a corporate local- or wide-area IP network.

Typically, users will have an IP phone that plugs directly into an Ethernet RJ45 wall jack instead of the traditional RJ11 telephone wall jack. However, those who use the phone extensively in conjunction with a PC can use handsets or headsets that plug into their PC. This arrangement allows both the phone and PC to be connected to the LAN with a single RJ45 connection.

The IP phone resembles a normal digital PBX electronic set and offers single-button access to line appearances and features. Administration and button configuration is accomplished through a Web browser. Built-in encryption protects the privacy of voice conversations.

The phone is characteristic of a PC in that it can operate as a standard IP device and has its own IP address. Because it is completely H.323 compatible, it can talk to any other H.323 device, such as a computer running Microsoft's NetMeeting for application sharing, video, chat, and whiteboarding.

A key attribute of the IP phone is its ability to handle compression in conjunction with CallManager. This gives the phone the ability to switch compression schemes on demand. For example, when a call is routed over a "skinny" IP pipe, CallManager instructs the IP phone to use a lower bit-rate audio compression, such as G.723, which provides 5.3 Kbps and 6.3 Kbps compression to another phone. For calls toward the PSTN, CallManager instructs the phones to use G.711 at 64 Kbps, which is the compression rate used over the circuit-switched public network.

For some users, a software PC phone (i.e., "virtual phone") is more desirable. In such cases, Microsoft's NetMeeting is used. The software is installed on a desktop or laptop PC equipped with sound drivers and a microphone. This arrangement combines the features of the IP phone with a number of desktop automation features that leverage all the benefits of the graphical user interface for common applications like video or whiteboarding.

When the virtual phone works in conjunction with an IP phone, the user gets the benefits of call mobility. For example, a laptop running the virtual phone gets the same line appearances and features of the IP phone on their desk. When calls arrive at the desktop IP phone, the same line appearances also ring on the virtual phone as well, enabling the user to answer either device, then transfer the caller to other parties if necessary. This call mobility works regardless of where the laptop is located—at home or at the office.

As noted, Cisco offers two models of IP phone. The 12 SP+ model is designed for business professionals and office workers. It supports 12 programmable line and feature buttons and has an internal two-way speakerphone with microphone mute. An LED associated with each of the 12 feature and line buttons provides feature and line status. The phone's large LCD display provides call status and identification.

The 30 VIP model is the full-featured IP telephone for executives and managers. It provides 30 programmable line and feature buttons, an internal two-way speakerphone with microphone mute, and a transfer feature button. An LED associated with each of the 30 feature and line buttons provides feature and line status. The phone's large 40-character LCD display provides information such as date and time, calling party name, calling party number, and digits dialed.

Both IP phone models use an extra pair of wires in the 4-pair Category 5 cable, which provides power to permit the phone's operation in the event of a building power failure. Power can be supplied near the phone or from a shared 48V power source near the Ethernet switch. Using a shared power source with an uninterruptible power supply (UPS) allows the phone to remain operational in the event of a power failure.

IP address assignment and configuration may be accomplished manually via the IP phone's keypad or the DHCP. The latter method makes IP phone setup and address management virtually automatic. DHCP is a standard in the data network environment for configuring and managing various IP devices.

For managers of large IP networks, DHCP reduces the work necessary to administer a large number of IP addresses. It does this by automatically assigning IP addresses to clients as they log on to the network. DHCP also reclaims unused IP addresses and maintains a pool of reusable addresses. DHCP also lets IP devices, including phones, to be moved and plugged in anywhere on the IP network (local or remote ports) with no manual database adjustments or wiring changes. The phones can be unplugged from one RJ45 wall jack and plugged into another location anywhere on the enterprise IP network. They will automatically boot and reregister with CallManager.

8.3.2 CallManager

CallManager is a software application that provides the intelligence necessary for PBX-like features. The application is installed on a Windows NT server and provides basic call processing, signaling, and connection services to IP phones and virtual phones, voice-over-IP gateways, and other local and remote devices (see Figure 8.3). This includes the management and control of various signaling protocols such as Q.931 for ISDN WAN control and H.225/H.245 for IP packet control. With total switch and network independence, administrators can create a virtual campus environment utilizing the familiar interface of a Web browser (see Figure 8.4).

CallManager manages the resources of the IP-PBX by signaling and coordinating call control activities. It sets up a call by instructing the calling party to set up a Real-Time Transfer Protocol (RTP) audio stream to the other device, either telephone or gateway. Once the audio stream is set up between the two devices, the CallManager is out of the picture until a new request is made, such as transfer or disconnect. Should the CallManager fail during a call, the two parties stay connected and can complete their call.

CallManager supports the RSVP—which, combined with routers that support the standard—reserves network resources and prioritizes voice traffic to guarantee QoS over the IP network (discussed later).

Supplementary and enhanced services such as hold, transfer, forward, conference, multiple line appearances, automatic route selection, speed dial, last-number redial, and other features are extended by Call-Manager to IP phones and gateways via parameters stored in a configuration database. Microsoft's Internet information server (IIS) is installed at the CallManager server to provide a browser interface to this database. With administrator approval, users can be granted access to configure their own phones, and any user with appropriate access privileges on the

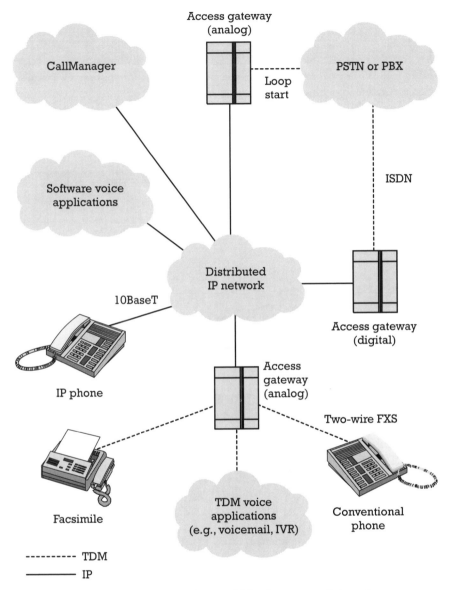

Figure 8.3 CallManager's place within the network topology.

IIS can administer the CallManager from any location through a Web browser.

An optional suite of integrated voice applications can be added to CallManager to provide basic voice messaging, voice conferencing,

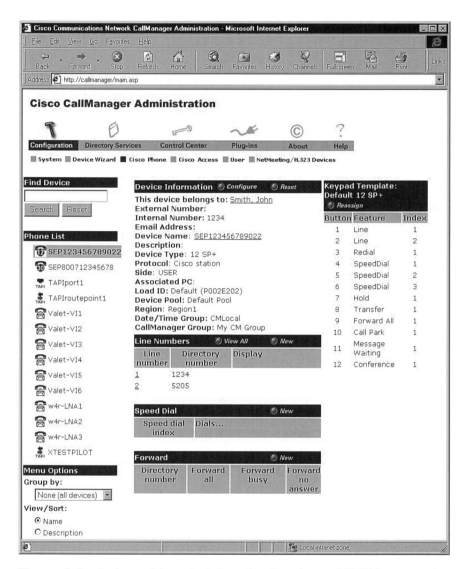

Figure 8.4   A view of the administrative interface of CallManager via the Web browser.

manual attendant console, click-to-call, and other functions. Because CallManager is a software application, these and other options and enhancements can be loaded for quick implementation, eliminating expensive hardware upgrade costs and installation delays.

CallManager supports the industry standard simplified message desk interface (SMDI) to provide connectivity to various voice mail and IVR systems, along with call detail record reporting for tracking call activity and billing. Call records are kept in a standard CDR database from which a variety of summary and detail reports can be generated.

At this writing, CallManager is implemented on a single server, but the architecture allows for a future scalable network of multiple, redundant CallManager servers. This will enable any IP phone to receive CallManager services from a primary, secondary, or tertiary server. In the event that communications between the active CallManager and the IP phones is disrupted, the phones will register with the backup CallManagers for their call processing service.

8.3.3 WAN/IP gateways

Like the IP/PSTN gateways of other vendors, Cisco's gateway products operate as the trunk interface to circuit-switched components such as PBXs and Class 5 central office switches. They convert voice from the packet domain to the circuit-switched domain. Specifically, this type of device converts packetized voice that has been placed into an Ethernet frame into the format that can be accepted by the PSTN. Gateways also pass signaling information, including dial tones and network signaling, like ISDN. Since the PSTN supports both digital and analog voice, the gateways must also be capable of supporting digital and analog traffic. Cisco offers both types of access gateways, the choice of which depends on the type of circuit-switched interface presented. Each model is managed and maintained through CallManager.

The digital gateway supports G.711 audio encoding, offers Ethernet access, and provides integrated DSP functions. The DSP interface contains an RJ48C connector for attachment to digital TDM devices. The digital gateway supports ISDN PRI at the T1 rate of 1.544 Mbps. Each interface card supports 24 channels with line echo cancellation and packet-to-circuit conversion for voice or fax calls. In addition to H.323 compliance, which allows interoperability with other H.323 client applications and gateways, the digital gateway also supports supplementary services such as call forward, transfer, and hold. The gateway is configured using a Web browser.

Cisco's analog gateway supports G.711 or G.723 audio compression, and comes with integrated DSPs and modular analog circuit-switched interfaces. It connects to analog circuit-switched interfaces through two-, four-, or eight-wire RJ11 connectors. The analog system not only connects to local analog telephone company lines, but also provides connectivity to devices such as fax machines, voicemail, and analog phones. It connects up to eight analog devices or trunks to a 10BaseT LAN.

8.4   Multiservice access routers

With companies spending billions of dollars each year on internal phone calls and faxes between their own offices, there is ample incentive to reduce these costs by integrating voice, fax, and data onto a single network infrastructure. With voice and fax over IP, companies, as well as service providers, can deploy integrated, scalable networks without sacrificing voice and fax quality. In addition, the deployment of these multiservice capabilities can be done without changing the way phone calls are made or the way faxes are sent.

8.4.1   Mid-range access routers

Multiservice access routers, such as the mid-range Cisco 2600 series or Cisco 3600 series, enable traditional telephony traffic such as voice and fax to be integrated with traditional data traffic such as IP, IPX, and SNA. This integration is achieved with traditional telephony interfaces so that PBXs, key systems, traditional phones, IP phones, fax machines, and even the PSTN can physically connect to the router. Once these connections are established, the voice and fax traffic are processed by the DSPs and placed into IP packets or frame relay cells for transfer to other network locations. In keeping as much voice and fax traffic as possible on the data network, toll charges normally paid for intracompany calls can be eliminated for substantial cost savings.

8.4.2   Remote access routers

Cisco 800 series remote access routers address the needs of small offices and telecommuters. Small offices and telecommuters not only can reduce telephone line charges by combining their data, telephone, and fax communications on a single ISDN line, they can simultaneously use the Internet, receive a fax, or make a phone call while reducing the expense of multiple telephone lines. In addition, with IOS features, these users can take advantage of advanced applications such as secure Internet access, managed network services, VPNs, and electronic commerce.

There are four router models in the series. The 801 model provides an ISDN basic rate interface (BRI) for use worldwide. The 802 model adds the integrated NT1 network termination device for use in North America. The 803 and 804 models add a four-port Ethernet hub and two RJ11 interfaces for telephones, fax machines, and modems. They also support supplementary telephone services such as caller ID, call waiting, call-waiting cancel, call hold, call retrieve, call transfer, and three-way conferencing.

The routers support X.25 service over the ISDN D channel. This feature supports transaction-oriented applications such as point-of-sale (POS) authorizations. It also allows users to take advantage of unused capacity on the 16-Kbps D channel. The Always On/Dynamic ISDN (AO/DI) technology, implemented in the IOS, allows users to minimize call charges by initiating the lower-bandwidth D channel in "always-on" mode. AO/DI activates the more costly B channel only when a specific transmission requires more bandwidth. This feature also includes priority queuing to improve the responsiveness of traffic over the link.

Security features, implemented through the IOS, protect the privacy of company communications and commerce transactions over the Internet. The IOS also provides the means to build custom security solutions, including standard and extended access control lists (ACLs), dynamic ACLs, router and route authentication, and generic routing encapsulation (GRE) for tunneling. Perimeter security features control traffic entry and exit between private networks, intranets, or the Internet. To protect the corporate LAN from unauthorized access, the routers also support token cards, Password Authentication Protocol (PAP), Challenge Handshake Authentication Protocol (CHAP), and other security features available through an optional firewall.

Workstations on the remote LAN can be assigned IP addresses dynamically by either the central network or the routers using DHCP. The units also support network address translation (NAT), which effectively creates a "private network" that is invisible to the outside world. NAT enables network administrators to assign IP addresses normally reserved for the Internet to a remote LAN. For businesses that want to allow select access to the network, NAT can be configured to allow only certain types of data requests, such as Web browsing, e-mail, or file transfers.

A range of platforms can be used to manage the routers, including Cisco ConfigMaker and Cisco Fast Step (Windows 95, 98, and NT 4.0), and CiscoView (UNIX). These applications provide configuration and security management, as well as performance and fault monitoring. Centralized administration and management can be applied via SNMP, Telnet, or local management through the router's console port.

8.5 Multiservice concentrator

Cisco offers a family of multiservice access concentrators that enables organizations to integrate all traffic—legacy data, LAN traffic, voice, fax, and video—over a single network backbone composed of frame relay, ATM, or leased lines at speeds up to T1/E1. The Cisco 3800 series routers can be deployed over private or public networks to reduce equipment and connection costs, simplify network management, and improve application performance. Through the IOS, these systems perform multiprotocol routing and bridging. They are also tightly integrated with Cisco's multiservice edge switch.

8.5.1 Capabilities

One multiservice concentrator, the MC3810, is a wire-speed T1/E1 router and serial data device that has voice, video, and ATM capabilities. It includes the extensive Ethernet LAN and data capabilities familiar to users of other Cisco IOS software-based devices, including a rich IP and SNA suite. In addition to providing 24/30 channels of voice through the T1/E1 port, the MC3810 provides echo cancellation for all voice channels and achieves further cost savings through the use of voice activity detection (VAD). This feature halts voice traffic during the silent periods of a conversation, allowing the idle bandwidth to be used for data. Further bandwidth efficiencies can be achieved with voice compression at 8 Kbps (G.729, G.729a) or 32 Kbps (ADPCM).

The MC3810 connects to any standard PBX switch, key system, or telephone. At small sites, the MC3810 can be used for local voice, which can possibly obviate the need for Centrex, key, or PBX switching. It offers an ISDN BRI voice interface and supports call-handling capabilities for voice connections. It can be used in tie-line and ring-down modes. It can also support DTMF digit-based per-call switching, using dialed digits to select destination sites and network calls.

The MC3810 supports transparent common channel signaling (CCS) and the Q.SIG voice signaling protocol. Q.SIG is a form of CCS that is based on ISDN Q.931, the signaling method used by the D channel for call setup and tear-down. Q.SIG provides transparent support for supplementary PBX services so that proprietary PBX features are not lost when connecting PBXs to networks comprised of MC3810 routers.

In addition to voice calls, the MC3810 also supports both circuit and packet mode video. Circuit video is transported bit-by-bit through circuit emulation over a constant bit rate (CBR) ATM connection. Packet video can be supported over a variable bit rate (VBR) ATM connection or over the LAN, through the router engine, and over an unspecified bit rate (UBR) ATM connection. Videoconferencing is supported on the MC3810 with an H.323 compliant gatekeeper. The Multimedia Conference Manager and proxy functions are integrated into the MC3810 to ensure QoS, interoperability, and bandwidth management for the video session on the MC3810.

The MC3810 is also compatible with the drop-and-insert capability of digital cross-connect systems (DCS) used on the PSTN. *Drop and insert* refers to the software-defined capability of the DCS to exchange channels from one digital facility to another, either to implement appropriate routing of the traffic, reroute traffic around failed facilities, or to increase the efficiency of all the available digital facilities. Accordingly, the MC3810 allows some time slots of a T1 facility to be used for on-net traffic and services, while the rest can be dropped/inserted off-net for transport over the PSTN when necessary.

8.5.2 Voice manager

The MC3810 is managed by Cisco Voice Manager, an end-to-end voice management solution that provides call-quality management and call-detail reporting, enabling network administrators to install, control, and lower the overall cost of owning corporate voice networks built on VoIP technology. It comes with tools to troubleshoot network problems, and provides call-history information that can be accessed from an SQL database through a Java-enabled Web interface.

Voice Manager automatically detects voice-supported products and establishes connections to VoIP-capable devices and gateways via an IP network. Communication between Voice Manager's embedded Web server and the Web browser clients occurs through HTTP, while communication between the Voice Manager server and the voice-supported devices occurs via SNMP and Telnet sessions.

8.5.3 Multimedia conference manager

Multimedia conferences through the MC3810 are implemented with Cisco's Multimedia Conference Manager, an application that provides network administrators with mechanisms to apply QoS and policy management to H.323 audio (i.e., VoIP) and videoconferences. Network administrators can allocate bandwidth and QoS to H.323 calls, while protecting mission-critical applications by limiting conferences within the LAN or across the WAN. QoS is provided by IP Precedence or RSVP. The management software is implemented in Cisco's IOS and runs on access routers and concentrators.

IP Precedence uses three bits in the packet header's precedence/type of service (ToS) field to provide eight different precedence levels, which are indicated in terms of 0 to 7, with 0 indicating normal and 7 providing the highest priority. ToS uses 4 bits to indicate quality of service in terms of a packet's delay, reliability, throughput, cost or security. The queuing technologies throughout the network can then use this information to provide the appropriate handling for the application. For example, a router that checks a packet header whose reliability bit is set to 1 will interpret this to mean that the packet is less eligible for discard than a packet whose reliability bit is set to 0.

RSVP enforces QoS in a different way—it runs on top of IP to provide receiver-initiated setup of resource reservations on behalf of an application data stream. When an application requests a specific QoS for its data stream, RSVP is used to deliver the request to each router along the path the data stream will take and maintain router and host states to support the requested level of service. In this way, RSVP essentially allows a router-based network to mimic a circuit-switched network on a best-effort basis.

At each node, the RSVP program applies a local decision procedure, called admission control, to determine if it can supply the requested

QoS. If admission control succeeds, the RSVP program in each router passes incoming data packets to a packet classifier that determines the route and the QoS for each packet. The packets are then queued as necessary in a packet scheduler that allocates resources for transmission on a particular link. If admission control fails at any node, the RSVP program returns an error indication to the application that originated the request.

For security, user authorization is implemented through a remote access dial-in user service (RADIUS) or terminal access controller access control system+ (TACACS+) account. Of the two, RADIUS is the more popular. Users are authenticated through a series of communications between the client and the server. When the client initiates a connection, the communications server puts the name and password into a data packet called the authentication request, which also includes information identifying the specific server sending the authentication request, the port that is being used for the connection. For added protection, the communications server, acting as a RADIUS client, encrypts the password before passing it on to the authentication server.

When an authentication request is received, the authentication server validates the request and decrypts the data packet to access the user name and password information. If the user name and password are correct, the authentication server sends back an authentication acknowledgment that includes information on the user's network system and service requirements. The acknowledgment can even contain filtering information to limit the user's access to specific network resources.

The older security system is TACACS, which has been updated by Cisco into a version called TACACS+. Although the protocols are different, the proprietary TACACS+ offers many of the same features as RADIUS, but is used mainly on networks consisting of Cisco remote access servers and related products. Companies with mixed-vendor environments tend to prefer use the more open RADIUS.

In addition to voice and videoconferencing services, the infrastructure built with the MC3810 multiservice concentrator platform accommodates streaming video via Cisco's IP/TV 3400 series of servers, which stream video programs to PC desktops over the enterprise network. The servers provide control, broadcast, and archive functions for the delivery of live video, scheduled video, and video on demand (VoD). In addition, they offer management functions that support a range of corporate applications that include training, internal employee communications, distance learning, and business TV.

8.6   Multiservice LAN switches

Cisco's Catalyst family of multiservice LAN switches implement telephony in the LAN and provide seamless integration with campus and WAN systems. In addition to a range of connectivity options and network services, the switches also provide an interoperable IP telephone and call manager.

8.6.1   Capabilities

The Catalyst family includes the 8500, 6000, 5000, and 4000 series of products. Together, they enable organizations to build corporate intranets for multicast, mission-critical, and voice applications. These systems also offer redundancy and topology resiliency for high availability, and Gigabit Ethernet and ATM interfacing for high performance. In addition, they reduce complexity with application awareness and policy classification, which eliminates the need for network managers and administrators to engage in detailed configuration of QoS parameters.

Cisco's top-of-the-line multiservice switch router is the Catalyst 8500, which can be deployed as one of two distinct editions, depending on which set of switch and router processors are purchased and installed. The Catalyst 8500 campus switch router (CSR) edition provides Layer 3 routing capabilities over Fast and Gigabit Ethernets, OC-3/OC-12 ATM, and packet over SONET (PoS) uplinks.

The Catalyst 8500 Multiservice ATM Switch Router (MSR) provides ATM switching capabilities from T1 through OC-48 speeds, in addition to the functionality available in the CSR edition. The ATM switching capabilities provide campus backbones with the means to integrate data, voice and video traffic over ATM LAN Emulation (LANE), Multi-Protocol Over ATM (MPOA), or Multi-Protocol Labeling Switching (MPLS)/tag switching networks.

Developed by the ATM Forum, LANE is a Layer 2 bridging protocol that causes a connection-oriented ATM network to appear to higher-layer protocols and applications as just another connectionless LAN segment. As such, LANE provides a means to migrate today's legacy networks toward ATM networks without requiring that the existing protocols and applications be modified. The scheme supports backbone implementations, directly attached ATM servers and hosts, and high performance, scalable computing workgroups. By defining multiple emulated LANs across an ATM network, switched virtual LANs can be created for improved security and greater configuration flexibility. Additional benefits include minimal latency for real-time applications and QoS for emulated LANs.

MPOA is another protocol developed by the ATM Forum. It preserves the benefits of LANE, while allowing inter-subnet, internetwork communication over ATM virtual circuits without requiring routers in the data path. This framework synthesizes bridging and routing with ATM in an environment of diverse protocols, network technologies, and IEEE 802.1 Virtual LANs. MPOA is capable of using both routing and bridging information to locate the optimal exit from the ATM cloud. It allows the physical separation of internetwork layer route calculation and forwarding, a technique known as virtual routing. This separation not only allows efficient intersubnet communication, it increases manageability by decreasing the number of devices that must be configured to perform internetwork layer route calculation. It also increases scalability by reducing the number of devices participating in internetwork layer route calculation and reduces the complexity of edge devices by eliminating the need for them to perform internetwork layer route calculation.

MPLS, based on Cisco's Tag Switching, is an IETF standard for IP service delivery. MPLS labels or "tags" provide the ability to differentiate service classes for individual data flows. The tags work like address labels on packages in an express delivery system—they expedite packet delivery on large corporate enterprise networks, allowing for the creation of faster, lower latency intranets that can effectively support data, voice, and video on a common network infrastructure.

8.6.2 Policy manager

Configuring and deploying QoS policies is achieved with Cisco's QoS Policy Manager. The Policy Manager's graphical user interface enables network administrators to define traffic classification and QoS enforcement policies. This system includes a rules-based policy builder, integrated policy validation and error reporting, and a policy distribution manager that downloads policies to network devices running Cisco's IOS.

Relying on IP Precedence to enforce QoS policy end to end, Policy Manager enables network administrators to quickly apply a mix of QoS policy objectives that protect business-critical application performance.

Policy Manager eliminates tedious and error prone device-by-device configuration of QoS parameters in switches and routers, giving network administrators a more efficient way to deploy QoS policies that achieve application service-level differentiation. In the process, organizations can more easily make the transition from unconstrained bandwidth utilization toward more consistent application performance over currently available bandwidth. This is not only increasingly necessary to eliminate unpredictable performance for business-critical applications running in corporate networks, but it will also be required to integrate data, voice, and video over a common network infrastructure.

8.7   Voice packet gateway

A voice packet gateway provides connectivity between PSTN and packet-based networks. Cisco's voice packet gateway is the AS5300, which is part of the AS5x00 product family of multiservice universal access servers.

8.7.1   Capabilities

As a data communications platform designed to support the growing volume of dial-in and voice connections to service providers and corporate networks, the AS5300 provides the functions of an access server, router, and digital modem in a single modular chassis. The access server is intended for ISPs, telecommunications carriers, and other service providers that offer managed Internet connections, as well as medium- to large-sized sites that provide both digital and analog access to users on an enterprise network.

The gateway can even be equipped to exchange routing and authorization information between carriers. The Internet roaming alliance of GRIC Communications, for example, offers Open Settlement Protocol (OSP) interoperability with the AS5300 voice packet gateway. OSP facilitates the exchange of routing and authorization information between multiple service carriers. This allows interdomain call detail records to be forwarded to the GRIC settlement system, processed, and then transmitted as settled records to the various participant's billing system platforms.

Cisco's AS5300 includes an access server; a voice/fax feature card; voice gateway application software, integrated into Cisco IOS software; and voice manager software. The system's T1/E1/PRI interfaces enable its connection to PBXs and PSTN digital switches. The system is H.323 compliant and features a coder that supports G.711, G.729a, and G.723.1 voice compression standards. Echo cancellation enhances the quality of voice and silence suppression enables more efficient transmission of voice and data. The system includes an integrated IVR capability and supports two-stage (account and PIN number) direct inward dialing (DID). The system performs multiprotocol routing and QoS, and supports RADIUS for security.

The voice/fax feature is coprocessor-enabled with a RISC engine and dedicated DSPs for each voice channel to ensure predictable, real-time voice processing. The coprocessor is coupled with direct access to the AS5300 routing engine for streamlined packet forwarding. Each voice/fax feature card provides 24 digital voice connections for a T1 or 30 digital voice connections for an E1. Since each AS5300 can accept two voice/fax feature cards, the system can actually scale up to 48/60 voice connections.

8.7.2 Management

The AS5300 is managed with CiscoView, which provides a graphical user interface that allows network managers to monitor the real-time performance and call status of the system, including the modems. Multiple displays, including a physical and logical view, provide in-progress call status information for both ISDN and analog modem calls as well as PRI, channelized line, and Ethernet interface statistics. Administrative tools include the ability to capture individual caller statistics and information such as IP address per port and remote phone number—all monitored and reported via call-information screens. Each modem can be directly accessed at any time—including when the unit is connected and online—to capture statistics, force a reset, or check a configuration.

8.7.3 Unified communications strategy

Cisco's AS5x00 product family is also the platform for implementing its Unified Communications strategy, which enables service providers to offer their customers consolidated voice, fax, and e-mail services on a single IP network. In conjunction with third-party applications providers, the scalable infrastructure allows service providers to derive revenue through new services aimed at multiple market segments including SOHOs, mobile work forces, traveling executives, and high-income households.

Examples of Unified Communications services that can be offered include:

) Integrated voicemail, fax, and e-mail;

) Voice, fax, and e-mail retrieval by phone;

) Integration of electronic documents with faxes;

) Personal messaging agents;

) Never busy fax lines;

) Broadcast fax;

) Voice over IP.

With millions of AS5x00 ports installed worldwide, this infrastructure can be leveraged further as the backbone for Universal Communications, allowing service providers to extend their existing network infrastructure investments to offer new services.

8.8 Multiservice router

The 7200 series of high-speed routers from Cisco provides organizations with the flexibility to meet the constantly changing requirements at the core and distribution points of their internetworks.

8.8.1 Capabilities

These routers deliver up to 1 Gbps of throughput over a multiservice interexchange (MIX) midplane that provides the ability to switch DS0 time slots between multichannel T1 and E1 interfaces, much like a TDM multiplexer. At the same time, the router provides digital voice connectivity via an ATM circuit emulation service (CES) module. Together, these capabilities allow the 7200 routers to be connected to an ATM network on one side and to the TDM network on the other side.

These devices run the IOS networking software to provide routing and bridging functions for a wide variety of protocols and network media, including any combination of Ethernet, Fast Ethernet, Token Ring, FDDI, ATM, serial, ISDN, and high-speed serial interface (HSSI). Port and service adapters are connected to the router's peripheral component interconnect (PCI) buses, enabling connection to external networks.

8.8.2 Reliability features

Designed for uninterrupted operation in support of mission-critical applications, the 7200-series of routers feature automated internetwork installation and management. IOS feature sets can be customized as well as upgraded to meet changing requirements.

Software-defined configuration changes take effect without rebooting or interrupting network applications and services. Port adapters and service adapters can be inserted and removed while the system is online. An automatic reconfiguration capability enables seamless upgrades to higher density and new port adapters without the need for rebooting, taking the system offline, or intervening manually. Dual hot-swappable, load-sharing power supplies provide system power redundancy; if one power supply or power source fails or is taken offline, the other power supply maintains system power without interruption. Alerts are issued when potentially problematic system fluctuations occur before they become critical, thereby enabling resolution while the system remains online.

8.9 Multiservice edge switch

Cisco's IGX 8400 series of multiservice edge switches provide data, voice, and video integration over the wide area. They extend the WAN backbone to branch offices, providing high service levels regardless of location.

Internetworking between existing routers and LAN switches provides seamless traffic flow between LAN/campus and WAN environments.

8.9.1   Capabilities

By combining ATM's dynamic bandwidth management with queuing techniques, the Cisco IGX 8400 series of edge switches minimizes recurring WAN bandwidth costs, while ensuring fairness and high QoS for individual applications. The switches support legacy applications, large-scale packet voice, frame relay, and ATM, as well as provide integrated LAN interfaces. Internetworking and management functions between the edge switch and LAN switching, routing, and branch-office devices improve end-to-end network performance and QoS.

Port interfaces ranging from 1.2 Kbps to 155 Mbps OC-3c/STM-1 are supported, as are network or trunk interfaces ranging from T1/E1 to OC-3c/STM-1. Advanced traffic management features and multiplexing techniques deliver high levels of bandwidth utilization and efficiency. The voice compression, silence suppression, and fax relay capabilities deliver additional bandwidth savings of up to 80% for voice traffic, according to Cisco.

Not only are LAN and WAN interfaces integrated on a single platform, in being able to consolidate multiple WAN infrastructures, organizations have the flexibility to deploy VoIP, voice over ATM, or voice over frame relay. Full control over network resources is exercised in a variety of ways—with per-virtual circuit (VC) queuing, per-VC rate scheduling, and multiple CoS—guaranteeing QoS levels for the individual applications. This enables all applications to be supported, according to their specific requirements, using advanced traffic management and CoS features.

Access devices and the IGX 8400 backbone can be connected through a leased line or though a public frame relay or ATM WAN (see Figure 8.1). Depending upon the modules used to connect to the access device, different levels of internetworking are possible. Data applications can be based on Ethernet or token ring LANs, as well as frame relay, ATM, and legacy protocols. Voice applications can use different transport technologies. In addition to VoIP, the edge switch supports voice over frame relay (VoFR) and voice transport over ATM (VtoA). Dial-up frame relay, SNA networking, and frame forwarding are also supported. For locations that need

ATM broadband speeds, the systems offer frame relay to ATM service internetworking.

8.9.2 Frame relay

Only the IGX 8400 frame relay port needs to be configured; the frame relay virtual connection parameters (e.g., committed information rate, committed burst, and excess burst) are automatically sent and set on the Cisco router ports. This enables the edge switch and router to dynamically exchange traffic shaping information for various permanent virtual connections on the interface. The routers can then dynamically configure their traffic shaping parameters, thereby reducing operational effort and eliminating potential errors that may be caused by a mismatch in the manual configuration of the router and WAN switches.

Frame relay-to-ATM service internetworking is accomplished by the edge switch by segmenting and mapping variable-length frame relay frames into fixed-length ATM cells. This enables the switch to provide transparent connectivity between large ATM and small frame relay locations, independent of the wide-area access protocol. Small sites that do not have enough traffic to justify a dedicated access connection can connect to a frame relay network on an as-needed basis using switched (dial-up) lines.

Through frame forwarding, the edge switch can transport frame-based protocols, such as SDLC, X.25, or any other HDLC-based protocol, at speeds ranging from 9.6 Kbps to 16 Mbps. With frame forwarding, all valid HDLC frames are forwarded (i.e., tunneled) from one frame relay port to another port without frame relay header processing or local management interface (LMI) control. This method of transport results in efficient bandwidth usage and low latency on the corporate network.

8.9.3 ATM

The edge switch provides voice connectivity between PBXs across the wide-area ATM backbone. Using compression, silence suppression, and cell switching, the switch can provide up to 240 active voice channels on a single T1/E1 line. The switch manages the priority, delay, delay variation, and echo cancellation requirements that are essential for voice networks, providing toll-quality voice across the ATM network.

To maximize the efficiency of available bandwidth, voice interfaces support a variety of alternatives to 64-Kbps PCM encoding that reduce bandwidth requirements to 32 Kbps, 16 Kbps, and 8 Kbps. Silence suppression further reduces bandwidth usage by transmitting cells onto the network only if voice is detected. This constant bit-rate stream is then converted into a variable bit rate type of connection, resulting in an additional 2:1 compression over that already achieved by voice compression.

The edge switch also supports common channel signaling (CCS) and channel associated signaling (CAS) between PBXs on both T1 and E1 interfaces.

With CCS, one channel is used exclusively to transport private signaling protocol information. The switch treats the channel as a transparent channel and transmits the signaling information across the network for delivery to the remote circuit line.

With CAS, up to four signaling bits are transported per voice channel using a separate channel in an E1 environment, or using robbed bits in a T1 environment. The switch monitors the state of the signaling bits associated with each voice channel and can transparently deliver the signaling bits to the correct destination PBX. CAS signaling can be used in conjunction with Cisco's VNS to dynamically set up connections across an ATM network.

The switch's T1/E1 interface automatically detects incoming voice-band data and fax calls coming from a PBX. The switch disables compression and echo cancellation on that channel and then adjusts its allocated bandwidth to ensure effective transport. The channel can be upgraded to 64 Kbps for high-speed data and fax calls, or 32 Kbps can be used for low-speed reception. To further reduce bandwidth consumption by data or fax calls, the switch can demodulate a Group III fax and remodulate it to a 9.6-Kbps stream across the network.

8.9.4 Voice network switching

With VNS, the switch receives signaling from all attached PBXs. It then interprets the signal and dynamically establishes voice or data connections between the source and destination PBXs. The network dynamically routes each call request on a per-call basis, and extends PBX features (e.g., transfer, caller ID, and camp-on) across the ATM WAN. The advantages include enhanced voice quality, higher degree of compression, reduced delay, improved bandwidth utilization, and lower equipment costs.

VNS reduces the number of T1/E1 trunks and interface cards required to interconnect PBXs, and enables the replacement or re-deployment of existing tandem PBXs. Adding VNS to the ATM network also improves efficiency by enabling single-hop routing across the network and by eliminating multiple compression/decompression cycles.

8.9.5 Circuit data services

Circuit data services are provided by the edge switch for transport of asynchronous or synchronous circuit data or video, which is transparently carried through a fixed-delay, fixed-throughput, zero-discard, point-to-point data connection across the WAN. This capability enables the transport of both legacy and TDM traffic and facilitates the migration to future ATM networks. Circuit data features include:

‣ Super-rate data provides up to 8×64 Kbps connectivity between two end devices connecting to the network via a digital T1/E1 interface.

‣ Sub-rate data supports standard data rates (2.4 Kbps, 4.8 Kbps, 9.6 Kbps) in which each circuit occupies one time slot of a digital T1/E1 interface.

‣ TDM transport transfers data, voice, and video traffic, and provides for a seamless migration path from existing TDM WAN equipment to an ATM-based multiservice WAN.

‣ Network connectivity offers extensive flexibility with a variety of trunking options at varying data rates, from narrowband to broadband.

The reliability of the IGX 8400 series switches is enhanced by common equipment that can be configured for redundancy. New software releases can be remotely downloaded onto the redundant processor for background installation while traffic continues to run. An advanced distributed intelligence algorithm enables the network to automatically route new connections and, if necessary, reroute traffic around failures in network facilities.

8.10   Analysis

Although this chapter has focused on multi-service networking for the enterprise, competitive local exchange carriers (CLECs) face similar challenges as they attempt to compete with the well-established Bell operating companies (BOCs). To facilitate competition in accordance with the Telecommunications Act of 1996, the Federal Communications Commission (FCC) has clarified its rules regarding the collocation of competitors' equipment in the central offices (COs) of the incumbent local exchange carriers (ILECs). The incumbents must not only provide floor space for the equipment of competitors, but they must allow them to establish cross-connects to the collocated equipment of other competing carriers. To the extent that such arrangements are technically feasible, they must be implemented expeditiously by the incumbent.

The intent of equipment collocation is to lower the cost of market entry and speed the rollout of alternative services and, in the process, give consumers more choice. However, would-be competitors cannot always afford to buy large switches to deploy in the CO. Consequently, they are turning to smaller, inexpensive carrier-class switches that can provide a range of voice and data services. These integrated access devices are modular, permitting CLECs to offer services with high market demand and expand capacity economically as the customer base grows.

Although important, many CLECs are not just looking to arrange equipment collocation at the CO and gain access to the local loop—real competition in the new era of voice-data convergence requires that they plan for the day when packet and cell-based data traffic catches and surpasses circuit-switched calls. This new paradigm entails the implementation of a multi-service platform capable of handling a mix of traffic formats—IP, frame relay, ATM, circuit emulation—simultaneously on the same ports. There are carrier-class platforms available that define in software the traffic handling capabilities of each port. This gives CLECs the same traffic handling capabilities of established carriers, but at much less cost, while giving them the flexibility they need to meet the demands of a diverse customer base.

For CLECs that want to offer customers economical IP-based voice services with hooks into the PSTN, there are a number of solutions available. Lucent Technologies, for example, offers a new software switch developed by Bell Labs that combines the reliability and features that customers expect from the public telephone network with the cost-effectiveness and flexibility of IP technology.

With the Lucent Softswitch, CLECs can provide a full range of IP-based communication services that are indistinguishable in quality and ease of use from services on traditional circuit-switched voice networks. Customers will be able to get the same quality of service, but at lower cost, and they can further benefit from the continuously improving price-performance of Internet technology.

Customers simply dial the desired telephone number, as they do normally. But unlike other voice services that are currently offered over Internet networks, there are no added numbers to dial, no extended call setup times, or the requirement for special equipment and software.

Lucent's Softswitch technology connects the signaling and database systems that operate the public telephone network to the service provider's IP network. The IP network translates circuit-based messages from customers to IP for low-cost transport and then, if required, converts them back to circuit-based messages for delivery over the public telephone network. If reliability on the IP network drops below the quality of the traditional network, the traffic is automatically switched over to the traditional network.

The Softswitch does not require expensive investments in CO hardware. It runs on general-purpose computers, such as Sparc workstations running the Solaris operating system. The platform can be located within the CO or at another location and is SNMP manageable with support from commercial network management tools. Multiple distributed systems offer load-balancing to enhance network performance and reliability. The platform interconnects with the PSTN through trunking arrangements with the local carrier. Call setup and routing between the IP network and PSTN is performed by SS7.

8.11 Conclusion

Although it appears to be the direction in which everyone is headed, it might be naïve to suppose that IP will alone provide the solution to all enterprise communications needs any time soon. Cisco has taken a practical approach in this regard. The company's value proposition is migration not replacement; integration rather than logical separation; and interoperability with existing network architectures all the way down to legacy systems. Its intelligent multiservice networks provide connectivity for all kinds of traffic, from the desktop to the WAN and to the core of the service provider. This approach provides the flexibility needed to adapt to changing business and application requirements. It results in performance improvement and cost savings, and globally, it reduces network complexity and increases the availability of corporate communications. The broad array of interoperable multiservice systems Cisco is able to bring to bear for effective integration of data, voice, and video provides choices of scale and capability across ATM, frame relay, and IP infrastructure technologies.

More information

The following Web pages contain more information about the topics discussed in this chapter:

3Com Corp.: http://www.3com.com/
Ascend Communications: http://www.ascend.com/
ATM Forum: http://www.atmforum.com/
Bay Networks (see Nortel Networks)
Cabletron: http://www.cabletron.com/
Cisco Systems: http://www.cisco.com/
Hewlett Packard: http://www.hp.com/
IBM: http://www.ibm.com/
Intel: http://www.intel.com/
Internet Engineering Task Force: http://www.ietf.org/
Lucent Technologies: http://www.lucent.com/
Microsoft: http://www.microsoft.com/
Nortel Networks: http://www.nortel.com/

Current regulatory environment

Certain characteristics of the Internet have ramifications for the development of regulations. Clearly, the same regulations that set the rules for traditional telephone service would be difficult or impossible to apply and enforce for Internet services.[1] In the United States, for example, it would not be feasible for each of the 50 states to claim regulatory authority over Internet traffic that traverses their jurisdiction in the same way that they do for traditional telephone traffic. In the circuit-switched world, the architecture makes it possible to set boundaries.

To complicate matters, many different communications can share the same physical facilities simultaneously (see Figure 9.1). Not even service providers that carry such services know what type of data packets are passing through their networks at any given

---

1. Instead of reiterating the history of telecom regulation in the United States—from the Communications Act of 1934 to AT&T's divestiture of the Bell Operating Companies in 1984 and noteworthy events thereafter—this chapter focuses on the Telecommunications Act of 1996, particularly those aspects that lay the ground rules for competition in the telecommunications industry and relate to voice-data convergence over IP.

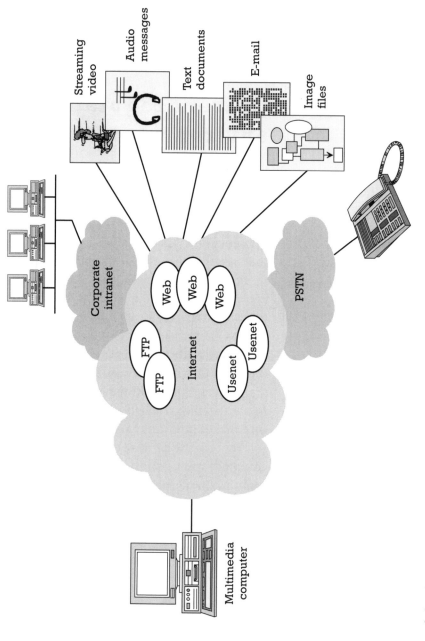

Figure 9.1   A user may open multiple voice-data-video sessions through the same Internet connection.

moment. The use of packets allows a user to engage in a voice conversation with a friend, check for mail, participate in a newsgroup, and browse through Web pages—all at the same time. And since a dedicated end-to-end channel need not be established for each communication, the notion of a discrete location for senders and receivers also breaks down.

Since a voice call originates and terminates at two known points, it can be assigned a jurisdictional category such as local, intraLATA toll, interLATA intrastate, interLATA interstate, intraLATA interstate, and international.[2] The fact that users dial ten or eleven digits instead of seven for calls outside their area code—and in some cases a two-digit country code—provides some indication of the categorization of a particular call. Because the Internet is a dynamically routed, packet-switched network, it can be difficult to identify the source and destination of a "call." Users generally do not open Internet connections to call a discrete recipient, but access various Internet sites during the course of a single connection. As noted, the same connection may support both data and voice, which traditionally have been regulated differently.

The Internet has evolved so fast that governments have not had a chance to seize control and limit user activity through regulation. Although there have been widely publicized attempts by China to curb internal Internet growth and impose access and content restrictions, the fact is that the Internet's connectionless, "adaptive" routing system makes it difficult for governments to exercise strict control. Unlike connection-oriented networks where traffic is sent over a specific circuit, Internet traffic is split into packets that are routed dynamically between multiple points based on the most efficient route at any given moment. When packets are prevented from reaching one location, the routers send them to another location until eventually they all arrive at their destination. This fluidity makes it difficult to rely on the traditional concept of geographic borders to stop the flow of communication.

In the United States, much of the debate over the Internet has centered around whether ISPs that transport voice calls over the Internet,

---

2. Local access and transport areas (LATAs) were an outcome of AT&T's divestiture agreement with the U.S. Justice Department in 1984. At that time, not only did AT&T divest itself of the Bell operating companies (BOC), about 200 LATAs were created to define the service boundaries of the local exchange carriers. Calls placed to destinations outside these boundaries are handled by long-distance carriers. When the telephone companies get permission to enter the long-distance market, they will be able to handle these types of calls.

or any part of it, should be treated as telecommunications carriers with respect to such things as universal service obligations, reciprocal compensation, access charges, and rate regulation. Traditional carriers contend that VoIP service providers have an unfair competitive advantage in that they are not subject to the same regulatory and cost burdens. ISPs and IP-based next-generation service providers claim that being forced to shoulder these burdens would snuff out any chance they have of ever competing against the larger, well-established telecommunications carriers. Besides, they say, such regulation would go against the intent of the Telecommunications Act of 1996 to create a more competitive telecommunications industry.

Despite IP's use as a medium for messaging and information sharing—and now voice services—the ISPs and next-generation carriers have not been considered telephone companies and have not been treated as such by the FCC, which has regulatory authority over telecommunications in the United States. For the most part, the FCC and other federal agencies have abided by the Clinton Administration's "hands-off" approach toward regulating any aspect of the Internet.[3] However, nothing prevents interest groups from putting pressure on Congress and the FCC to enact rules on their behalf, but which are couched in terms of the public interest.

However, the application of old rules to new technologies creates problems. Attempts to clarify old rules to make them fit today's environment also create problems. And both the old rules and the clarifications are subject to different interpretations intended to benefit one or another of the parties seeking competitive advantage. Many lawsuits and appeals have been filed to the federal court system in an attempt to sort through all the claims and counterclaims about the meaning of the

---

3. At the time President Clinton issued the "hands off" directive, no less than 18 government agencies had been looking at regulating various parts of the Internet, including the Department of Commerce, which favored taxing electronic transactions over the Internet. In 1998, Congress proposed a three-year moratorium on taxing electronic commerce, which was signed into law by President Clinton. The Clinton Administration even achieved an agreement with the World Trade Organization for the Internet to be free of customs tax and duty. But Congress is not above passing legislation to control other aspects of the Internet, the most controversial being the Communications Decency Act of 1996, which was intended to restrict access to pornography by children. Eventually, the Supreme Court ruled that the Act improperly restricted the free-speech rights of adults. Since then, numerous content filtering schemes have been proposed in the House and Senate.

Telecommunications Act of 1996 and about the authority of the FCC to make certain rules. However, the Federal Courts in different jurisdictions have themselves issued conflicting rulings.

These issues have enormous ramifications with respect to voice-data convergence. For example, at what point does an "information" provider become a "telecommunications" provider? If information providers are suddenly classified as telecommunications providers, how will this impact competition, rates for consumers, and continued innovation in products and services? The purpose of this chapter is to ferret out the key issues. The solutions will likely take a while longer to evolve.

9.1 The core issue

The Telecommunications Act of 1996—which became law on February 8, 1996—establishes a procompetitive, deregulatory framework for telecommunications in the United States. The intent of this law is to rapidly accelerate private sector deployment of advanced telecommunications and information technologies and services to all Americans by opening all telecommunications markets to competition.

For years, competitive service providers have wanted to offer local telecommunications services, a $110 billion market dominated by the traditional monopolistic telephone companies under the former Bell system (i.e., Bell operating companies, or BOCs). At the same time, the local telephone companies have wanted to expand into long-distance services—a $90 billion market. Among the objectives of the Telecommunications Act of 1996 is to encourage competition in both markets by all types of carriers, provided that a level playing field can be established for all participants.

Accordingly, Section 271 of the Telecommunications Act provides a 14-point checklist that is meant to ensure a competitive market for local and long-distance services (see Appendix A). State regulators and the FCC would have to review this checklist for full compliance before any former BOC would be allowed to provide long-distance telephone service. This is meant to ensure that they do not impose burdensome restrictions on would-be competitors, even as they themselves branch out into new markets.

By late 1999, more than three years after the Telecommunications Act became law, no former BOC had been able to demonstrate full compliance with the checklist. Although the incumbent telephone companies have submitted many applications for long-distance entry in the belief that they had met all of the requirements, they still have not been given permission to enter that market. The applications have been consistently denied because, in many cases, the incumbent telephone companies had only articulated a "plan" to meet the checklist requirements and had not actually demonstrated that real and sustained competition exists in their local serving areas.

These application denials set off a long legal fight in federal courts and appeals to Congress to curb the FCC's authority, which culminated in a decision by the Supreme Court. The arguments of the incumbent telephone companies focused on whether the FCC had the authority to:

▶ Set the prices competitors must pay incumbents for access to their local phone networks;

▶ Require incumbents to allow new entrants to "pick and choose" provisions of existing interconnection agreements and use them in their own contracts;

▶ Force incumbents to recombine unbundled network elements (UNEs) for competitors.

In October 1998, the Supreme Court heard oral arguments from all sides and issued its decision in January 1999. The high court upheld the FCC's authority to set price rules for providers who want to start local phone service, a right the states had claimed for themselves.

Pricing rules stipulate how expensive it is for competitors to interconnect with the prevailing local provider, which ultimately decides whether consumers have a choice for local phone service. Previously, incumbent carriers had priced interconnection so high that it discouraged local competition. Instead of letting the states set prices, which would have forced competitors to grapple with 50 sets of rules, the FCC can implement a single standard that sets interconnection charges at a level that will encourage competition.

After issuing the ruling upholding the authority of the FCC, the Supreme Court declined to review a lower court's decision upholding the constitutionality of the Telecommunications Act itself, which the Bells had challenged on the basis that it singled them out for punishment.

The high court, however, upheld the challenge to FCC bundling rules that required the incumbents to separate previously combined elements of their networks and let competitors pick and choose from among those elements for recombination in new leasing contracts. The regional Bells successfully argued that it was unfair that new entrants could pick and choose network elements to lease, since the Bells made the network investment. Unbundling would make it unprofitable for them to lease their services to competitors who will offer them for resale at lower prices. The FCC lost its argument that separating network elements would help jumpstart competition.

As these issues were being hashed out, improvements in IP telephony made it possible for the first time to offer commercial telephone service. Some IP telephony providers discovered they could offer phone-to-phone long-distance service without necessarily paying access charges to the incumbent telephone companies for originating and terminating the calls.

9.2 Information versus telecommunication

The potential competitive ramifications of IP telephony were apparent to the telephone companies in 1995. As noted in Chapter 2, at about this time improvements in microprocessors, DSP, codec technology, and routing protocols all came together to make feasible products for mainstream use. While industry analysts were still proclaiming that IP telephony would not progress beyond the gadget stage for hobbyists,[4] America's Carriers Telecommunication Association (ACTA)—a trade association of 130 competitive long-distance carriers—clearly understood that IP telephony could seriously cut into telco revenues if the FCC did not put a stop to it.

Accordingly, ACTA petitioned the FCC in March 1996 to stop companies from selling software and hardware products that enable use of the

---

4. All new technologies seem to go through a multi-stage acceptance cycle. First, the technology is pooh-poohed as "unworkable," ostensibly because there are too many problems. Second, the technology becomes "feasible," but only for niche applications. Third, the technology becomes worthy of "consideration" for enterprise deployment, but users should be careful anyway. Finally, the technology is "hot, hot, hot"—if companies do not climb aboard now, they risk losing competitive position.

Internet for long-distance voice services, creating the ability to bypass the access charge and reciprocal settlement mechanisms of local, long-distance, and international carriers and permitting calls to be made for virtually no cost.[5] ACTA made compelling arguments that continued bypass of traditional carriers for placing phone calls will result in serious economic hardship on all existing participants in the long-distance marketplace and the public. Accordingly, ACTA requested the FCC to exercise jurisdiction over the use of the Internet for unregulated interstate and international telecommunications services. ACTA noted that long-distance and international carriers must be approved by the FCC to operate in the United States and must file tariffs before both the FCC and state public service commissions. All of these requirements are stipulated in the Communications Act of 1934 and the Telecommunications Act of 1996.

To bolster its arguments, ACTA argued that providers of Internet "phoneware" do not contribute to the universal service fund[6] or make other contributions to extend telephone service to remote areas, libraries and schools. Under the Telecommunications Act of 1996, this obligation had been extended to all communications carriers, not just the large ones. Left unchecked, the growth of Internet telephony and improvements in the technology could damage the telecommunications infrastructure of the country, which requires huge capital investments for construction and maintenance, according to ACTA.

In response to ACTA's effort to have the FCC crack down on telephone calling over the Internet, a user group called Voice On Net (VON) organized a coalition of users and software vendors to present its views to

---

5. See Appendix B for the original full-length petition submitted to the FCC by ACTA, as well as Appendixes C and D for responses by the National Telecommunications Industry Association (NTIA) and the Voice On Net (VON) Coalition, respectively.

6. The "universal service" system was originally designed to make local telephone service available to low-income consumers who might not otherwise be able to afford it. In many cases, universal service policies have required that rates for certain services be set above the cost of providing those services in order to generate a subsidy to be used for this purpose. The Telecommunications Act of 1996 updates the traditional universal service system, expanding both the base of companies that contribute to the fund and the category of customers who benefit from service discounts. In addition to helping low-income consumers, the universal service fund is now used to help schools, libraries, and rural health care providers obtain affordable telephone service. The universal service charges are passed on to consumers and now appear as a separate item on most telephone bills. The charge is about $1 per month.

the FCC. VON argued that the continued unrestricted growth and development of the Internet and IP telephony are in the public interest and could lead to entirely new opportunities for personal and business communications as well as new education, health care, and entertainment services. Accordingly, the FCC should encourage and not stifle this innovation with regulation.

The National Telecommunications Industry Association (NTIA), representing the views of the Clinton Administration, also chimed in on the issue, advising the FCC to deny the ACTA petition. The NTIA claimed that ACTA had misinterpreted existing law and that its arguments reflected a fundamental misunderstanding of the way in which the Internet operates and of the services making use of the Internet.

The FCC did not rule on the ACTA petition, claiming that the issue is a complex one and deserving of all the time necessary to reach a decision. This allowed IP telephony to progress to the point where it is today—able to compete with established long-distance carriers. In fact, the technology is evolving so rapidly that ACTA may have presented the FCC with a fait accompli. Not only has a new class of next-generation carriers emerged, which use IP to carry voice over high-speed fiber nets from phone to phone, but a growing number of ISPs of every type and size are offering some form of VoIP service as well.

In early 1999, 43% of local and regional ISPs surveyed by Infonetics were already offering some form of VoIP service. Among competitive local exchange carriers, 10% were offering VoIP service in 1998, and 50% said they plan to offer it in 1999. And among national ISPs, 54% indicated that they planned to offer IP telephony in 1999. ISPs tend to limit voice offerings to their own networks, where high-quality connections are easier to maintain, thus slowing their evolution to full-fledged telephone companies. With partnerships that link the well-managed networks of ISPs, however, even this barrier is falling.

The FCC has always contended that an information service should not be classified as a telecommunications service just because it could also be used for telephony. However, with IP voice phone-to-phone, the information component is no longer present, which removes the FCC's original justification for not treating such activity as a telecommunications service.

Since the ACTA petition, and in light of technological advancements, the FCC has clarified its position on this and related matters.

9.3 The FCC weighs in

Since 1995, the FCC had always denied that it had any intention of regulating the Internet, even in cases where it carried voice as a commercial service. Yet, the FCC appears to leave the door open for such regulation in the future.

In April 1998, the FCC submitted a report to Congress on the status of the universal service fund, which also enumerated its position on IP telephony.[7] The report revisited many of the FCC's major decisions related to the implementation of the universal service provisions of the Telecommunications Act of 1996. The commission reiterated its commitment toward ensuring that low-income and rural consumers have access to local telephone service at affordable rates and that an evolving level of telecommunications services is available and affordable for all Americans. At the same time, the report reaffirmed the commission's commitment to encouraging the continued development of new services and technologies related to the Internet.

The commission analyzed several definitions in the Telecommunications Act and the impact of its interpretations of those definitions on the current and future provision of universal service. The commission concluded that the categories of "telecommunications service" and "information service" in the Telecommunications Act are mutually exclusive and consistent with pre-existing definitions. The commission found generally that Congress intended to maintain a regime in which information service providers are not subject to regulation as common carriers merely because they provide their services via telecommunications. The commission also reaffirmed that information service providers are not subject to universal service obligations, the access charges paid by long-distance providers, or rate regulation.

In its report, the commission stated that the provision of transmission capacity to Internet service providers constitutes the provision of "telecommunications." As a result, telecommunications providers offering leased lines to Internet service providers must contribute to the universal service fund. The commission noted that at least some leased-line providers understand this and comply with that requirement, and the prices paid by Internet service providers for their leased lines reflect that universal service obligation. As Internet-based services grow, Internet service

---

7. Report No. CC 98-9, Common Carrier Action, April 10, 1998.

providers will have a greater need to lease lines. The payments for those additional leased lines will in turn lead to increased universal service contributions by leased-line providers.

In cases where an Internet service provider uses its own transmission facilities to provide an information service, the commission's rules do not require Internet service providers to contribute to the universal service fund. However, the commission stated that, as a "theoretical matter," it may exercise its discretion under the statute to require such providers to contribute to universal service.

The commission also examined the application of the statutory definitions to various new services, particularly Internet telephony. Some forms of Internet telephony involve the use of ordinary telephone handsets and employ IP in routing calls to their destination. The commission noted that certain forms of phone-to-phone IP telephony lack the characteristics that would render them "information services" within the meaning of the statute, and instead bear the characteristics of "telecommunications services." The commission, however, did not find it appropriate to make any definitive pronouncements in the absence of a more complete record focused on individual IP service offerings.

In the intervening year since the FCC issued this report, there has been no attempt by the FCC to regulate the Internet, or take steps to treat VoIP as a telecommunications service. As late as March 1999, FCC Chairman William E. Kennard provided even more reassurance on the matter, saying "as long as I am chairman of the FCC, we will not regulate the Internet."[8] However, some telephone companies—notably BellSouth and U S West—have cited the FCC's report to Congress as justification for making companies that use the local phone network and Internet Protocol to complete long-distance phone calls pay access charges just like all other companies providing long-distance phone service (discussed later).

9.4 Asserting jurisdiction over the Internet

The FCC's jurisdiction over the Internet had been debated for several years. In fact, the states had claimed jurisdiction over their piece of the Internet. The FCC itself never emphatically claimed this authority until

---

8. Remarks delivered by William E. Kennard before a national investors' conference sponsored by Legg Mason in Washington, D.C., on March 11, 1999.

February 1999. The opportunity came when it moved to address the payment of reciprocal compensation for Internet-bound traffic and, in the process, established that it alone has jurisdiction over Internet traffic. The FCC's reasoning: most traffic does not begin and end locally, but ends up at sites far beyond the local server. This makes it interstate traffic, which puts it firmly under the FCC's jurisdiction.

In response to requests by carriers that the commission clarify how local telephone companies should compensate one another for delivering traffic to Internet service providers, the FCC concluded that carriers are bound by their existing interconnection agreements, as interpreted by state commissions, and thus are subject to the reciprocal compensation obligations contained in those agreements.

The FCC declared that Internet traffic is jurisdictionally mixed and appears to be largely interstate in nature, putting the commission in agreement with the large telephone companies, which have always contended that such traffic should be categorized as interstate. But the decision preserves the rule that exempts the Internet and other information services from interstate access charges. This means that those consumers who continue to access the Internet by dialing a seven-digit number will not incur long-distance charges when they do so.

Specifically, the commission had been asked by industry participants to determine whether local telephone companies are entitled to receive reciprocal compensation for delivering calls to their ISP customers. Generally, new entrants to the local telephone business contend that calls to ISPs are local traffic and, therefore, subject to reciprocal compensation. Incumbent local telephone companies, on the other hand, generally contend that calls to ISPs are interstate in nature and that they are, therefore, beyond the scope of reciprocal compensation agreements.

The FCC noted that it traditionally has determined the jurisdictional nature of communications by the end points of the communication. Accordingly, it concluded that the calls to the ISP's local servers do not terminate there, but continue to their ultimate destinations, specifically at Web sites that are often located in other states or countries. As a result, the commission found that, although some Internet traffic is intrastate, a substantial portion of Internet traffic is interstate and, therefore, subject to federal jurisdiction.

This decision does not, however, determine whether calls to ISPs are subject to reciprocal compensation in any particular instance. The commission noted that parties may have agreed that ISP-bound traffic should be subject to reciprocal compensation, or a state commission, in the exercise of its statutory authority under Sections 251 and 252 of the Telecommunications Act to arbitrate interconnection disputes, may have imposed reciprocal compensation obligations for this traffic. In either case, the commission noted that carriers are bound by their existing interconnection contracts, as interpreted by state commissions.

The FCC also stated that adopting a federal rule to govern reciprocal compensation in the future would serve the public interest, but tentatively concluded that commercial negotiations are the ideal means of establishing the terms of interconnection contracts, and reciprocal compensation agreements in particular.

Critics of the FCC's decision claim that asserting jurisdiction over the Internet is the first step toward regulating it. The FCC denies this, claiming that its action has reconfirmed the so-called Internet exemption and that consumers will see no new charges on their Internet or phone bills in either the short run or the long run. By the FCC asserting its jurisdiction over Internet traffic, the states have no power to impose long-distance charges for local calls to the Internet. Thus, the FCC's action protects consumers and its continued hands-off policy toward the Internet will allow it to keep growing.

As viewed by the FCC, the reciprocal compensation order not only allows competition to flourish on the Internet, but it takes the industry closer to the type of competitive and open telecommunications marketplace that Congress envisioned when it wrote the Telecommunications Act—a market where the relationships between carriers are governed by contracts, not government regulation.

9.5 Access charges

Access charges are paid by long-distance providers to support the cost of maintaining the local phone network and ensure universal service. At this writing, long-distance carriers must pass on roughly six cents per minute in access charges for every interstate call, which is split between the local telephone companies that originate and terminate the call. Access charges not only help provide support to universal service, they help keep local rates affordable for consumers.

Although more companies are providing long-distance service via the Internet, or using IP technology, they still depend upon the local phone network to receive and deliver the calls. The telephone companies contend that if these companies are allowed to get by without paying access charges, it could put universal service at risk and cause local rates to go higher to offset the losses. These losses could be substantial. According to BellSouth, the amount of IP telephone traffic that escapes access charges could amount to 13% in the year 2000, and escalate thereafter, if allowed to continue.

ISPs currently enjoy a special status as "enhanced service providers," by which the FCC exempts them from having to pay access charges. However, this exemption applies only to data-oriented or information services and is transitional, which means it will one day come to an end. It is up to the FCC to decide exactly when this will happen.

The issue today concerns the voice traffic that these ISPs carry, especially the kind that occurs phone-to-phone, which means it lacks the information component. In the view of the telephone companies, if it functions like a phone and rings like a phone, it is a phone, and long-distance calls that originate and terminate on the local exchange network are subject to access charges, just like the calls carried by any other long-distance company.

As noted earlier, some telephone companies have seized on the FCC's distinction between information service providers and telecommunications service providers to start implementing access charges on long distance competitors that use the Internet Protocol for phone-to-phone IP telephony.

Further justification comes from the same FCC report to Congress, in which the commission states:

> Specifically, when an IP telephone provider deploys a gateway within the network to enable phone-to-phone service, it creates a virtual transmission path between points on the public switched telephone network. ... From a functional standpoint, users of these services obtain only voice transmission, rather than information services such as access to stored files. Routing and protocol conversion within the network does not change this conclusion, because from the user's standpoint there is no net change in form or content.

As if this were not enough, the telephone companies point to the Telecommunications Act itself, which defines "telecommunications" as the:

> transmission, between or among points specified by the user, of information of the user's choosing, without change in the form or content of the information as sent and received.

9.5.1 Unilateral implementation

The first telephone company to justify access charges on these grounds was BellSouth. In September 1998, BellSouth informed customers that it would begin applying access charges on telephone-to-telephone long-distance calls placed using IP technology. BellSouth even cited the portion of the FCC report to Congress in April of that year, which stated that "certain IP telephony services lack the characteristics that would render them 'information services' and instead bear the characteristics of telecommunications services."

BellSouth concluded that such carriers should be subject to appropriate state and interstate access charges and transitioned their service to circuits used to carry long-distance calls. Use of these circuits requires the assessment of access charges. BellSouth noted that the same access charges currently are paid by all other long-distance carriers, per FCC rules, and that the Telecommunications Act is "technology indifferent" when it comes to what facilities are used to offer common carrier telephone service.

As BellSouth sees it, as the amount of long-distance traffic attributable to Internet telephony continues to grow, a substantial amount of access support would otherwise go uncollected, forcing other carriers and their customers to subsidize the business plans of IP telephony providers.

With regard to the universal service fund, BellSouth agrees that as long as ISPs are defined as information service providers they are exempt from both access charges and the requirement to contribute to the support of universal service. However, as telecommunications service providers, companies offering long-distance telephone service over the Internet should not only pay access charges, but they should contribute to the universal service fund as well.

366 IP Convergence: The Next Revolution in Telecommunications

Later the same month, U S West followed BellSouth's lead. It notified companies that use the local phone network and Internet Protocol to complete long-distance phone calls they would have to pay access charges just like all other companies providing long-distance phone service. U S West justified its action using the same arguments as BellSouth and citing the same passages of the FCC's report to Congress in April 1998.

9.5.2 Bilateral implementation

While some regional telephone companies have unilaterally imposed access charges on IP telephony service providers, Bell Atlantic has recognized the opportunities in linking up with IP telephony carriers and negotiating termination agreements, which is the way the FCC prefers that such arrangements be handled.

In 1998, Bell Atlantic became the first regional telephone company to provide a gateway service to carriers for the completion of phone calls made over IP data networks. The gateways provide translation services between the traditional circuit-switched voice network and new IP data networks that provide telephone service. Bell Atlantic's first gateway customer is ITXC Corp., which relies on the service for "last-mile" connections for international telephone calls coming into the New York City metropolitan area over the Internet. ITXC oversees the largest network of operators of Internet telephony gateways, and provides routing and clearinghouse services to the carriers of IP telephony services.

As a member of ITXC's WWeXchange network, the originating carrier hands the call over to ITXC which then routes all New York-bound calls over an IP network to Bell Atlantic. Bell Atlantic, in turn, translates the data call back to voice and sends it over its local network to the person being called. WWeXchange provides intermediary functions needed to link multiple networks.

Bell Atlantic's wholesale gateway service may ultimately eliminate the need for carriers to build and operate their own facilities and provide access throughout the company's East Coast region. Bell Atlantic is rapidly evolving from a voice telecommunications company that carries data into a data company that also transports voice. The company is aggressively modernizing its network with the latest in data transmission and switching technology, including ATM and IP. Bell Atlantic also is building state-of-the-art, next-generation data networks from the ground up, as it prepares to enter the long-distance market.

Although most calls to the East Coast travel the last mile on Bell Atlantic's network already, the regional telephone companies previously accepted calls only in traditional circuit-switch format. Under the agreement with ITXC, Bell Atlantic will accept calls in IP format. The termination agreement with ITXC for IP traffic is much like the agreements Bell Atlantic already has to terminate circuit-switched traffic from traditional long-distance carriers like AT&T, MCI WorldCom, and Sprint.

Internet callers served by ITXC affiliates around the world place their calls using standard telephones. The calls are routed by prepaid calling card companies, call-back companies, and other resellers to PSTN/IP gateways, which convert the calls from circuit-switched form to IP. The gateway operator hands the calls off to ITXC for delivery to their destination.

ITXC routes some calls over the public Internet or over a private data network to Bell Atlantic, for termination in its territory. Terminating gateways operated by Bell Atlantic then translate the calls back into traditional voice format for completion over the Bell Atlantic local network. Bell Atlantic, as an affiliate in the ITXC WWeXchange Service, is paid to terminate calls routed to it by ITXC just as it is paid to terminate traditional long-distance and international calls by traditional carriers.

9.6 The international scene

The many challenges the Internet generates impact the United States more rapidly and acutely not only because it has the largest percentage of the Internet's infrastructure and traffic, but because it is more deeply concerned with fostering competition in the telecommunications market than almost any other country. In fact, countries that are concerned at all with similar issues frequently look to the United States, specifically the FCC, for guidance on how to approach their policy issues.

In most countries, there are no formal regulatory barriers to using the Internet for telephone calls. The telecom laws of most countries were written long before there was an Internet, and even today IP telephony is still considered a hobbyist's toy that does not pose a significant threat to established carriers. However, there are some countries that have laws that could be interpreted as outlawing the practice.

The Netherlands, for example, allows only telephone companies to handle real-time speech over public networks; presumably, this includes the Internet. In addition, since local call charges are usage-based, it could be costly to run an IP/PSTN gateway that distributes calls locally.

There are only isolated instances of strict prohibitions against the transmission of voice over the Internet. In Pakistan, Internet telephony is banned and it is possible that the use of other streaming audio and video products can also be restricted. Section 3.3 of the contracts required by ISPs in Pakistan reads as follows:

"Voice transmission of any sort is strictly prohibited: Violation of this clause shall lead to prosecution according to the Telephone and Telegraph Act of 1885 by the relevant authority. In such a case the Service Provider shall terminate the service without notice and with no refunds of any sort."

Countries that maintain a liberal stance on Internet issues will likely take note of the FCC's inclination to make a distinction between "information" and "telecommunication" services. If the decision of some carriers in the United States to impose access charges on phone-to-phone IP traffic withstands regulatory scrutiny, this could be used as justification by other countries to start putting surcharges on services that travel the Internet or use IP technology in ways that compete with conventional carrier-provided services.

9.7 Regulation of advanced services

As noted, one of the fundamental goals of the Telecommunications Act of 1996 is to promote innovation and investment by all participants in the telecommunications marketplace in an effort to stimulate competition. These goals apply to advanced services such as those delivered over broadband DSL technologies. Today, both incumbent carriers and new entrants are in the early stages of developing and deploying DSL for advanced voice and data services, and the market is ripe for competition to develop in a robust fashion.

To ensure that this happens, the FCC is looking at ways to remove barriers to competition so that new entrants can compete effectively with incumbent carriers and their affiliates in the provision of advanced services. At the same time, the FCC wants to ensure that incumbent carriers are able to make their decisions to invest in, and deploy, advanced telecommunications services based on market demand and their own strategic business plans, rather than on regulatory requirements.

DSL technology, coupled with packet-switched networks, addresses the performance constraints of the current local loop. With DSL technology, two modems are attached to each telephone loop: one at the subscriber's premises and the other at the telephone company's central office. The use of DSL modems allows transmission of data over the copper loop at vastly higher speeds than can be achieved with analog data transmission. Moreover, combining DSL technology with packet switching permits more efficient use of the network because information generated by multiple users can be sent over a telecommunications facility that in a circuit-switched environment may be dedicated to only one customer for the duration of a call. In addition, the customer can make ordinary voice calls over the public switched network while using the same line for high-speed data transmission.

When the DSL-equipped line carries both ordinary telephone service and data channels, the carrier separates the two streams when they reach the telephone company's central office. This is generally done by a device known as a digital subscriber line access multiplexer, or DSLAM. The DSLAM and central office DSL modem send the customer's POTS traffic to the public, circuit-switched telephone network. The DSLAM sends the customer's data traffic (combined with that of other DSL users) to a packet-switched data network. Thus, the data traffic, after traversing the local loop, avoids the circuit-switched telephone network altogether.

Once on the packet-switched network, the data traffic is routed to a location selected by the customer, such as a corporate LAN or an Internet service provider. That location may itself be a gateway to a new packet-switched network or set of networks, like the Internet, a corporate VPN, or private intranet.

9.7.1 Collocation

To facilitate the development of competition in the advanced services market, the FCC has strengthened its collocation rules to reduce the costs and delays faced by competitors that seek to collocate equipment in

370 IP Convergence: The Next Revolution in Telecommunications

an incumbent's central office. For example, the commission requires incumbents to make available to requesting competitors space for their equipment. Moreover, when collocation space is exhausted at a particular location, the FCC requires incumbents to permit collocation in adjacent controlled environmental vaults or similar structures to the extent technically feasible.

A collocation method used by one incumbent or mandated by a state commission is deemed technically feasible for any other incumbent and must therefore be offered to competitors as well. Incumbents may not require a competitor's equipment to meet more stringent safety requirements than those the incumbent imposes on its own equipment.

To compete effectively in the advanced services marketplace, competitive service providers must be permitted to collocate integrated equipment that lowers costs and increases the services they can offer their customers. Accordingly, incumbents must permit competitors to collocate all equipment used for interconnection and/or access to unbundled network elements (UNEs)—even if it includes a switching or routing function. Furthermore, incumbents cannot require that the switching or routing functions of integrated equipment be disengaged.[9]

Incumbents are further required to allow competing carriers to establish cross-connects to the collocated equipment of other competing carriers at the incumbent's premises. The competitors may even construct their own cross-connect facilities between collocated equipment without having to purchase any equipment or cross-connect capabilities from the incumbent at tariffed rates. Often this is as simple as running a copper or fiber transmission facility from one collocation rack to an adjacent rack. Even where competitive equipment is collocated in the same room as the incumbent's equipment, the FCC requires the incumbent to permit the new entrant to construct its own cross-connect facilities.

The FCC rules also allow competitors to tour the incumbent's entire central office in cases where the incumbent has denied the competitor collocation space. Incumbents must provide a list of all offices in which there is no more space. They must also remove obsolete, unused equipment, in order to facilitate the creation of additional collocation space within a central office.

---

9. The FCC does not require incumbent local service providers to permit the collocation of equipment that is not necessary for either access to UNEs or for interconnection.

These collocation rules for facilitating the provision of advanced services by competitors serve as minimum standards, and permit any state to adopt additional requirements. In issuing these rules, the FCC was mindful of the Supreme Court's opinion issued in January 1999, which vacated its previous rules delineating what network elements incumbents must make available to competitors. At this writing, the FCC was still seeking comments on the issue of whether network elements used in the provision of advanced services should be unbundled.

9.7.2 Spectrum compatibility

The FCC has also issued rules regarding spectrum compatibility, allowing competitive providers to deploy innovative advanced services technology in a timely manner. Spectrum compatibility refers to the ability of various loop technologies to reside and operate in close proximity while not significantly degrading each other's performance. According to the FCC, any loop technology that complies with existing industry standards, has been successfully deployed by any carrier without significantly degrading the performance of other services, or has been approved by the FCC, any state commission, or an industry standards body is presumed acceptable for deployment. An incumbent may not deny a carrier's request to deploy technology that is presumed acceptable for deployment, unless it can demonstrate that the particular technology will significantly degrade the performance of other services.

Although these rules focus on the provision of advanced services, the FCC has emphasized that they apply to all telecommunications services, whether traditional voice services or advanced data services.

9.7.3 Line sharing

While there is nothing to stop incumbent carriers from offering competitors space on their loops for advanced services, a future possibility is that the FCC may mandate line sharing for the delivery of advanced services by multiple service providers. This arrangement is entirely feasible with various DSL technologies, which allows the use of different frequencies to transport voice or data over that line. For example, ADSL technology allows a high-speed data channel to run on higher frequencies above the frequency used for delivery of analog voice signals. By separating the line into a voice channel and an advanced services channel, such a line can carry both voice and advanced services traffic simultaneously, and potentially, each service could be provided by a different carrier. The issue is whether competitors should have the right to run high frequency data signals, or other advanced services, over the same line as the incumbent's voice signal.

Shared line access makes it possible for a competing carrier to offer advanced services over the same line that a consumer uses for voice service without requiring the competing carrier to take over responsibility for providing the voice service. Such shared line access would enable new entrants to focus solely on the advanced services market without having to acquire the resources or the expertise to provide other types of telecommunications services, such as analog voice service. Shared line access could also remove any cost disadvantage that a provider of advanced services might face if it had to use a standalone line. A competitor, therefore, may want to take advantage of the ability of advanced services technology, such as ADSL, to run on the frequency above the analog voice channel by providing only high-speed data service, without voice service, over a loop.

The FCC believes each customer should be able to choose from a broad array of services and from whom to obtain these services—just as today they can choose one carrier to provide local service, another to provide long distance, and a third to provide Internet access over the same line. A customer should, therefore, have the right to purchase voice service from one carrier and advanced services from another over the same line. In particular, the FCC believes that allowing consumers to keep their voice service provider, while allowing them to obtain advanced services on the same line from a different provider, will foster consumer choice and promote innovation and competitive deployment of advanced services.

9.8 Conclusion

In passing the Telecommunications Act of 1996, Congress expressed its intent to implement a procompetitive deregulatory national communications policy. Congress recognized that the Internet provides both a space for innovative new services, as well as potential competition for existing communications technologies. The FCC's role has been to ensure a level playing field where efficiency and market forces drive competition.

Clearly, the federal government's efforts to avoid burdening the Internet with regulation have contributed to its rapid development as a useful tool that virtually anybody can use for practical benefit—including schools and libraries.

However, the continued industry turmoil concerning the Internet stems from the difficulty in applying existing service boundaries and regulatory classifications to Internet-based services. This turmoil is especially contentious with regard to the convergence of the telephony and Internet worlds. Many in government and the telecom industry have called for regulatory action for the Internet to match that of the telephone companies. But many of the reasons for regulating the telecommunications industry may no longer exist, and the Internet's power and ubiquity may actually provide an opportunity to deregulate telecommunications even further.

In pursuing the opposite course—eliminating regulatory roadblocks and other disincentives to investment for new entrants—the FCC has sought to encourage both incumbents and new entrants to develop innovative solutions that go beyond the capabilities of the existing network. What makes the Internet such a catalyst for innovation is that it offers a low-cost means of providing access to all members of the supply chain, as well as a standards-based platform for creating applications that span diverse industries. This has eliminated many of the complexities of doing business with multiple suppliers and also created a more level playing field for new and smaller competitors.

With the FCC's authority to act in these matters reaffirmed by the Supreme Court, many of the obstacles to fulfilling the intent of the Telecommunications Act of 1996 have been removed. Nevertheless, the goals of the Telecommunications Act will take many more years to be fully realized.

In an effort to get sound technical advice for developing policies regarding the difficult technical issues in such areas as spectrum management, convergence, and the Internet, the FCC has created a Technical Advisory Council. The council—comprised of recognized technical experts in diverse fields—will help provide the expertise the commission needs to stay abreast of innovations and new developments in the communications industry so that its policies will be more effective in fulfilling the public interest during this era of exceptionally rapid technological change.

More information

The following Web pages contain more information about the topics discussed in this chapter:

AT&T: http://www.att.com/
Bell Atlantic: http://www.bellatlantic.com/
BellSouth: http://www.bellsouth.com/
Federal Communications Commission: http://www.fcc.gov/
ITXC: http://www.itxc.com/
MCI WorldCom: http://www.mciworldcom.com/
Sprint: http://www.sprint.com/
U S West: http://www.uswest.com/

Appendix A

Summary of 14-point checklist

Section 271 of the Telecommunications Act of 1996 provides a 14-point checklist that is meant to ensure a competitive market for local and long distance services.

A.1 Checklist item 1

The Act requires, "Interconnection in accordance with the requirements of Sections 251(c)(2) and 252(d)(1)."

This means that all incumbent local exchange carriers (LECs) must allow interconnection to their networks: (1) for exchange service and exchange access; (2) at any technically feasible point; (3) that is at least equal in quality to what the local exchange carrier gives itself, its affiliates, or anyone else; and (4) on rates terms and conditions that are just, reasonable, and nondiscriminatory. (251(c)(2))

Any interconnection, service, or network element provided under an approved agreement shall be made available to any other requesting telecommunications carrier upon the same terms and conditions as those provided in the agreement. (252(i)) Prices for interconnection shall be based on cost (without reference to any rate-based proceeding) and be nondiscriminatory, and may include a reasonable profit. (252(d)(1))

A.2 Checklist item 2

The Act requires, "Nondiscriminatory access to network elements in accordance with the requirements of sections 251(c)(3) and 252(d)(1)."

This means that all incumbent local exchange carriers must provide, to any requesting telecommunications carrier for the provision of a telecommunications service, nondiscriminatory access to network elements on an unbundled basis at any technically feasible point on rates, terms, and conditions that are just, reasonable, and nondiscriminatory. These unbundled network elements will be provided in a manner that allows carriers to combine the elements in order to provide the telecommunications service. (251(c)(3)) A network element is a facility or equipment used in the provision of a telecommunication service, including features, functions, and capabilities such as subscriber numbers, databases, signaling systems, and information sufficient for billing and collection, or used in transmission, routing, or provision of a telecommunications service. (3(a)(45))

In determining which network elements will be made available, the FCC shall consider, at a minimum, whether (A) access to network elements that are proprietary is necessary and (B) whether failure to provide access to these network elements would impair the ability of a carrier to provide the services it wishes. (251(d)(2)) Prices shall be based on cost (without reference to any rate-based proceeding) and be nondiscriminatory, and may include a reasonable profit. (252(d)(1)) As part of their competitive checklist, BOCs are required to unbundle loop transmission, trunk side local transport, and local switching. (271(c)(2)(B)(iv)-(vi))

A.3 Checklist item 3

The Act requires, "Nondiscriminatory access to the poles, ducts, conduits, and rights-of-way owned or controlled by the Bell operating company at just and reasonable rates in accordance with the requirements of section 224."

This means that each local exchange carrier is to afford nondiscriminatory access to the poles, ducts, conduits, and rights-of-way to competing providers of telecommunications services, but they may deny access for reasons of safety, reliability, and generally applicable engineering purposes. (251(b)(4), 224(f)) Within two years, the FCC must prescribe regulations for charges for pole attachments used by telecommunications carriers (not incumbent local exchange carriers) to provide telecommunications services, when the parties fail to agree. Charges must be just, reasonable, and nondiscriminatory. (224(a)(5), (e)(1)) Pole attachment charges shall include costs of usable space and other space. (224(d)(1)-(3), (e)(2)) Duct and conduit charges shall be no greater than the average cost of duct or conduit space. (224(d)(1)) A utility must impute and charge affiliates its pole attachment rates. (224(g))

A.4 Checklist item 4

The Act requires, "Local loop transmission from the central office to the customer's premises, unbundled from local switching or other services."

This means that BOCs must unbundle loop transmission. (271(c)(2)(B)(iv)) This is to be provided at any technically feasible point and in a way that is nondiscriminatory, including rates, terms, and conditions that are just, reasonable, and nondiscriminatory. Unbundled network elements will be provided in a manner that allows carriers to combine the elements in order to provide the telecommunications service. (251(c)(3)) In determining which network elements will be made available, the FCC shall consider, at a minimum, whether (A) access to network elements that are proprietary is necessary and (B) whether failure to provide access to these network elements would impair the ability of a carrier to provide the services it wishes. (251(d)(2)) Prices shall be based on cost (without reference to any rate-based proceeding) and be nondiscriminatory, and may include a reasonable profit. (252(d))

A.5 Checklist item 5

The Act requires, "Local transport from the trunk side of a wireline local exchange carrier switch unbundled from switching or other services."

This means that BOCs must unbundle trunk side local transport. (271(c)(2)(B)(v)) This is to be provided at any technically feasible point and in a way that is nondiscriminatory, including rates, terms, and conditions that are just, reasonable, and nondiscriminatory. Unbundled network elements will be provided in a manner that allows carriers to combine the elements in order to provide the telecommunications service. (251(c)(3)) In determining which network elements will be made available, the FCC shall consider, at a minimum, whether (A) access to network elements that are proprietary is necessary and (B) whether failure to provide access to these network elements would impair the ability of a carrier to provide the services it wishes. (251(d)(2)) Prices shall be based on cost (without reference to any rate-based proceeding) and be nondiscriminatory, and may include a reasonable profit. (252(d))

A.6   Checklist item 6

The Act requires, "Local switching unbundled from transport, local loop transmission, or other services."

This means that BOCs must unbundle local switching. (271(c)(2)(B)(vi)) This is to be provided at any technically feasible point and in a way that is nondiscriminatory, including rates, terms, and conditions that are just, reasonable, and nondiscriminatory. Unbundled network elements will be provided in a manner that allows carriers to combine the elements in order to provide the telecommunications service. (251(c)(3)) In determining which network elements will be made available, the FCC shall consider, at a minimum, whether (A) access to network elements that are proprietary is necessary and (B) whether failure to provide access to these network elements would impair the ability of a carrier to provide the services it wishes. (251(d)(2)) Prices shall be based on cost (without reference to any rate-based proceeding) and be nondiscriminatory, and may include a reasonable profit. (252(d))

A.7 Checklist item 7

The Act requires, "Nondiscriminatory access to: (I) 911 and E911 services; (II) directory assistance services to allow the other carrier's customers to obtain telephone numbers; and (III) operator call completion services."

A.8 Checklist item 8

The Act requires, "White pages directory listings for customers of the other carrier's telephone exchange service."

This means that access or interconnection provided or generally offered by a BOC to other telecommunication carriers must include white pages directory listings for customers of the other carrier's telephone exchange service. (271(c)(2)(B)(viii))

A.9 Checklist item 9

The Act requires, "Until the date by which telecommunications numbering administration guidelines, plan, or rules are established, nondiscriminatory access to telephone numbers for assignment to the other carrier's telephone exchange service customers. After that date, compliance with such guidelines, plan, or rules."

This means the FCC must create or designate one or more impartial entities to administer telecommunications numbering and to make numbers available on an equitable basis. The FCC has exclusive jurisdiction over the U.S. portion of the North American Number Plan, but may delegate any or all jurisdiction to State commissions or other entities. (251(e)(1)) BOCs are required to provide nondiscriminatory access to telephone numbers for assignment by other carriers until telecommunications numbering administration guidelines, plan, or rules are established. Once these guidelines, plans, or rules are established, BOCs must comply with them. (271(c)(2)(B)(ix))

A.10 Checklist item 10

The Act requires, "Nondiscriminatory access to databases and associated signaling necessary for call routing and completion."

This means that access or interconnection provided or generally offered by a BOC to other telecommunication carriers shall include nondiscriminatory access to databases and associated signaling necessary for call routing and completion. (271(c)(2)(B)(x)) In determining which of these network elements will be made available, the FCC shall consider, at a minimum, whether (A) access to network elements that are proprietary is necessary and (B) whether failure to provide access to these network elements would impair the ability of a carrier to provide the services it wishes. (251(d)(2)) Prices of network elements shall be based on cost (without reference to any rate-based proceeding) and be nondiscriminatory, and may include a reasonable profit. (252(d)(1))

A.11 Checklist item 11

The Act requires, "Until the date by which the Commission issues regulations pursuant to section 251 to require number portability, interim telecommunications number portability through remote call forwarding, direct inward dialing trunks, or other comparable arrangements, with as little impairment of functioning, quality, reliability, and convenience as possible. After that date, full compliance with such regulations."

This means that all local exchange carriers must provide number portability, to the extent feasible, and in accordance with the FCC's requirements. (251(b)(2)) Number portability allows customers to retain, at the same location, their existing telecommunications numbers without impairment of quality, reliability, or convenience when switching from one telecommunications carrier to another. (3(a)(46)) Until the date that the FCC establishes for number portability, BOCs are required to provide interim number portability through remote call forwarding, direct inward dialing trunks, or other comparable arrangements, with as little impairment of functioning, quality, reliability, and convenience as possible. BOCs must fully comply with all FCC number portability regulations. (271(c)(2)(B)(xi))

A.12 Checklist item 12

The Act requires, "Nondiscriminatory access to such services or information as are necessary to allow the requesting carrier to implement local dialing parity in accordance with the requirements of section 251(b)(3)."

This means that access or interconnection provided or generally offered by a BOC to other telecommunication carriers shall include non-discriminatory access to such services or information as are necessary to allow the requesting carrier to implement local dialing parity in accordance with 251(b)(3). (271(c)(2)(B)(xii)) All local exchange carriers have the duty to provide dialing parity to competing providers of telephone exchange service and telephone toll service, and the duty to permit all such providers to have nondiscriminatory access to telephone numbers, operator services, directory assistance, and directory listing, with no unreasonable dialing delays. (251(b)(3))

A.13 Checklist item 13

The Act requires, "Reciprocal compensation arrangements in accordance with the requirements of section 252(d)(2)."

This means that all local exchange carriers must establish reciprocal compensation arrangements for transport and termination of telecommunications traffic. (251(b)(5)) The terms and conditions shall allow each carrier to cover its additional costs of terminating the traffic, including offsetting of reciprocal obligations such as bill-and-keep. Commissions may not engage in any rate proceedings nor require record keeping to determine the additional costs of the calls. (252(d)(2))

A.14 Checklist item 14

The Act requires, "Telecommunications services are available for resale in accordance with the requirements of sections 251(c)(4) and 252(d)(3)."

This means that all local exchange carriers must not prohibit and not impose unreasonable or discriminatory restrictions on resale. (251(b)(1)) Incumbent local exchange carriers must offer wholesale rates for any telecommunications service that is provided at retail to customers who are not telecommunications carriers. (251(c)(4)(A)) Wholesale prices shall be based on retail prices less the marketing, billing, collection, and other costs that will be avoided by selling the service at wholesale. (252(d)(3)) State commissions may, to the extent permitted by the FCC, prohibit a reseller from buying a service available only to one category of customers and reselling it to different category of customers. (251(c)(4)(B))

Appendix B

Transcript of ACTA petition against telephone calls over the Internet

BEFORE THE FEDERAL COMMUNICATIONS COMMISSION
Washington, D.C. 20554

In the Matter of

THE PROVISION OF INTERSTATE AND INTERNATIONAL IN-
TEREXCHANGE TELECOMMUNICATIONS SERVICE VIA THE
"INTERNET" BY NON-TARIFFED, UNCERTIFIED ENTITIES

AMERICA'S CARRIERS TELECOMMUNICATION ASSOCIATION
("ACTA"),

Petitioner

PETITION FOR DECLARATORY RULING, SPECIAL RELIEF, AND INSTITUTION OF RULEMAKING AGAINST:

VocalTec, Inc.; Internet Telephone Company; Third Planet Publishing Inc.; Camelot Corporation; Quarterdeck Corporation; and Other Providers of Non-tariffed, and Uncertified Interexchange Telecommunications Services, Respondents.

To the Commission:

SUMMARY OF FILING

America's Carriers Telecommunication Association ("ACTA"), a trade association of interexchange telecommunications companies, submits this Petition for Declaratory Ruling, for Special Relief, and for Institution of Rulemaking Proceedings. This petition concerns a new technology: a computer software product that enables a computer with Internet access to be used as a long distance telephone, carrying voice transmissions, at virtually no charge for the call.

ACTA submits that the providers of this software are telecommunications carriers and, as such, should be subject to FCC regulation like all telecommunications carriers. ACTA also submits that the FCC has the authority to regulate the Internet.

ACTA submits that it is not in the public interest to permit long distance service to be given away, depriving those who must maintain the telecommunications infrastructure of the revenue to do so, and nor is it in the public interest for these select telecommunications carriers to operate outside the regulatory requirements applicable to all other carriers.

ACTA asks the Commission to issue a declaratory ruling confirming its authority over interstate and international telecommunications services using the Internet.

ACTA asks the Commission, as special relief, to order the Respondents to immediately stop their unauthorized provisioning of telecommunications services pending their compliance with 47 U.S.C. Sections 203 and 214. and in order to give the Commission time for appropriate rulemaking.

ACTA asks the Commission to institute rulemaking to govern the use of the Internet for providing telecommunications services.

PETITION FOR DECLARATORY RULING, SPECIAL RELIEF, AND
INSTITUTION OF RULEMAKING
America's Carriers Telecommunication Association ("ACTA"), by its
attorneys, submits this Petition for Declaratory Ruling, for Special Relief,
and for Institution of Rulemaking Proceedings. In support of this petition,
the following is shown.

STANDING
ACTA is a national trade association of competitive interexchange, non-
dominant telecommunications companies. Its members provide interexchange telecommunications services on an intrastate, interstate and
international basis to the public at large.

Some of its members also act as underlying (or wholesale) carriers
providing network facilities, equipment and service to other member
carriers which permits telecommunications services to be resold to the
public. Other ACTA members supply facilities and equipment to member
and non-member wholesale and resale carriers.

ACTA's carrier members must be certificated and tariffed before the
FCC and most state regulatory commissions in order to render their
telecommunications service to the public. In addition, ACTA carrier
members are subject to the requirements of the Communications Act of
1934, as amended (the "Act"), and various state laws and regulations
which prohibit engaging in unreasonable practices and/or unduly discriminatory conduct.

ACTA carrier members are required to pay, directly, or indirectly,
various fees and charges in order to render their services to the public.
Filing fees and annual fees are levied by the FCC and most states.

In addition, the FCC and most states require interexchange carriers to
assess and collect from the using public specific charges to support various
regulatory policies and programs used to sustain and advance national
and state goals for telecommunications.

Entities, like those which are described hereinafter, which do not
comply with or operate subject to the same statutory and regulatory
requirements as ACTA's carrier members, distort the economic and public interest environment in which ACTA carrier members and nonmembers must operate. Continuing to allow such entities to operate without
complying with or being subject to the same legal and regulatory

requirements as ACTA carrier members threatens the continued viability of ACTA's members and their ability to serve the public and acquit their public interest obligations under federal and state laws.

As the appointed representative of its members charged with advancing their economic interests and assisting in achieving and maintaining their legal and regulatory compliance, ACTA has standing to file and prosecute these petitions.

STATEMENT OF FACTS AND BACKGROUND

A growing number of companies are selling software for the specific purpose of allowing users of the Internet to make free or next to free local, interexchange (intraLATA, interLATA) and international telephone calls using the user's computer (Attachment 1). One of the Respondents, VocalTec, Inc., advertises the ability of its software called "Internet Phone," to connect any user of "Internet Phone" with any other user of "Internet Phone" anywhere in the world. The software enables users to audibly talk with one another in real-time. Respondents make a one-time charge for the software, but users incur no other charges for making local or long distance telephone calls to any other "Internet Phone" user in the world (except for whatever the user already pays monthly to whomever provides them Internet access).

ASSERTION AND ENFORCEMENT OF JURISDICTION

ACTA submits that it is incumbent upon the Commission to exercise jurisdiction over the use of the Internet for unregulated interstate and international telecommunications services. As a first step, ACTA submits that the Commission may deem it appropriate to issue a declaratory ruling officially establishing its interest in and authority over interstate and international telecommunications services using the Internet.

Secondly, ACTA submits that the Commission has an obligation, heightened by the recent enactment of the Telecommunications Act of 1996, to address on a focused basis the on-going, unregulated and unauthorized provisioning of telecommunications services. The Commission should, as special relief, issue an order to the Respondents to immediately stop arranging for, implementing, and marketing non-tariffed, uncertified telecommunications services without complying with applicable provisions of the Act, particularly Sections 203 and 214, codified at 47 U.S.C. Sections 203 and 214.

Further, ACTA submits that it is incumbent upon the Commission to examine and adopt rules, policies and regulations governing the uses of the Internet for the provisioning of telecommunications services. The use of the Internet to provide telecommunications services has an impact on the traditional means, methods, systems, providers, and users of telecommunications services. The unfair competition created by the current unregulated bypass of the traditional means by which long distance services are sold could, if left unchecked, eventually create serious economic hardship on all existing participants in the long distance marketplace and the public which is served by those participants. Ignored, such unregulated operations will rapidly grow and create a far more significant and difficult to control "private" operational enclave of telecommunications providers and users. Such development will clearly be detrimental to the health of the nation's telecommunications industry and the maintenance of the nation's telecommunications infrastructure.

ARGUMENT

Commission's Authority to Regulate the Internet. ACTA submits that the Commission has the authority to regulate the Internet under the provisions of 47 U.S.C. Section 151, which created the Commission:

> ... for the purpose of regulating interstate and foreign commerce in communication by wire and radio so as to make available, so far as possible, to all the people of the United States a rapid, efficient, Nation-wide, and world-wide wire and radio communication service with adequate facilities at reasonable charges, for the purpose of the national defense, for the purpose of promoting safety of life and property, through the use of wire and radio communication ...

The Internet is a unique form of wire communication. It is a resource whose benefits are still being explored and whose value is not fully realized. Its capacity is not, however, infinite. The misuse of the Internet as a way to bypass the traditional means of obtaining long distance service could result in a significant reduction of the Internet's ability to handle the customary types of Internet traffic. The Commission has historically protected the public interest by allocating finite communications resources/frequencies and organizing communications traffic. ACTA submits that here also it would be in the public interest for the

Commission to define the type of permissible communications which may be effected over the Internet.

Commission's Authority to Regulate Respondents as Interstate Telecommunications Carriers. ACTA submits that by both established precedents defining "common carriage" or public utility type of operations for purposes of regulatory jurisdiction, and by statutory enactment, the Respondents, as purveyors of Internet long distance services, are interstate telecommunications carriers, subject to federal regulation. Section 3 of the new "Telecommunications Act of 1996," Pub. L. No. 104-104, 110 Stat. 56 (1996), to be codified at 47 U.S.C. Section 153, includes the following definitions:

> (48) Telecommunications.—The term "telecommunications" means the transmission, between or among points specified by the user, of information of the user's choosing, without change in the form or content of the information as sent and received.

> (49) Telecommunications Carrier.—The term "telecommunications carrier" means any provider of telecommunications services, except that such term does not include aggregators of telecommunications services (as defined in section 226). A telecommunications carrier shall be treated as a common carrier under this Act only to the extent that it is engaged in providing telecommunications services, except that the Commission shall determine whether the provision of fixed and mobile satellite service shall be treated as common carriage.

> (51) Telecommunications Service.—The term "telecommunications service" means the offering of telecommunications for a fee directly to the public, or to such classes of users as to be effectively available directly to the public, regardless of the facilities used.

It would appear that Respondents are currently operating without having complied with the requirements of the Communications Act of 1934, as amended, applicable to providing interstate and international telecommunications services. e.g., Sections 203 and 213, codified at 47 U.S.C. Sections 203 and 214.

Case law also supports the Commission's authority to regulate the Respondents. In 1968, the Supreme Court was presented the issue of the Commission's authority to regulate the cable television industry, or CATV, then still in its infancy but growing quickly. In United States v. Southwestern Cable Co., 392 U.S. 157 (1968), the Supreme Court had to decide whether the Federal Communications Commission (1) had the authority under the Communications Act of 1934, as amended, to regulate CATV systems, a new technology and therefore not specifically discussed in the Act, and (2) if the Commission had such authority, whether it also had the authority to issue the particular prohibitory order that it had: one designed generally to preserve the status quo pending further investigation and proceedings, and not issued pursuant to the cease and desist rules of Section 312 of the Act (47 U.S.C. Section 312).

The Supreme Court answered both questions in the affirmative. The Supreme Court stated that "the [Federal Communications] Commission has reasonably concluded that regulatory authority over CATV [was] imperative if it [was] to perform with appropriate effectiveness certain of its other responsibilities." Id. at 173. At that time, cable television characteristically neither produced its own programming nor paid producers or broadcasters for use of the programming which CATV redistributed. Id. at 162.

The Court noted the Commission's concern that competition by CATV might destroy or degrade the service offered by local broadcasters and exacerbate the financial difficulties of UHF and educational television broadcasters.

Commission's Authority to Grant Special Relief to Maintain the Status Quo. With regard to the procedural issue, the Court in Southwestern Cable upheld the authority of the Commission to issue an order to maintain the status quo. The argument was made that the Commission could only issue prohibitory orders under the Act's Section 312 cease and desist provisions which, the Court assumed without finding, were only proper after a hearing or the waiver of the right to a hearing. The Court rejected that argument stating:

The Commission's order was thus not, in form or function, a cease-and-desist order that must issue under Sections 312(b), (c). The Commission has acknowledged that, in this area of rapid and significant change, there may be situations in which its generalized regulations are inadequate, and special or additional forms of relief are imperative. It has found that the present case may prove to be such a situation, and that the public interest demands "interim relief limiting further expansion," pending hearings to determine appropriate Commission action.

Such orders do not exceed the Commission's authority. This Court has recognized that "the administrative process [must] possess sufficient flexibility to adjust itself" to the "dynamic aspects of radio transmission," F.C.C. v. Pottsville Broadcasting Co., supra, at 138, and that it was precisely for that reason that Congress declined to "stereotype the powers of the Commission to specific details ... National Broadcasting Co. v. United States, supra, at 219."

The Commission should take the same action in 1996 with regard to the new technology of long distance calling via Internet as it did thirty years ago in 1966 with regard to the then-new technology of cable television: grant special relief to maintain the status quo so that it might carefully consider what rules are required to best protect the public interest and to carry out its statutory duties.

Other Issues Necessitating the Commission's Regulation of Long Distance via the Internet. The Commission has a duty to oversee and effect the Telecommunications Act of 1996 as well as its longstanding duties under 47 U.S.C. Section 151. The Commission should take action in order to preserve fair competition and the health of the Nation's telecommunications industry. Absent a healthy industry, with users paying telecommunications companies a fair price for telecommunications services, the Commission's duty to effectively promote universal service cannot be achieved.

Absent action by the Commission, the new technology could be used to circumvent restrictions traditionally found in tariffs concerning unlawful uses, such as gambling, obscenity, prostitution, drug traffic, and other illegal acts.

INFORMATION REGARDING RESPONDENTS

ACTA does not possess a listing of all the companies providing free long distance calls via computer software. However, Attachment I contains some information regarding the following Internet telephone software companies and products:

a.   Company: VocalTec, Inc.
                    157 Veterans Drive
                    Northvale, NJ 07647
                    Telephone: (201) 768-9400

      Product: Internet Phone

      Distributors: VocalTec, Inc.; and Ventana Communications Group, Research Triangle Park, NC

b.   Company: Internet Telephone Company
                    Boca Raton, FL
                    Telephone: (407) 989-8503

      Product: WebPhone

c.   Company: Third Planet Publishing Inc.,
                    a division of Camelot Corporation

      Product: Digiphone

d.   Company: Quarterdeck Corporation
                    13160 Mindanao Way, 3rd Floor
                    Marina Del Ray, CA 90292
                    Telephone: (310) 309-3700

      Product: WebTalk

e.   Company: Unknown
      Product: CyberPhone

CONCLUSION

Permitting long distance service to be given away is not in the public interest. Therefore, ACTA urges the Federal Communications Commission ("the Commission") to exercise its jurisdiction in this matter and: issue a declaratory ruling establishing its authority over interstate and international telecommunications services using the Internet; grant special relief to maintain the status quo by immediately stopping the sale of this software; and institute rulemaking proceedings defining permissible communications over the Internet.

Respectfully submitted,

AMERICA'S CARRIERS TELECOMMUNICATIONS ASSOCIATION

Charles H. Helein
General Counsel

Of Counsel:
Helein & Associates, P.C.
8180 Greensboro Drive
Suite 700
McLean, Virginia 22102
(703) 714-1300 (Telephone)
(703) 714-1330 (Facsimile)

Dated: March 4, 1996

Footnotes

[1]   47 U.S.C. 201 et seq.

[2]   The user must hook up a microphone to his computer and either a headset
       or speakers.

[3]   ACTA asserts that Respondents are also intrastate telecommunications
       carriers, subject to regulation by state public utility commissions.

[4]   The Commission had ordered that respondents, a cable company, generally
       restrict their carriage of Los Angeles signals to areas served by them on
       February 14, 1966, pending hearings to determine whether the carriage of
       such signals into San Diego contravened the public interest. The order did
       not prohibit the addition of new subscribers within areas served by
       respondents on February 15, 1966; it did not prevent service to other
       subscribers who began receiving service or who submitted an accepted
       subscription request between February 15, 1966, and the date of the
       Commission's order; and it did not preclude the carriage of San Diego and
       Tijuana, Mexico, signals to subscribers in new areas of service. United States
       v. Southwestern Cable Co., 392 U.S. 157, 180 (1968).

[5]   Id. at 180.

Source: Federal Communications Commission, Washington, D.C.
(Attatchments references in this Appendix are not included.)

Appendix C

Transcript of NTIA letter to the FCC addressing ACTA's petition for rulemaking against telephone calls over the Internet

May 8, 1996

The Honorable Reed Hundt
Chairman
Federal Communication Commission
Room 814
1919 M Street, N.W.
Washington D.C. 20554

Re: RM 8775

Dear Chairman Hundt:

This letter addresses the petition for rulemaking filed before the Commission by America's Carriers Telecommunication Association (ACTA) in March 1996. ACTA asks the Commission to: (1) order Internet software providers to "immediately stop their unauthorized provisioning of telecommunications software"; 2) confirm the Commission's authority over interstate and international telecommunications services offered over the Internet; and 3) institute rules to govern the use of the Internet for providing telecommunications services.

On behalf of the Administration, NTIA strongly urges the Commission to deny the ACTA Petition. The Petition not only mischaracterizes the existing law, but also reflects a fundamental misunderstanding of the way in which the Internet operates and of the services now making use of the Internet.

ACTA requests that the Commission stop firms such as the Respondents from selling software that enables "a computer with Internet access to be used as a long distance telephone, carrying voice transmissions, at virtually no charge for the call" [ACTA Petition at i]. ACTA asserts that such firms are common carrier providers of telecommunications services that should not be allowed to operate without first obtaining a certificate from the Commission [ACTA Petition at 6-7].

That argument is wrong. The Respondents provide their customers with goods, not services. Although the software that those firms sell does enable individuals to originate voice communications, all of the actions needed to initiate such communications are performed by the software users, rather than the vendors. At no time do the Respondents engage in the "transmission" of information, which, according to the Telecommunications Act of 1996, is the sine qua non of both a telecommunications service and a telecommunications carrier. [See Telecommunications Act of 1996, Pub. L. No. 104-104, 110 Stat. 56, 3(a) amending Section 153 of the Communications Act of 1934 to add new definitions of "telecommunications," "telecommunications service," and "telecommunications carrier."] In that critical sense, the Respondents are no more providing telecommunications services than are the vendors of the telephone handsets, fax machines, and other customer premises equipment that make communications possible.

ACTA also asks the Commission for a declaratory ruling "confirming its authority over interstate and international telecommunications services using the Internet." [ACTA Petition at 6. While ACTA claims the Commission has jurisdiction to regulate the Internet pursuant to Section 1 of the Communications Act, citing United States v.

Southwestern Cable Co., 392 U.S. 157 (1968), ACTA also acknowl-
edges that such jurisdiction is limited to actions ancillary to the effective
performance of its specific responsibilities under other parts of the Act.
Id. at 5, 7-8. ACTA suggests that unregulated growth of the Internet
presents "unfair competition" to Title II regulated interexchange carriers that "could, if left unchecked, eventually create serious economic
hardship on all existing participants in the long distance marketplace"
and could be "detrimental to the health of the nation's telecommunications industry and the maintenance of the nation's telecommunications
infrastructure." Id. at 4, 5. Voice telephony via the Internet, however, is
still a limited and cumbersome capability: both parties to the call need
computers and must have compatible software. Moreover, there is no
assurance that a call placed will be completed or not interrupted. While
the technology involved may improve rapidly, presently there is no
credible evidence to justify Commission regulation of the Internet.] In
fact, as the Federal Networking Council pointed out in comments filed
on May 4, there are no telecommunications services currently being
offered via the Internet. The services that now involve the Internet are
more likely to be "enhanced," or information services over which the
Commission has disclaimed jurisdiction under the Communications
Act. The Commission decision in the 1980's not to regulate enhanced
services was a wise one that has conferred substantial benefits on
American consumers. The Telecommunications Act of 1996 in no way
requires a change in that decision.

The Internet now connects more than 10 million computers, tens
of millions of users, and is growing at a rate of 10–15 percent a month.
This growth has created opportunities for entrepreneurs to develop
new services and applications such as videoconferencing, multicast-
ing, electronic payments, networked virtual reality, and intelligent
agents. Perhaps more importantly, it creates a growing number of
opportunities for consumers to identify new communication and
information needs and to meet those needs. The Commission should
not risk stifling the growth and use of this vibrant technology in order
to prevent some undemonstrated harm to long distance service providers. If Internet-based services eventually develop to an extent that
raises concerns about harm to consumers or the public interest, the
Commission would have ample time to more fully address the issue.
Now is not that time.

NTIA, therefore, urges the Commission to reject the ACTA petition without delay.

Larry Irving
Assistant Secretary for Communications and Information

cc: The Honorable James H. Quello
    The Honorable Rachelle B. Chong
    The Honorable Susan Ness

Source: Federal Communications Commission, Washington, D.C.

Appendix D

Statement to the FCC from the VON Coalition

FEDERAL COMMUNICATIONS COMMISSION
Washington, D.C. 20554

In the Matter of	)	
	)	
The Petition of America's Carriers	)	RM-8775
Telecommunication Association for	)	
Declaratory Ruling, Special Relief,	)	
and Institution of Rulemaking	)	

OPPOSITION OF THE VON COALITION

Summary

The VON Coalition urges the Commission not to regulate the Internet or ban VON products. The unfettered growth and development of the

Internet and VON are in the public interest. Both the Internet and VON are used to facilitate communications among millions of people in innovative and often very efficient and cost-effective ways that open entirely new opportunities for personal and business communications, education, health care, and entertainment. The Commission should encourage and not stifle this innovation.

In its most basic form, VON is no different from other forms of data communication using the Internet or other communications networks. As such, its development and use is inseparable from the development and use of other multimedia software and applications.

There is no apparent reason for government regulation that might upset the dynamic nature of the development of the Internet, including VON. ACTA has failed completely to show that the Internet or VON have produced any harm to the public interest or even to the narrow self-interest of ACTA's members. We are unaware of any present commercial use of VON software to connect to standard telephones that are part of the public switched telephone network. But, even if there were such uses, the logical government response should be to welcome and encourage the development of new services that add to competition and consumer choice.

ACTA's legal argument that the Commission must regulate the Internet and VON because VON software developers are "Telecommunications Carriers" flies in the face of the plain language of the Communications Act, years of Commission rulings, and common sense. The companies that ACTA attacks in its Petition offer software that customers install on their computers and use over various computer networks, including in many cases proprietary networks. ACTA's argument that the offering of this software turns these companies into "carriers" is completely unreasonable and should be rejected by the Commission.

Background

The VON Coalition

The VON Coalition (http://www.von.org) was formed as a direct result of ACTA's attempt to get the Commission to regulate computer software and Internet services. Since its formation in early March, over four hundred individuals and 81 corporations have joined. The corporate membership includes many of the software companies that are developing and marketing VON products. These include: FreeTel Communications, Inc. (http://www.freetel.com); Netspeak Corporation (http://www.netspeak.com); VDONet Corporation (http://www.vdolive.com); VocalTec, Inc. (http://www.vocaltec.com); Voxware Inc. (http://www.voxware.com); White Pine Software, Inc. (http://www.wpine.com); and Xing Technology, Inc. (http://www.xingtech.com). Information about these companies and their VON products is attached as Exhibit A.

The initial force in organizing the coalition was Jeff Pulver, a computer professional, long-time amateur radio enthusiast, and Internet activist who was an early experimenter with VON products. Among Jeff's projects is Free World Dial-up ("FWD"), a non-commercial experiment in using VON in ways that amateur radio has been used. A description of FWD is attached as Exhibit B. Other members of the coalition include many professionals who use VON products in their work. For instance, Dr. Takeshi Utsumi is a well-known innovator in the use of computer networks for distance learning.

The VON Coalition's goal is to educate the public and government officials regarding the use of voice and video on communications networks. The VON Coalition seeks to preserve the Internet as a place for emerging technologies and business. A copy of the Coalition's initial press release is attached as Exhibit C.

Voice on the Net

In its most basic form, VON is not much different from other forms of data communication using the Internet.[1] VON uses the same basic transport lines, routers, and servers as are used, for instance, for Internet e-mail. From the user's perspective, VON requires the same equipment and configuration as other Internet communications, with a few additional items. Altogether, the user needs a computer with certain minimum capability (in the case of a PC, at least a 486), a modem with at least 14.4 kbps transmission capability, a sound card, speakers or a headset, a microphone, and special software that processes the sound of the user's voice into small digital packets that are sent through the Internet. Data packets from a computer user at another point on the Internet are similarly processed into a sound that is transmitted through the computer's speakers. Some VON products are limited to half duplex operation, others permit full duplex conversations. Some operate through an Internet Relay Chat server, others are designed to contact a user's Internet Protocol address directly.

The compression algorithms typically used for VON make it quite efficient. An Internet user typically uses sophisticated signal processing algorithms to first compress his speech, then send this compressed signal over the Internet. The compression algorithms used vary, and are constantly being improved. They typically compress speech to between 2.4 and 14.4 kbps, and only transmit when the user is actively speaking, which is less than half of the duration of the call. This compares with 64 kbps for a telephone conversation using the PSTN dial up network.

The economics of VON involve not only an investment in the necessary computer hardware and software, but also the cost of accessing the Internet. This cost varies widely, depending on the user. Some users have Internet accounts that provide them with a flat monthly rate for unlimited access at a particular bandwidth or data rate (e.g., 28.8 kbps or 1.5 mbps), others pay a flat rate for several hours of access at a particular bandwidth or data rate and a per-minute price for additional access. The Internet is not designed with sophisticated accounting and billing mechanisms. Users are never billed based on any distinction between "local" and "long distance" packets.

There is nothing about VON products that limits their use to the Internet. The same equipment and software can be used to provide voice communications over other properly equipped networks, such as corporate and institutional computer Local Area Networks. There has been a tremendous increase in the use of these and other internal networks in the recent past and that trend is expected to continue with the growth of what are referred to as "intranets." VON is expected to play an important role in the growth of these networks.

The VON Coalition estimates that there are currently at least 18 companies offering their versions of VON software. This software has been acquired by over a million computer users and Netscape Corporation recently released a beta version of its browser software that contains VON capability. Most of the current use is by hobbyists who are experimenting with the technology. As the software becomes more widely distributed and there is a greater critical mass of potential users, the VON Coalition expects its use to increase substantially.

The likely uses of VON include a number of personal, business and educational applications. In general, VON can be expected to be widely used whenever there is communication among people with computers. Thus, for instance, companies that operate World Wide Web sites to promote their products and services may make available "live" operators to answer questions from customers who are visiting their website.

The distance learning applications of voice and video on the Internet are particularly exciting. Anyone who has witnessed the use of CU-SeeMe technology over the Internet can immediately grasp the educational potential for multimedia applications as a way to inexpensively expand the classrooms of our finest teachers. Early informal comments filed by Educom, an organization that represents the information technology interests of over 600 colleges and universities, eloquently describe the developing role of the Internet and VON products in education:

Institutions of higher education have been deeply involved with computer networking technology since its inception some twenty five years ago. Our participation has included fundamental research, applied development, and precommercial deployment of products and services. In the course of these activities, campuses have partnered with federal research agencies and with private industry on many occasions to achieve common networking objectives.

Today, higher education is a major user of the worldwide facilities of the Internet, which have become an essential component of our teaching, research and public service missions. In the United States, fifteen million students and several million faculty and staff at more than three thousand accredited institutions are served by 1.9 million Internet host computers, a number which has doubled in the last year and is still growing rapidly.

The Internet is not only important today, but its successful continuing growth is an integral part of the strategic planning efforts within higher education to enable greater access by students throughout the nation, to reach out to the primary and secondary school community, and to forge new ties with employers of our graduates.

The availability of fully digital, multimedia Internet services has enormous future value to the realization of many teaching and learning goals, especially in connection with distance education. Many of the well known shortcomings of "talking head" forms of projecting classrooms to off campus sites based on switched or off the air analog television will be ameliorated by a fully interactive, multimedia, wireline and wireless Internet. In addition to its "anyone, anywhere" capability, the future multimedia Internet will facilitate the transition of instruction from "teacher centered" to "learner centered" forms of instruction.

Additionally, the flexible provision of multimedia capability will allow the development of new learning tools and systems that support the needs of individuals with learning impairments such as blindness, deafness, limited muscular control, dyslexia, etc.

Comments of Educom (April 3, 1996)

VON software in its current form has certain limitations. Beyond the obvious limitations that use requires a relatively expensive computer and a minimal amount of technical competence, there are also often problems with the quality of the communications, due to the nature of the Internet as it is presently structured and operated. This can mean that calls do not always go through, sound quality can be poor, or there can be annoying delays within a conversation. There is no independent power supply. Thus, if the called party's computer is not operating or no one is nearby, the call may go unanswered. Another limitation is that there is at present no single standard for compressing audio into data packets. As a result, in many cases users of one software product cannot communicate with users of other products.

Solving these problems, to make VON less costly, more convenient, and of better quality is a matter of ongoing effort that the coalition expects will be successful, but it is impossible at this time to predict how much time and money will be required to produce results. For instance, there are new routing protocols, such as one recently introduced by Cisco Systems called "RSVP" which should help to reduce the latency between packets. In addition, the Internet Engineering Task Force is working to develop a voice standard protocol and many of the leading companies have endorsed that process, but such an effort may take substantial time to complete.

Some of the great deal of publicity in the general press about VON software gives the impression that it is being used to provide a commercial service that connects the user to the PSTN to terminate communications at a standard telephone. Our understanding is that any such description at least at the present time is inaccurate. As described above and in Exhibit B, the Free World Dial-up project was strictly a non-commercial venture by a volunteer group of hobbyists interested in promoting technological innovation and communications, in the spirit of amateur radio. [2]

In the future commercial applications may develop that involve terminating communications at standard telephones but, as discussed below, such a development should not be relevant to the Commission's decision regarding either the regulation of the Internet or the regulation of VON software.

VON software is used for more than just two-way point-to-point communications. Many radio stations are beginning to use the Internet to transmit programming. The MBONE, an experimental portion of the Internet that was critical in the development of VON, is being used to experiment with audio and video-conferencing, the broadcast of concerts, and other "non-traditional" uses of computer networks. A further description of the MBONE is attached as Exhibit D.

All of the bits that are transmitted through the Internet, whether voice conversations, e-mail, pictures of comets, or Rolling Stones concerts, appear the same and are handled the same by the hardware and software that comprise the Internet. It may be theoretically possible for Internet service providers to try to identify customers using VON products, but it is practically impossible to do so. The only possible way to distinguish VON from other packets is by the particular port that is typically selected by the software provider for that product and encoded into the software. As a practical matter, if entities attempted to block packets from those ports, it would be relatively easy for users to simply change the designated port. Moreover, attempting to block the use of particular ports might inadvertently block other software and applications that happen to designate the same port.

The ACTA Petition

ACTA, which describes itself as a trade association of interexchange telecommunications companies, filed its Petition on March 4, 1996. ACTA contends that the providers of VON software are "telecommunications carriers" under the new Telecommunications Act and, as such, should be subject to FCC regulation like all telecommunications carriers. ACTA also asks the Commission to order certain software companies that it names as respondents to immediately stop "arranging for, implementing, and marketing nontariffed, uncertified telecommunications services" that, according to ACTA, do not comply with the Communications Act, §§ 203 and 214 [3].

ACTA contends that providers of VON software, because they are not subject to the same statutory and regulatory requirements as ACTA's carrier members, "distort the economic and public interest environment in which ACTA carrier members and nonmembers must operate." According to ACTA, unless the Commission takes the requested action, the "continued viability" of ACTA's members and their ability to "acquit their public interest obligations" under federal and state laws is threatened.

ACTA also contends that the FCC should exercise broad authority to regulate the Internet. ACTA asks the Commission to issue a declaratory ruling establishing its authority over interstate and international telecommunications services using the Internet and begin a rulemaking to govern the use of the Internet for providing telecommunications services. ACTA expresses its concern that without action by the Commission, "the new technology could be used to circumvent restrictions traditionally found in tariffs concerning unlawful uses, such as gambling, obscenity, prostitution, drug traffic, and other illegal acts."

ACTA seems to claim that its petition is intended to help the development of the Internet, stating that "the misuse of the Internet as a way to bypass the traditional means of obtaining long distance service could result in a significant reduction of the Internet's ability to handle the customary types of Internet traffic." ACTA therefore submits that it would be in the public interest for the Commission to "define the type of permissible communications which may be effected over the Internet."

Discussion

As discussed below, the evidence is overwhelming that the Commission should not regulate the Internet or ban VON products. [4] The unfettered growth and development of the Internet and VON are in the public interest, providing important new services in a way that is economical and efficient. There is no apparent reason for government regulation that might upset the dynamic nature of their development. ACTA certainly has failed to show that the Internet or VON have produced any harm to the public interest or even to the interests of ACTA's members, let alone the kind of harm that can or should be cured by government regulation. To the extent that the Internet provides competition to the public switched telephone network, that competition should be encouraged, not stifled.

ACTA's legal arguments are similarly misguided. Its claim that VON software providers are "Telecommunications Carriers" flies in the face of the plain language of the Communications Act, years of Commission rulings, and common sense. VON software providers develop and market software that customers install on their computers and use over various computer networks, including in many cases proprietary networks. To argue, as ACTA does, that the offering of this software turns these companies into "carriers" is, frankly, absurd.

I. The Continued Unfettered Development of the Internet and VON Is in the Public Interest

An objective look at the Internet and VON demonstrates beyond any doubt that they have many significant benefits and do not cause any apparent harm. Both the Internet and VON can be used to facilitate communications among millions of people in innovative ways that open entirely new opportunities for personal and business communications, entertainment, education, health care, and other uses.

Chairman Hundt described the importance of networking classrooms with computer resources in a speech that he made over a year ago:

> I have seen networked classrooms, and they work. I saw this future in the Ralph Bunche Elementary school—P.S. 125—in Harlem, New York. Two classrooms were connected to the phone network. I watched fifth and sixth graders share a lesson with kids in Nova Scotia and Hawaii. They use the CIA World Fact Book to conduct science projects. They electronically questioned researchers in Australia.

> Networks carry those kids to their classrooms, beyond the walls of P.S. 125, outside of Harlem, and around the globe. Networks show them the way to a brighter future. Networks bring them education resources that no school district could otherwise physically import.

> ... I'm talking about real learning. Social scientists have repeatedly proved that education over networks captures students' imaginations

and calls forth a greater willingness to learn. Test scores go up when learning occurs over networks. Self esteem rises. Fluency in self expression increases.

... if we get every school and classroom networked, it won't matter whether the school is in a rich or a poor state. All information will be equally accessible to every child. Communications technology may not be a panacea for all the challenges of education. But it can be the great equalizer of opportunity.

Prepared Remarks at a Meeting of Kidsnet (August 22, 1995)

One of the most important things about VON is what it represents as a product of the experimentation that characterizes the Internet. That experimentation makes the Internet a dynamic testing ground for the kinds of innovative computer and communications concepts that bring tremendous benefit to society, in terms of both new services and more efficient versions of existing services. The fact that much of this innovation is taking place in the United States is a testament both to the United States' preeminence in these fields and to recognition of the importance of encouraging innovation if the United States is to maintain that leadership role.

As discussed above, it also appears unlikely that any attempt to ban VON would be practically enforceable. The packets that are transmitted are indistinguishable from e-mail or other packets and the software has many applications that go well beyond its use on the Internet, so the Commission could not justify a wholesale ban on the sale of the software. Moreover, much of the software is distributed without charge.

ACTA alleges that VON is harmful, but it fails completely to make a showing that this is the case. ACTA makes no showing that VON is used for illegal activities or, more to the point, that existing criminal laws would be inadequate to deal with such activities. ACTA also fails to make any showing to support its argument that the use of VON will harm the Internet itself. The evidence, moreover, is to the contrary, that the increased utility of the Internet provided by VON and other innovations will lead to an increase in the facilities committed to support the network, thus relieving any congestion. The Internet Network Engineering Task Force is also engaged in discussions to develop ways of prioritizing packets to reduce congestion for applications that require faster transmission.

Even ACTA's blatantly protectionist claim of harm to its own members, which presumably is the heart of its petition and the area in which ACTA is best able to supply evidence, is totally conclusory. ACTA does not present a shred of evidence that any of its members has lost one customer or one minute of traffic due to the existence of VON. In this regard, it is interesting to note the rejection of ACTA's position by the largest interexchange carriers. [5] ACTA's claim that its members suffer from being unfairly subjected to regulatory requirements is also unsubstantiated. For the most part, as discussed below, the Commission's policies, codified by the passage of the Telecommunications Act, have been to deregulate non-dominant carriers such as ACTA's members. As a result, these entities are not generally subject to burdensome regulation. [6]

As discussed above, at present there is no commercial use of VON to connect to standard telephones that are part of the public switched telephone network. Even assuming for the sake of argument, however, that VON will be used in the near future to provide such a service or will otherwise be competitive with the interexchange services provided by ACTA's members, it does not necessarily follow that such a service would be harmful to the public interest. Quite to the contrary, such a new service would add to consumer choice. Thus, absent a compelling reason to regulate, the logical government response should be to welcome and encourage the development of the new service. [7] The largest savings from any such service would come in those instances in which the existing pricing for telephony is uneconomical, as is particularly the case with respect to charges for certain international calls. As James Clark of Netscape has said, "If you can bypass the tariffs and it works, then there must be something wrong with the tariffs." Communications Daily (April 8, 1996). In those cases, just as with International Callback services, the Commission's policies should be one of forbearance. [8]

Incumbent providers such as ACTA's members also should welcome the new competition, at least insofar as it would provide the basis for their arguing that the new competition means less need for government regulation of their own activities. Moreover, there is nothing preventing ACTA's members from themselves providing the services that they claim have an unfair advantage.

II. The Communications Act and FCC Policies Support the Continued Unfettered Development of VON

A. The Internet is not a "Telecommunications Service" and VON software companies are not "Telecommunications Carriers."

Without explanation, ACTA contends that Congress has conferred authority on the FCC to regulate the Internet as a "Telecommunications Service" and VON software companies as "Telecommunications Carriers." No more than perfunctory examination of the issue is necessary to show that these definitions are inapplicable to the Internet and VON.

The new Telecommunications Act, Section 3, defines these terms as follows:

> (48) Telecommunications. — The term "telecommunications" means the transmission, between or among points specified by the user, of information of the user's choosing, without change in the form or content of the information as sent and received.

> (49) Telecommunications Carrier. — The term "telecommunications carrier" means any provider of telecommunications services, except that such term does not include aggregators of telecommunications services (as defined in section 226). A telecommunications carrier shall be treated as a common carrier under this Act only to the extent that it is engaged in providing telecommunications services, except that the Commission shall determine whether the provision of fixed and mobile satellite service shall be treated as common carriage.

> (51) Telecommunications Service. — The term "telecommunications service" means the offering of telecommunications for a fee directly to the public, or to such classes of users as to be effectively available directly to the public, regardless of the facilities used.

47 U.S.C. §§ 153(48), (49), and (51)

The definitions of "Telecommunications," "Telecommunications Service," and "Telecommunications Carrier" are based almost verbatim on the existing Commission definition of "Basic Services."[9] Those Commission decisions have never classified services provided over computer networks such as the Internet as "basic services." The Telecommunications Act codifies the policy of the Commission regulating only basic services. By extension, since the Internet is not a Telecommunications Service, the software used on such networks cannot possibly require the classification of the software developed as Telecommunications Carriers. [10]

Moreover, VON software developers clearly do not offer "telecommunications ... directly to the public" for purposes of the Communications Act. They do not offer any service "for a fee," but rather merely offer software for a one-time purchase price. Accordingly, because VON software developers do not provide Telecommunications Service, by extension they cannot meet the definition of Telecommunications Carrier for purposes of the Communications Act. ACTA's assertion that these companies should be required to apply for a Section 214 authorization or file tariffs is completely misplaced.

The ACTA proposal is also directly contrary to the theme of the Telecommunications Act, which fundamentally favors competition over regulation. Indeed, the Act's caption is:

> [a]n Act to promote competition and reduce regulation in order to secure lower prices and higher quality services for American telecommunications consumers and encourage the rapid development of new telecommunications technologies.

B. Other specific provisions of the Act confirm that the Internet and VON should remain unregulated.

The Telecommunications Act creates relevant new definitions for "Access Software Providers" and "Interactive Computer Services." An Access Software Provider is defined as a provider of software (including client or server software), or enabling tools that do any one or more of the following:

(A) filter, screen, allow, or disallow content;

(B) pick, choose, analyze, or digest content; or

(C) transmit, receive, display, forward, cache, search, subset, organize, reorganize, or translate content.

47 U.S.C. § 230(e)(4)

As such, it would appear that VON software manufacturers might be considered to be Access Software Providers. An Interactive Computer Service is defined as any information service, system, or access software provider that provides or enables computer access by multiple users to a computer server, including specifically a service or system that provides access to the Internet and such systems operated or services offered by libraries or educational institutions.

47 U.S.C. § 230(e)(2)

These definitions are relevant because the Telecommunications Act makes no attempt to regulate these services, but rather makes clear that the absence of regulation heretofore has been a good thing and it is U.S. policy to leave them unregulated. Specifically, Section 230(a)(4) contains a finding that "[t]he Internet and interactive computer services have flourished, to the benefit of all Americans, with a minimum of government regulation." Section 230(b)(2) goes on to add that "[I]t is the policy of the United States ... to preserve the vibrant and competitive free market that presently exists for the Internet and other interactive computer services, unfettered by Federal or State regulation...." Similarly, Section 223(e)(6) clarifies that "[n]othing in this section [concerning restrictions on obscene communications via telephone facilities or interactive computer services] shall be construed to treat interactive computer services as common carriers or telecommunications carriers." 47 U.S.C.§ 223(e)(6).

VON may also be seen as resembling Customer Premises Equipment ("CPE") as defined in the new Telecommunications Act, [11] although CPE is defined as applying to equipment involved in Telecommunications, which makes it only imperfectly applicable to VON. In any event,

VON software, like CPE, does not independently provide a user with the means to transmit or receive communications. Rather, the VON software can only function where the following five prerequisites are met: (i) each user participating in the voice communications over the Internet must have a computer meeting certain performance specifications; (ii) each such user must have compatible VON software installed on their computer; (iii) each user must install a web browser; (iv) each user must subscribe to a local exchange carrier or other Telecommunications Service provider that provides dial tone (and thereby, access to Internet access providers); and (v) each such user must subscribe to an Internet access provider.

Manufacturers of CPE are not regulated as interexchange carriers or other Telecommunications Carriers. [12] In Carterfone, the FCC held that any form of CPE could be attached to the telephone network provided that the interconnection does not adversely affect "the telephone company's operations or the telephone system's utility to others." Subsequently in 1975, the Commission decided that the "customer's right to interconnect" should not be curtailed "merely because the device he seeks to interconnect can be defined to constitute a substitution for telephone system equipment." [13] To implement this policy, the Commission adopted Part 68 of the Commission's Rules which allows users to connect any type of CPE to the telephone network provided that either the equipment is connected through protective circuitry registered with the Commission or that the equipment is itself registered with the Commission. [14] In addition, the Commission's decision to not regulate CPE except in the context of its registration program preempts any state regulation of CPE. [15]

Conclusion

The Internet and VON have flourished without regulation by the Commission, providing new and innovative products and services for millions of people around the world, and generating thousands of new jobs for Americans and billions of dollars of value to the U.S. and world economies. In contrast to this clear record of benefit, ACTA submits totally unsubstantiated claims of harm, even with respect to its own members, and seeks to misapply the new Telecommunications Act to establish broad new regulations, including a total ban on computer software, something the Commission has never before regulated. As such, and based on the foregoing discussion, the VON Coalition urges the Commission to reject ACTA's petition.

Respectfully submitted,
THE VON COALITION
/ s /
Jeffrey L. Pulver
Chairman
http://www.von.org
  / s /
Bruce D. Jacobs
Stephen J. Berman
Fisher Wayland Cooper Leader & Zaragoza L.L.P.
2001 Pennsylvania Avenue, N.W.
Suite 400
Washington, D.C. 20006-1851
(202) 659-3494
Attorneys for the VON Coalition
May 8, 1996

Footnotes

[1]   VON was first introduced in the early 1990's with the advent of the VAT20 program on the MBONE. Cornell University introduced CU-SeeMe for the Macintosh platform in 1993, which provided two-way videoconferencing capability. With CU SeeMe and projects like Netphone from the Electric Magic Company in 1994, people began using the Internet for the first time to transmit voice and video.

[2]   The FCC's rules permit "autopatches" by amateur radio operators between their facilities and the public telephone network, as long as those operations are non-commercial. See § 97.113; Amendment of Part 97 of the Commission's Rules to Relax Restrictions on the Scope of Permissible Communications in the Amateur Service, PR Docket No. 92 136, 8 FCC Rcd 5072, 1993 FCC LEXIS 3852, 73 RR 2d 679 (1993).

[3]   Section 203 of the Act establishes the requirement that common carriers file tariffs. Section 214 requires carriers to obtain approval from the Commission for extensions of lines.

[4] In these comments, the VON Coalition does not address ACTA's assertion that the Commission could legally assert jurisdiction over the Internet, although such an assertion is clearly questionable (particularly with respect to Internet software) and the coalition reserves the right to discuss this issue at greater length in the future. The more relevant questions from the VON Coalition's perspective are whether the Commission, even if it has jurisdiction, must or should exercise its authority to regulate either the Internet or VON. The answers to those questions quite clearly are that the FCC has discretion not to regulate the Internet and VON products and that, consistent with the Communications Act and Commission policies, it should exercise that discretion by forbearing from regulating either one.

[5] See, e.g., AT&T's recent announcement that AT&T WorldNet will help develop and market VON products. Interactive Week Online (May 1, 1996). AT&T WorldNet Vice President Tom Evslin said AT&T is opposed to the ACTA petition. Id.

[6] See, e.g., Notice of Proposed Rulemaking in CC Docket No. 96-61, FCC 96-123 (proposing to apply mandatory forbearance to all interstate long-distance carriers). The obligation to support universal service, which ACTA alludes to, is the subject of a Commission rulemaking proceeding that seeks to reform the current structure of support. CC Docket No. 96-45. See, e.g., Comments of Netscape (April 12, 1996) (the Communications Act does not permit the Commission to impose universal service support obligations on Internet service providers, since as "information service providers" they are not subject to an obligation imposed only on "telecommunications carriers"). The VON Coalition urges the Commission to limit consideration of universal service support issues to that rulemaking.

[7] The recent announcement of the Swiss Office of Communications reflects this approach. A copy of that decision, along with an informal translation of the announcement is attached as Exhibit E.

[8] See Order on Reconsideration, File Nos. I-T-C-93-031 and I-T-C-93-050, 10 FCC Rcd 9540 (1995).

[9] See Amendment to Section 64.702 of the Commission's Rules and Regulations, Report and Order, 77 FCC 2d 384 ("Computer II"), modified on recon. 84 FCC 2d 50 (1980), further modified on recon., 88 FCC 2d 512, aff'd sub nom. Computer & Communications Indus. Assn. v. FCC, 693 F.2d 198 (D.C. Cir. 1982). That decision is widely regarded as having been a wise decision that permitted the robust development of the U.S. computer industry. There is no reason to change that policy now, and absolutely no reason to take the unprecedented action of going so far as to ban a particular type of computer software.

[10] Moreover, even if the Commission were to consider VON to be a Telecommunications Service, Section 401 of the new Act gives the FCC discretion to forbear from regulating certain common carrier services. 47 U.S.C. § 160.

[11] Specifically, Section 3 of the Telecommunications Act, 47 U.S.C. § 153(38), defines "Customer Premises Equipment" as follows:

> (38) CUSTOMER PREMISES EQUIPMENT.— The term 'customer premises equipment' means equipment employed on the premises of a person (other than a carrier) to originate, route, or terminate telecommunications.

[12] See Use of the Carterfone Device in Message Toll Telephone Services, 13 FCC 2d 420, recon. denied, 14 FCC 2d 571, 572 (1968) ("Carterphone").

[13] AT&T Co.'s Proposed Tariff Revisions in Tariff F.C.C. No. 263 Exempting Mebane Home Telephone Co. of North Carolina from the Obligations to Afford Customers the Option of Interconnecting Customer-Provided Equipment to Mebane's Facilities; AT&T Transmittal No. 12321, 53 FCC 2d 473, 476-477 (1975) ("Mebane").

[14] 47 C.F.R. Part 68. Initially, the Registration Program did not apply to PBXs, key telephone systems, regular handsets, or coin telephones. In 1976, however, the Commission broadened the Part 68 rules to apply to all forms of CPE. Proposal for New or Revised Classes of Interstate and Foreign Message Toll Telephone Service (MTS) and Wide Area Telephone Service (WATS), Second Report and Order, 58 FCC 2d 736 (1976).

[15] Computer and Communications v. FCC, 693 F.2d 198, 210 (D.C. Cir. 1982), cert. denied, 461 U.S. 938 (1983).

Note: The exhibits referred to in this document are not included with this Appendix.

Source: Federal Communications Commission, Washington, D.C.

List of acronyms and abbreviations

3G   third generation

AAL   ATM adaptation layer

ABR   available bit rate

AC   access control

AC   address copied

AC   alternating current

AC   authentication center

ACC   analog control channel

ACD   automatic call distributor

ACELP   algebraic code excited linear predictive

ACK   acknowledge

ACL   access control list

ACL   asynchronous connectionless

ACP   access control point

ACPI   advanced configuration and power interface

ACTA   America's Carriers Telecommunication Association

ADA   Americans with Disabilities Act

ADC   analog to digital converter

ADCR   alternate destination call routing (AT&T)

ADM   add-drop multiplexer

ADN   advanced digital network (Pacific Bell)

ADPCM   adaptive differential pulse code modulation

ADS   Active Directory Service (Novell, Inc.)

ADSL   asymmetrical digital subscriber line

AFLC   adaptive frame loss concealment

AFP   Apple file protocol

AGC   automatic gain control

AGRAS   air-ground radiotelephone automated service

AIOD   automatic identification of outward dialed calls

AIN   advanced intelligent network

AJBM   automatic jitter buffer management

ALI   automatic location information

AM   amplitude modulation

AMI   alternate mark inversion

AMPS   Advanced Mobile Phone Service

ANI   automatic number identification

ANR   automatic network routing (IBM Corp.)

ANSI   American National Standards Institute

ANT   ADSL network terminator

AO/DI   Always On/Dynamic ISDN

AOL   America Online

AP   access point

APC   access protection capability (AT&T)

APD   avalanche photodiode

API   application programming interface

APPC   advanced program-to-program communications (IBM Corp.)

APPN   advanced peer-to-peer network (IBM Corp.)

APC   automatic protection switching

ARCnet   attached resource computer network (Datapoint Corp.)

ARB   adaptive rate based (IBM Corp.)

ARP   address resolution protocol

ARPA   Advanced Research Projects Agency

ARQ   automatic repeat request

ARS   action request system (Remedy Systems Inc.)

AS   autonomous system

ASAI   Adjunct Switch Application Interface

ASCII   American Standard Code for Information Interchange

ASIC   application-specific integrated circuit

ASN.1   Abstract Syntax Notation 1

ASTN   alternate signaling transport network (AT&T)

AT&T   American Telephone & Telegraph

ATDMA   asynchronous time division multiplexing access

ATE   advanced television enhancement

ATIS   Alliance for Telecommunications Industry Solutions (formerly, ECSA)

ATM   Asynchronous Transfer Mode

ATSC   Advanced Television Systems Committee

ATVEF   Advanced Television Enhancement Forum

AUI   attachment unit interface

AVC   analog voice channel

AWG   American Wire Gauge

B   byte

B8ZS   binary eight zero substitution

BACP   bandwidth allocation control protocol

BBS   bulletin board system

BCC   block check character

BCCH   broadcast control channel

BDCS   broadband digital cross-connect system

BECN   backward explicit congestion notification

Bellcore   Bell Communications Research, Inc.

BER   bit error rate

BERT   bit error rate tester

BGP   border gateway protocol

BHCA   busy hour call attempts

BIB   backward indicator bit

BIOS   basic input-output system

BLSR   bi-directional line switched ring

BMAN   broadband metropolitan area networks

BMC   block multiplexer channel (IBM Corp.)

BMS-E   bandwidth management service-extended (AT&T)

BNC   bayonet nut connector

BOC   Bell operating company

BONDING   Bandwidth On Demand Interoperability Group

BootP   Boot Protocol

BPDU   bridge protocol data unit

BPS   bits per second

BPV   biploar violation

BRI   basic rate interface (ISDN)

BSA   basis serving arrangement

BSA   business software alliance

BSC   base station controller

BSC   binary synchronous communication (IBM Corp.)

BSD   Berkeley Software Distribution

BSE   basic service element

BSN   backward sequence number

BSSC   base station system controller

BT   British Telecom

BTS   base transceiver station

BUS   broadcast/unknown server

CA   communications assistant

CACH   call appearance call handling

CAD   computer aided design

CAM   computer aided manufacturing

CAN   campus area network

CAP   Carrierless Amplitude/Phase (modulation)

CAP   competitive access provider

CARP   Cache Array Routing Protocol

CARS   cable antenna relay services

CAS   channel associated signaling

CASE   computer aided software engineering

CATV   community antenna television

CB   citizen's band

CBR   constant bit rate

CBSC   centralized base site controller

CCC   clear channel capability

CCCH   common control channel

CCF   call control function

CCITT   Consultative Committee for International Telegraphy and Telephony

CCR   customer controlled reconfiguration

CCS   common channel signaling

CCSNC   common channel signaling network controller

CCSS 6   common channel signaling system 6

CCTV   closed circuit television

CCU   central control unit

CD   compact disk

CDCS   continuous dynamic channel selection

CDG   CDMA Development Group

CD-R   compact disk-recordable

CD-ROM   compact disk-read only memory

CDMA   code division multiple access

CDO   community dial office

CDPD   cellular digital packet data

CDR   call detail record

CDSL   consumer digital subscriber line

CEI   comparably efficient interconnection

CEMA   Consumer Electronics Manufacturers Association

CENTREX   central office exchange

CEPT   Conference of European Posts and Telegraphs

CES   circuit emulation service

CGI   Common Gateway Interface

CGSA   cellular geographic servicing areas

CHAP   Challenge Handshake Authentication Protocol

CICS   customer information control system (IBM Corp.)

CIF   common intermediate format

CIR   Committed Information Rate

CIX   commercial internet exchange

CLASS   custom local area signaling services

CLEC   competitive local exchange carrier

CLI   calling line identification

CLNP   Connectionless Network Protocol

CLP   cell loss priority

CMC   Common Messaging Call

CMI   cable microcell integrator

CMIP   Common Management Information Protocol

CMIS   common management information services

CMRS   commercial mobile radio service

CMS   Call Management System (Lucent Technologies)

CNR   customer network reconfiguration

CNS   complementary network service

CO   central office

CON   concentrator

COO   chief operations officer

COPS   common open policy service

CORBA   common object request broker architecture

CoS   class of service

COT   central office terminal

CP   communications processor

CP   connection point

CP   coordination processor

CPE   customer premises equipment

CPID   calling party identification

CPS   cycles per second (Hertz)

CPU   central processing unit

CRC   cyclic redundancy check

CRIMP   Connectivity Routing and Infrastructure Modeling Program
(Cablesoft, Inc.)

CSA   carrier serving area

CSM   communications services management

CSMA/CD   carrier sense multiple access with collision detection

CSU   channel service unit

CT   cordless telecommunications

CTI computer-telephony integration

CUG closed user group

CVSD continuously variable slope delta (modulation)

CWIX Cable and Wireless Internet Exchange

DA destination address

DAC data acquisition and control

DACS digital access and cross-connect system (AT&T)

DAC digital to analog converter

DAMPS Digital Advanced Mobile Phone System

DAP Demand Access Protocol

DAP Directory Access Protocol

DAS dual attached station

DASD direct access storage device (IBM Corp.)

DAT digital audio tape

dB decibel

DBA dynamic bandwidth allocation

DBMS data base management system

DBS direct broadcast satellite

DBU dial backup unit

DCCH digital control channel

DCE data communications equipment

DCE distributed computing environment

DCF data communication function

DCOM Distributed Component Object Model (Microsoft Corp.)

DCPT dynamic cell packet transport (3Com Corp.)

DCS   digital cross-connect system

DDS   digital data services

DDS/SC   digital data service with secondary channel

D/E   debt/equity (ratio)

DECT   digital enhanced (formerly, European) cordless telecommunication

DES   data encryption standard

DFB   distributed feedback

DFSMS   data facility storage management subsystem (IBM Corp.)

DHCP   Dynamic Host Configuration Protocol

DID   direct inward dialing

DIF   digital interface frame

diff-serv   differentiated services

DIP   dual inline package

DLL   data link layer

DLCI   data link connection identifier

DLCS   digital loop carrier system

DLL   dynamic link library

DLSw   data link switching (IBM Corp.)

DLU   digital line unit

DM   distributed management

DMA   direct memory access

DME   distributed management environment

DMI   desktop management interface

DMT   discrete multi tone

DMTF   Desktop Management Task Force

DNS   Domain Name Service or Domain Name Server

DOCSIS   data over cable service interface specification

DoD   Department of Defense (U.S.)

DOD   direct outward dialing

DOM   document object model

DOS   disk operating system

DOV   data over voice

DPL   digital power line

DQDB   distributed queue dual bus

DQPSK   differential quadrature phase-shift keying

DS0   digital signal-level 0 (64 Kbps)

DS1   digital signal-level 1 (1.544 Mbps)

DS1C   digital signal-level 1c (3.152 Mbps)

DS2   digital signal-level 2 (6.312 Mbps)

DS3   digital signal-level 3 (44.736 Mbps)

DS4   digital signal-level 4 (274.176 Mbps)

DSA   directory system agent

DSI   digital speech interpolation

DSI   directory service integration

DSL   digital subscriber line

DSLAM   digital subscriber line access multiplexer

DSMA   digital sense multiple access

DSN   defense switched network

DSP   digital signal processor

DSP   Directory System Protocol

DSS   decision support system

DSSS   direct sequence spread spectrum

DSU   data service unit

DSX1   digital systems cross-connect 1

DTC   digital traffic channel

DTE   data terminal equipment

DTMF   dual tone multifrequency

DTR   dedicated token ring

DTU   data transfer unit

DTV   digital television

DUA   directory user agent

DVD   digital video disk

DWDM   dense wavelength division multiplexer

DWMT   discrete wavelet multitone

DXI   data exchange interface

E&M   ear and mouth (a signaling method)

e-mail   electronic mail

E-TDMA   expanded time division multiple access

EAS   Expert Agent Software (Lucent Technologies)

EB   exabyte (1,000,000,000,000,000,000 bytes)

EBCDIC   extended binary code decimal interchange code (IBM Corp.)

ECMA   European Computer Manufacturers Association

ECSA   Exchange Carriers Standards Association

ED   ending delimiter

EDFA   erbium-doped fiber amplifier

EDGE   enhanced data for global evolution

EDI   electronic data interchange

EDIFACT   electronic data interchange for administration and transport

EDRO   enhanced diversity routing option (AT&T)

EEROM   electronically erasable read-only memory

EFF   Electronic Frontier Foundation

EFRC   enhanced full rate codec

EFS   encrypting file system (Microsoft Corp.)

EFT   electronic funds transfer

EGP   external gateway protocol

EHF   extremely high frequency (more than 30 GHz)

EIA   Electronic Industries Association

EIR   equipment identity register

EISA   extended industry standard architecture

EMA   Electronic Messaging Association

EMI   electromagnetic interference

EMS   element management system

EOC   embedded overhead channel

EOS   expanded originating service

EOT   end of transmission

EP   extension point

ERP   enterprise resource planning

ESCON   enterprise system connection (IBM Corp.)

ESD   electronic software distribution

ESF   extended super frame

ESMA   enterprise storage management architecture (Legato Systems)

ESMR   enhanced specialized mobile radio

ESMTP   Extended Simple Mail Transfer Protocol

ESN   electronic serial number

ETC   Exempt Telecommunication Companies

ETSI   European Telecommunication Standards Institute

EVRC   enhanced variable rate encoder

4GL   fourth-generation language

FACCH   fast associated control channel

FASB   financial accounting standards board

FASC   Fraud Analysis and Surveillance Center (AT&T)

FASTAR   fast automatic restoral (AT&T)

FAT   file allocation table

FBI   Federal Bureau of Investigation (U.S.)

FC   frame control

FC   fiber channel

FC-0   fiber channel—layer 0

FC-1   fiber channel—layer 1

FC-2   fiber channel—layer 2

FC-3   fiber channel—layer 3

FC-4   fiber channel—layer 4

FC-AL   fiber channel-arbitrated loop

FCC   Federal Communications Commission (U.S.)

FCS   frame check sequence

FDD   frequency division multiplex

FDDI   fiber distributed data interface

FDIC   Federal Deposit Insurance Corporation

FDL   facilities data link

FDMA   frequency division multiple access

FEC   forward error correction

FECN   forward explicit congestion notification

FEP   front-end processor

FHSS   frequency hopping spread spectrum

FIB   forward indicator bit

FIB   forwarding information base

FIFO   first-in/first-out

FIRST   flexible integrated radio systems technology

FITL   fiber-in-the-loop

FM   frequency modulation

FOCC   forward control channel

FOD   fax on demand

FPF   fraud protection feature

FRAD   frame relay access device

FRF   frame relay forum

FRS   family radio service

FS   frame status

FSN   forward sequence number

FTAM   file transfer, access, and management

FT1   fractional T1

FTP   File Transfer Protocol

FTS   Federal Telecommunications System

FTTB   fiber to the building

FTTC   fiber to the curb

FTTH   fiber to the home

FTTN   fiber to the neighborhood

FWA   fixed wireless access

FWD   Free World Dialup (project)

FWT   fixed wireless terminal

FX   foreign exchange (line)

FXO   foreign exchange office

FXS   foreign exchange station

GATT   general agreement on tariffs and trade

GB   gigabyte (1,000,000,000 bytes)

GDS   generic digital services

GEO   geostationary earth orbit

GFC   generic flow control

GFR   guaranteed frame rate

GHz   gigahertz (billions of cycles per second)

GIF   Graphics Interchange Format

GIS   geographic information systems

GloBanD   global bandwidth on demand

GMRS   general mobile radio service

GNAP   global network access point

GPI   general purpose interface

GPRS   general packet radio services

GPS   global positioning system

GRE   generic routing encapsulation

GSA   General Services Administration (U.S.)

GSM   Global System for Mobile Communications (formerly, Groupe Spéciale Mobile)

GSN   gigabit system network

GUI   graphical user interface

H0   high-capacity ISDN channel operating at 384 Kbps

H11   high-capacity ISDN channel operating at 1.536 Mbps

HAN   Home Area Network (Boca Research)

HDLC   high-level data link control

HDML   Handheld Device Markup Language

HDSL   high-bit-rate digital subscriber line

HDT   host digital terminal

HDTV   high definition television

HEC   header error check

HF   high frequency (3 MHz to 30 MHz)

HFC   hybrid fiber/coax

HIC   head-end interface converter

HIPPI   high performance parallel interface

HLR   home location register

HNF   high performance networking forum

HPNA   home phoneline networking alliance

HIPPI   high performance parallel interface

HomePNA   Home Phoneline Networking Alliance

HPR   high performance routing (IBM Corp.)

HSCSD   high-speed circuit switched data

HSM   hierarchical storage management

HST   helical scan tape

HSTR   high speed token ring

HSTRA   high speed token ring alliance

HTML   Hypertext Markup Language

HTTP   Hypertext Transfer Protocol

HVAC   heating, ventilation, and air conditioning

Hz   hertz (cycles per second)

I/O   input/output

IAB   Internet Architecture Board

IANA   Internet Assigned Numbers Authority

IAPP   Inter Access Point Protocol

ICC   Internet call center

ICI   interexchange carrier interface

ICMP   Internet Control Message Protocol

ICP   Internet Cache Protocol

ICR   intelligent call routing

ICS   intelligent calling system

ICSA   International Computer Security Association

ID   identification

IDDD   international direct dialing designator

IDPR   inter-domain policy routing

IDSL   ISDN digital subscriber line

IEC   International Electrotechnical Commission

IEEE   Institute of Electrical and Electronics Engineers

IESG   Internet Engineering Steering Group

IETF   Internet Engineering Task Force

IF   intermediate frequency

IGMP   Internet Group Management Protocol

IGP   Interior Gateway Protocol

IIS   Internet information server (Microsoft Corp.)

ILD   injection laser diode

ILEC   incumbent local exchange carrier

IMAP   Internet Mail Access Protocol

IMEI   international mobile equipment identity

IMS/VS   information management system/virtual storage (IBM Corp.)

IMSI   international mobile subscriber identity

IMT   intelligent multimode terminals

IMTS   improved mobile telephone service

IN   intelligent network

INC   Integrated Network Connect (AT&T)

INMARSAT   International Maritime Satellite Organization

INMS   integrated network management system

IOC   inter office channel

ION   Integrated On-Demand Network (Sprint)

IOS   Internetwork Operating System (Cisco Systems)

IP   Internet Protocol

IPH   integrated packet handler

IPI   intelligent peripheral interface

IPN   intelligent peripheral node

IPTC   IP Telephony Solution for Carriers

IPX   internetwork packet exchange

IR   infrared

IRC   Internet relay chat

IrDA   infrared data association

IrLAN   infrared LAN

IrLAP   Infrared Link Access Protocol

IrLMP   Infrared Link Management Protocol

IrPL   infrared physical layer

IRQ   interrupt request

IrTTP   Infrared Transport Protocol

IS   information system

IS   industry standard

IS-IS   intra-autonomous system to intra-autonomous system

ISA   industry standard architecture

ISAPI   Internet server applications programming interface

ISDL   ISDN subscriber digital line

ISDN   Integrated Services Digital Network

ISH   integrated service hub

ISM   industrial, scientific, and medical (frequency bands)

ISO   International Organization for Standardization

ISOC   Internet Society

ISP   Internet service provider

ISR   intermediate session routing (IBM Corp.)

ISSI   inter-switching systems interface

IT   information technology

ITFS   instructional television fixed service

ITG   Internet Telephony Gateway (Lucent Technologies)

ITR   intelligent text retrieval

ITU   International Telecommunications Union

IVR   interactive voice response

IXC   interexchange carrier

JCE   Java cryptography extensions

JDC   Japanese Digital Cellular

JEPI   Joint Electronic Payments Initiative

JIT   just in time

JPEG   Joint Photographic Experts Group

JRE   Java runtime environment

JTAPI   Java telephony application programming interface

JTC   joint technical committee

JVM   Java virtual machine

K (Kilo)   one thousand (e.g., Kbps)

KB   kilobyte (1,000 bytes)

KHz   Kilohertz (thousands of cycles per second)

KSU   key service unit

KTS   key telephone system

KVM   keyboard, video, mouse (a type of switch)

L2F   Layer 2 Forwarding

L2TP   Layer 2 Tunneling Protocol

LAN   local area network

LANCES   LAN resource extension and services (IBM Corp.)

LANE   Local Area Network Emulation

LAPB   link access procedure-balanced

LAT   local area transport (Digital Equipment Corp.)

LATA   local access and transport area

LBO   line build out

LCD   liquid crystal display

LCN   local channel number

LCP   Link Control Protocol

LD   laser diode

LDAP   Lightweight Directory Access Protocol

LEC   local exchange carrier

LED   light-emitting diode

LEO   low earth orbit

LF   low frequency (30 KHz to 300 KHz)

LI   length indicator

LIPS   lightweight Internet person schema

LLC   logical link control

LMDS   local multipoint distribution system

LMI   local management interface

LMS   location and monitoring service

LNP   local number portability

LOM   LAN on motherboard

LSAPI   licensing service application programming interface

LSI   large scale integration

LTG   line trunk group

LU   logical unit (IBM Corp.)

LVD   low voltage differential

M (Mega)   one million (e.g., Mbps)

ma   milliamp

MAC   media access control

MAC   moves, adds, changes

MAE   metropolitan area exchange

MAN   metropolitan area network

MAPI   messaging applications programming interface   (Microsoft Corp.)

MAS   multiple address system

MATV   metropolitan antenna television

MAU   multistation access unit

MB   megabyte (1,000,000 bytes)

MCA   micro channel architecture (IBM Corp.)

MCU   multipoint control unit

MD   mediation device

MDBS   mobile database system

MDF   main distribution frame

MDI   medium dependent interface

MDLP   Mobile Data-Link Layer Protocol

MDS   multipoint distribution service

MEO   middle earth orbit

MES   master earth station

MF   mediation function

MF   medium frequency (300 KHz to 3 MHz)

MF-TDMA   multi-frequency time division multiple access

MGCP   Multimedia Gateway Control Protocol

MHz   megahertz (millions of cycles per second)

MIB   management information base

MIC   Management Integration Consortium

MIF   management information format

MII   media independent interface

MIME   Multipurpose Internet Mail Extensions

MIN   mobile identification number

MIPS   millions of instructions per second

MIS   management information services

MISR   multiprotocol integrated switch-routing

MIX   multiservice interexchange (Cisco Systems)

MJU   multipoint junction unit

MLPPP   Multi-Link Point-to-Point Protocol

MM   mobility manager

MMDS   multichannel, multipoint distribution service

MMITS   modular multifunction information transfer system

MNLP   Mobile Network Location Protocol

MNRP   Mobile Network Registration Protocol

MO   magneto-optical

Modem   modulation/demodulation

MOS   mean opinion score

MOSPF	multicast open shortest path first

MPEG	Motion Pictures Experts Group

MPLS	Multi-Protocol Label Switching

MPOA	Multi-Protocol Over ATM

MPPP	Multilink Point-to-Point Protocol

MRI	magnetic resonance imaging

ms	millisecond (thousandths of a second)

MS	mobile station

MSC	mobile switching center

MSF	Multiservice Switching Forum

MSN	Microsoft Network

MSRN	mobile station roaming number

MSS	mobile satellite service

MTA	message transfer agent

MTA	multimedia terminal adapter

MTBF	mean time between failure

MTBSO	mean time between service outages

MTP	message transfer part

MTSO	Mobile Telephone Switching Office

MTTR	mean time to restore

MVC	multicast virtual circuit

MVDS	microwave video distribution system

MVL	Multiple Virtual Line

MVPRP	multi-vendor problem resolution process

mW	milliwatt

N-AMPS   Narrowband Advanced Mobile Phone Service

NAC   network applications consortium

NAK   negative acknowledge

NAM   numeric assignment module

NAP   network access point

NAT   network address translation

NAU   network addressable unit or network accessible unit (IBM Corp.)

NAUN   nearest active upstream neighbor

NC   network computer

NCP   network control program (IBM Corp.)

NCP   network control point

NCSA   National Center for Supercomputer Applications

NDIS   network driver interface specification

NDS   network directory service

NDS   Novell Directory Services (Novell, Inc.)

NE   network element

NEBS   new equipment building specifications

NECA   National Exchange Carrier Association

NetBIOS   network basic input/output system

NEF   network element function

NEI   network equipment identifier

NFS   network file system (or server)

NIC   network interface card

NiCd   nickel cadmium

NIF   network interconnection facilities

NiMH nickel-metal hydride

NIST National Institute of Standards and Technology

NLM netware loadable module (Novell Inc.)

nm nanometer

NM network manager

NMS NetWare management system (Novell, Inc.)

NMS network management system

NMT Nordic Mobile Telephone

NNM network node manager (Hewlett-Packard Co.)

NNTP Network News Transfer Protocol

NOC network operation center

NOS network operating system

NPC network protection capability (AT&T)

NPV net present value

NRC Network Reliability Council

NRIC Network Reliability and Interoperability Council

ns nanosecond

NSA National Security Agency

NSF National Science Foundation

NTIA National Telecommunications Industry Association

NTSA Networking Technical Support Alliance

NTSC National Television Standards Committee

OA&M operations, administration, management

OAM&P operations, administration, maintenance, and provisioning

OC optical carrier

OC-1   optical carrier signal-level 1 (51.84 Mbps)

OC-3   optical carrier signal-level 3 (155.52 Mbps)

OC-9   optical carrier signal-level 9 (466.56 Mbps)

OC-12   optical carrier signal-level 12 (622.08 Mbps)

OC-18   optical carrier signal-level 18 (933.12 Mbps)

OC-24   optical carrier signal-level 24 (1.244 Gbps)

OC-36   optical carrier signal-level 36 (1.866 Gbps)

OC-48   optical carrier signal-level 48 (2.488 Gbps)

OC-96   optical carrier signal-level 96 (4.976 Gbps)

OC-192   optical carrier signal-level 192 (9.952 Gbps)

OC-256   optical carrier signal-level 256 (13.271 Gbps)

OCC   Online Communications Center (Lucent Technologies)

OCR   optical character recognition

OCUDP   office channel unit data port

ODBC   Open Database Connectivity (Microsoft Corp.)

ODI   open datalink interface

ODS   operational data store

OEM   original equipment manufacturer

OFX   open financial exchange

OLAP   online analytical processing

OLE   object linking and embedding

OMA   object management architecture

OMAP   operations, maintenance, administration, and provisioning

OMC   operations and maintenance center

OMF   object management framework

OMG   object management group

ONU   optical network unit

OOP   object-oriented programming

OPX   off premises extension

ORB   object request broker

OS   operating system

OS/2   operating system/2 (IBM Corp.)

OSF   Open Software Foundation

OSF   operations systems function

OSI   Open Systems Interconnection

OSP   Open Settlement Protocol

OSS   operations support system

OTDR   optical time domain reflectometry

PA   preamble

PACS   personal access communications system

PAD   packet assembler-disassembler

PAL   Phase Alternating by Line

PAN   personal area network

PAP   Password Authentication Protocol

PAT   port address translation

PB   petabyte (1,000,000,000,000,000 bytes)

PBX   private branch exchange

PC   personal computer

PCB   printed circuit board

PCH   paging channel

PCI   peripheral component interconnect

PCLEC   packet competitive local exchange carrier

PCM   pulse code modulation

PCN   personal communications networks

PCS   personal communications services

PCT   private communication technology

PDA   personal digital assistant

PDF   portable document format (Adobe Systems)

PDN   packet data network

PDU   payload data unit

PEM   privacy enhanced mail

PERL   Practical Extraction and Reporting Language

PGP   pretty good privacy

PHS   personal handyphone system

PHY   physical layer

PIM   personal information management

PIN   personal identification number

PIN   positive-intrinsic-negative

PIP   picture-in-picture

PKE   public key encryption

PLMRS   private land mobile radio services

PMA   physical medium attachment

PMD   physical media dependent

PnP   plug and play

PoS   packet over SONET

POP   point of presence

POP   Post Office Protocol

POS   point of sale

POTS   plain old telephone service

PPN   policy powered network (3Com Corp.)

PPP   Point-to-Point Protocol

PPS   packets per second

PPTP   Point-to-Point Tunneling Protocol

PQ   priority queuing

PRI   primary rate interface (ISDN)

PSAP   public safety answering point

PSN   packet switched network

PSTN   public switched telephone network

PT   payload type

PTT   Post Telephone & Telegraph

PU   physical unit (IBM Corp.)

PUC   public utility commission

PUK   personal unblocking key

PVC   permanent virtual circuit

PWT   personal wireless telecommunications

PXE   preboot execution environment

QA   quality assurance

QAM   quadrature amplitude modulation

QCIF   quarter common intermediate format

QIC   quarter inch cartridge

QoS   quality of service

QPSK   quadrature phase shift keying

RAC   remote access concentrator

RACH   random access channel

RAD   remote antenna driver

RADIUS   remote access dial-in service

RAID   redundant array of inexpensive disks

RAM   random access memory

RAS   remote access server

RASDL   rate adaptive digital subscriber line

RASP   remote antenna signal processor

RBES   rule based expert systems

RBOC   regional Bell operating company

RCU   remote control unit

RDBMS   relational database management system

RDSS   radio determination satellite service

RED   random early detection

RECC   reverse control channel

RFC   request for comment

RF   radio frequency

RF   routing field

RFI   radio frequency interference

RFI   request for information

RFP   request for proposal

RFQ   request for quotation

RG/U   radio guide/utility

RI/RO   ring in/ring out

RIP   Routing Information Protocol

RISC   reduced instruction set computing

RJE   remote job entry

RJU   residential junction unit

RLL   run length limited

RMON   remote monitoring

ROI   return on investment

ROM   read only memory

RPC   remote procedure call

RSCN   registered state change notification

RSS   residential service system

RSVP   Resource Reservation Protocol

RT   remote terminal

RTNR   real-time network routing (AT&T)

RTP   Rapid Transfer Protocol (IBM Corp.)

RTP   Real-Time Transfer Protocol

RTT   radio transmission technology

RX   receive

SA   source address

SACCH   slow associated control channel

SAFER   split access flexible egress routing (AT&T)

SAN   storage area network

SAP   second audio program

SAP   service access point

SAP   Service Advertising Protocol

SAS   single attached station

SATAN   security administrator tool for analyzing networks

SBA   Small Business Administration (U.S.)

SBCCS   single byte command code set (IBM Corp.)

SC   subscriber channel

SC   system controller

SCC   Standards Coordinating Committee (IEEE)

SCF   service control function

SCO   synchronous connection oriented

SCP   service control point

SCSI   small computer systems interface

SD   starting delimiter

SDCCH   standalone dedicated control channel

SDF   service data function

SDH   synchronous digital hierarchy

SDK   software development kit

SDLC   synchronous data link control (IBM Corp.)

SDM   subrate data multiplexing

SDMA   space division multiple access

SDN   software defined network (AT&T)

SDP   service delivery point

SDR   software defined radio

SDSL   symmetric digital subscriber line

SET   secure electronic transaction

SFD   start frame delimiter

SGCP   Simple Gateway Control Protocol

SGML   standard generalized markup language

SHF   super high frequency (3 GHz to 30 GHz)

SHTTP   Secure Hypertext Transfer Protocol

SIF   signaling information field

SIF   SONET Interoperability Forum

SIM   Subscriber Identity Module

SIMA   simple integrated media access

SINA   static integrated network access

SIP   SMDS Interface Protocol

SLA   service level agreement

SLD   service level definition

SLIC   serial line interface coupler (IBM Corp.)

SLIP   Serial Line Internet Protocol

SLP   Service Location Protocol

SMDI   station message desk interface

SMDR   station message detail recording

SMDS   switched multimegabit data services

SMP   symmetric multi-processor

SMR   specialized mobile radio

SMS   service management system

SMS   short message service

SMT   station management

SMTP   Simple Mail Transfer Protocol

SN   switching network

SNA   Systems Network Architecture (IBM Corp.)

SNAGAS   SNA gateway access server

SNDCP   Subnetwork Dependent Convergence Protocol

SNI   subscriber network interface

SNMP   Simple Network Management Protocol

SNS   simple name service

SOHO   small office home office

SONET   synchronous optical network

SPA   Software Publishers Association

SPC   stored program control

SPI   service provider interface

SPID   service profile identifier

SPX   synchronous packet exchange (Novell, Inc.)

SQE   signal quality error

SQL   Structured Query Language

SRB   source route bridging

SS   switching system

SS7   signaling system 7

SSA   serial storage architecture (IBM Corp.)

SSCP   system services control point (IBM Corp.)

SSCP/PU   system services control point/physical unit (IBM Corp.)

SSF   service switching function

SSG   Service Selection Gateway (Cisco Systems)

SSL   Secure Sockets Layer

SSP   service switching point

SSP   Switch-to-Switch Protocol

ST   simple twist

STDM   statistical time division multiplexing

STP   shielded twisted pair

STP   signal transfer point

STP   spanning tree protocol

STS   shared telecommunications services

STS   synchronous transport signal

STX   start of transmission

SUBT   subscriber terminal

SVC   switched virtual circuit

SWAP   Shared Wireless Access Protocol

SWC   serving wire center

SYNTRAN   synchronous transmission

SYSGEN   system generation

T1   transmission service at the DS1 rate of 1.544 Mbps

T3   transmission service at the DS3 rate of 44.736 Mbps

TA   technical advisor

TA   technical advisory

TACACS   terminal access controller access control system

TACS   Total Access Communication System

TAG   technical advisory group

TAPI   telephony application programming interface (Microsoft Corp.)

TASI   time assigned speech interpolation

TB   terabyte (1,000,000,000,000 bytes)

TBOP   transparent bit oriented protocol

Tbps   terabit-per-second

TCAP   transaction capabilities applications part

TCL   Tool Command Language

TCO   total cost of ownership

TCP   Transmission Control Protocol

TD-CDMA   time division-code division multiple access

TDD   time division duplex

TDM   time division multiplexer

TDMA   time division multiple access

TDMA/TDD   time division multiple access with time division duplexing

TDR   time domain reflectometry

TFTP   trivial file transfer protocol

TIA   telecommunications industry association

TIB   tag information base

TIMS   transmission impairment measurement set

TL1   transaction language 1

TLS   transparent LAN service

TMN   Telecommunications Management Network

ToS   type of service

TRS   telecommunications relay services

TSAPI   telephony services application programming interface (Novell Inc.)

TSI   time slot interchange

TSR   terminal stay resident

TTAC   Telemetry, Tracking, and Control Center

TTRT   target token rotation time

TTY   text telephone

TV   television

TWX   teletypewriter exchange (also known as Telex)

TX   transmit

UART   universal asynchronous receiver/transmitter

UAWG   Universal ADSL Working Group

UBR   unspecified bit rate

UDP   User Datagram Protocol

UHF   ultra high frequency (300 MHz to 3 GHz)

UI   unit intervals

ULS   user location service

UMS   universal messaging system

UMTS   Universal Mobile Telecommunications System

UN   United Nations

UNE   unbundled network elements

UNI   user-network interface

UPS   uninterruptible power supply

USB   Universal Serial Bus

USDLA   United States Distance Learning Association

USNC   U.S. National Committee

UTP   unshielded twisted-pair

UTRA   UMTS terrestrial radio access

UWCC   Universal Wireless Communications Consortium

VAD   Voice Activity Detection

VAR   value added reseller

VBNS   Very High-Speed Backbone Network Service

VBR   variable bit rate

VBR-rt   variable bit rate real-time

VC   virtual circuit

VCI   virtual channel identifier

VCR   video cassette recorder

VCSEL   vertical-cavity surface-emitting laser

VDN   Vector Directory Number

VDSL   very high-speed digital subscriber line

VF   voice frequency

VFN   vendor feature node

VG   voice grade

VHF   very high frequency (30 MHz to 300 MHz)

VLF   very low frequency (less than 30 KHz)

VLR   visitor location register

VLSI   very large scale integration

VM   virtual machine

VMS   virtual machine system (Digital Equipment Corp.)

VNS   voice network server (3Com Corp.)

VNS   voice network switching (Cisco Systems)

VoD   video on demand

VoFR   voice over frame relay

VoIP   voice over Internet Protocol

VON   Voice On Net

VP   virtual path

VPI   virtual path identifier

VPIM   Voice Profile for Internet Mail

VPN   virtual private network

VSAT   very small aperture terminal

VT   virtual terminal

VT   virtual tributary

VTAM   virtual telecommunications access method (IBM Corp.)

VtoA   voice transport over ATM

W3C   World Wide Web Consortium

W-CDMA   Wideband Code Division Multiple Access

W-CDMA/NA   Wideband Code Division Multiple Access/North America

WACS   wireless access communications system

WAN   wide area network

WAP   Wireless Access Protocol

WATS   wide area telecommunications service

WBM   Web-based management

WCS   wireless communications service

WDCS   wideband digital cross-connect system

WDM   wave division multiplexer

WEP   wired equivalent privacy

WfM   wired for management

WFQ   weighted fair queuing

WGS   worldwide geodetic system

WIMS   wireless multimedia and messaging services

WINS   Windows Internet name service (Microsoft Corp.)

WLAN   wireless local area network

WLL   wireless local loop

WML   Wireless Markup Language

WORM   write once read many

WRC   World Radio Conference

WRED   weighted random early detection

WTO   World Trade Organization

WWW   World Wide Web

WWW3   World Wide Web consortium

XNS   Xerox Network System (Xerox Corp.)

Y2K   year 2000

YAG   yttrium-aluminum-garnet

YB   yottabyte (1,000,000,000,000,000,000,000,000 bytes)

ZB   zettabyte (1,000,000,000,000,000,000,000 bytes)

About the author

Nathan Muller is an independent consultant in Sterling, Virginia, specializing in advanced technology marketing, research, and education. In his 29 years of industry experience, he has written extensively on many aspects of computers and communications, having published 17 books—including three encyclopedias—and over 1,500 articles about computers and communications in 50 publications worldwide.

In addition, he is a regular contributor to the Gartner Group's Datapro Research Reports. He has participated in market research projects for Dataquest, Northern Business Information, and Faulkner Technical Reports. He also does custom projects for technology-oriented clients in the computer, telecommunications, and health care industries.

Muller has held numerous technical and marketing positions with such companies as Control Data Corporation, Planning Research Corporation, Cable & Wireless Communications, ITT Telecom, and General DataComm Inc. He has an M.A. in Social and Organizational Behavior from George Washington University.

He maintains *Strategic Information Resources* on the World Wide Web, which is located at http://www.ddx.com. He can be reached via e-mail at nmuller@ddx.com.

Index

A

ABR. *See* Available bit rate
ACC. *See* Analog control channel
Access charges, 363–67
Access control list, 333
Access experience information, 42
Access router, 332–34
Accrue Software Insight, 133–34
ACD. *See* Automatic call distributor
ACTA. *See* America's Carriers
    Telecommunication
    Association
Adaptive differential pulse code
    modulation, 334
Add-drop multiplexer, 172, 179
Adjustable volume control, 58
ADM. *See* Add-drop multiplexer
ADSL. *See* Asymmetric digital
    subscriber line
Advanced caller ID, 58
Advanced mobile phone
    service, 12, 180, 279, 284
Advanced phone book, 58
Advanced Television Enhancement
    Forum, 230–31

AGC. *See* Automatic gain control
AGIS, 252–53
All-optical network, 178–80
Always On/Dynamic ISDN, 151, 333
America's Carrier
    Telecommunication
    Association, 357–59,
    383–92, 403–9
Ameritech Cellular, 14
AMPS. *See* Advanced mobile phone
    service
Analog system, 22, 283, 332
APD. *See* Avalanche photodiode
API. *See* Application program
    interface
Application program interface, 17
Applications
    integration, 15–18
    protocols, 37–38
    reliability, 25–28
    sharing, 201–2
Artisoft, 210
AS5300 gateway, 340–342
Ascend Communications, 137
Ascend Online Services, 138–40

J

Japanese Digital Cellular, 279
Java, 111, 230
JDC. *See* Japanese Digital Cellular
Joint Photographic Experts
      Group, 17
JPEG. *See* Joint Photographic Experts
      Group

L

L2F. *See* Layer 2 forwarding
L2TP. *See* Layer 2 Tunneling Protocol
LAN. *See* Local area network
LANE. *See* ATM LAN emulation
LANtastic, 210
Last-party redial, 60
Latency, 26–27, 262
Layer 2 forwarding, 80
Layer 2 Tunneling Protocol, 80, 173
LDAP. *See* Lightweight Directory
      Access Protocol
Leased lines, 4
LED. *See* Light-emitting diode
LEO satellite. *See* Low earth orbit
      satellite
Level 3 Communications, 4, 260–61
Light-emitting diode, 174, 327
Lightweight Directory Access
      Protocol, 16, 85, 325
Line sharing, 371–72
Linskys, 217–18
Liquid crystal display, 233, 327
LNP. *See* Local number portability
Local access, 43–44
Local access and transport areas, 353
Local area network, 7
      modem, 212–14
      switch, 338–40
Local loop, 144, 154–69

Local management interface, 345
Local number portability, 193
Lotus cc:Mail, 234–35
Low earth orbit satellite, 15, 186–87
Lucent Technologies, 23, 71, 95,
      115, 315, 348–49
      Internet telephony
      services, 107–16

M

MAC. *See* Media access control
Macro Capacity Fiber Network, 28
MAE. *See* Metropolitan area
      exchange
Management, performance
      and, 30–31
Map feature, 61
Market drivers
      Internet telephony, 73–74
      phoneline networking, 218–20
MC3810 concentrator, 334–35
MCI WorldCom, 4, 106, 132, 236,
      244, 251, 262, 270–72
MCI WorldCom
      Click'nConnect, 129–30
MDBS. *See* Mobile data base system
MD-IS. *See* Mobile data-intermediate
      system
MDLP. *See* Mobile data-link layer
MDSL. *See* Multi-bit-rate digital
      subscriber line
Mean opinion score, 67n
Media access control, 185
Mediation, Internet protocol, 28–30
MEO satellite. *See* Middle-earth-
      orbit satellite
Meridian Data, 201
M-ES. *See* Mobile-end system
Message Care software, 112–14

Ubiquity, 36–37
ULS. *See* User location service
ULTRA. *See* UMTS terrestrial radio
    access
UMTS. *See* Universal mobile
    telecommunications system
UMTS terrestrial radio access, 298
Unbundled network
    element, 356, 370
Uni-directional access, 210–11
Unified Communications, 342
Uninterruptible power supply, 327
United States, 304–5
Universal ADSL Working
    Group, 154–55
Universal asymmetric digital
    subscriber line, 155
Universal digital subscriber line, 157
Universal messaging, 15–16
Universal mobile
    telecommunications
    system, 298–304, 312–13
Universal serial
    bus, 7, 158, 194, 226–27
Universal Wireless Communications
    Consortium, 305, 310
Unspecified bit rate, 335
USB. *See* Universal serial bus
User-defined group, 62
User location service, 62
Users, scalability and, 18–20
U S West, 5, 12, 44, 74*n*, 228, 366
Utility companies, 6–7, 191–93
UTRA, 302
Uunet Technologies Inc., 41
UWC-136, 310–11

V

Value-added network, 66–67
Value-added reseller, 321

VAN. *See* Value-added network
VAP. *See* Voice applications partner
VAR. *See* Value-added reseller
Variable bit rate, 29, 335
VBR. *See* Variable bit rate
VDN. *See* Vector directory number
Vector directory number, 111
Video feature, phoneware, 62
Video over Internet Protocol, 203
Vienna Systems, 6
Virtual circuit queuing, 344
Virtual private
    network, 4, 8, 18–19,
     66, 173
    Internet protocol-based, 80–82
    popularity, 33
    security, 32–33
VNS. *See* Voice name server
VocalTec Communications
    Ltd., 55, 65, 71
Voice activity detection, 334
Voice applications partner, 321
Voice funnel, 49
Voice mail, 62–63
Voice manager, 335–36
Voice name server, 83
Voice network switching, 346–47
Voice On Net, 358–59, 397–12
Voice over frame relay, 344
Voice-over-Internet
    Protocol, 6, 67, 83, 85,
     88–90, 203, 335–36
Voice packet gateway, 340–42
Voice Profile for Internet Mail, 16–17
Voice transport over ATM, 344
VoIP. *See* Voice-over-Internet
    Protocol
VON. *See* Voice On Net
VPIM. *See* Voice Profile for Internet
    Mail
VPN. *See* Virtual private network

Recent Titles in the Artech House Telecommunications Library

Vinton G. Cerf, Senior Series Editor

Telecommunications Deregulation, James Shaw

Telemetry Systems Design, Frank Carden

Teletraffic Technologies in ATM Networks, Hiroshi Saito

Understanding Modern Telecommunications and the Information Superhighway, John G. Nellist and Elliott M. Gilbert

Understanding Networking Technology: Concepts, Terms, and Trends, Second Edition, Mark Norris

Understanding Token Ring: Protocols and Standards, James T. Carlo, Robert D. Love, Michael S. Siegel, and Kenneth T. Wilson

Videoconferencing and Videotelephony: Technology and Standards, Second Edition, Richard Schaphorst

Visual Telephony, Edward A. Daly and Kathleen J. Hansell

Winning Telco Customers Using Marketing Databases, Rob Mattison

World-Class Telecommunications Service Development, Ellen P. Ward

For further information on these and other Artech House titles, including previously considered out-of-print books now available through our In-Print-Forever® (IPF®) program, contact:

Artech House
685 Canton Street
Norwood, MA 02062
Phone: 781-769-9750
Fax: 781-769-6334
e-mail: artech@artechhouse.com

Artech House
46 Gillingham Street
London SW1V 1AH UK
Phone: +44 (0)20 7596-8750
Fax: +44 (0)20 7630-0166
e-mail: artech-uk@artechhouse.com

Find us on the World Wide Web at:
www.artechhouse.com